MARKS'
MECHANICS
PROBLEM-SOLVING
COMPANION

MARKS'
MECHANICS
PROBLEM-SOLVING
COMPANION

Larry Silverberg
James P. Thrower

McGraw-Hill

New York   Chicago   San Francisco   Lisbon   London   Madrid
Mexico City   Milan   New Delhi   San Juan   Seoul
Singapore   Sydney   Toronto

Library of Congress Cataloging-in-Publication Data

Silverberg, L. (Larry)
    Marks' mechanics problem-solving companion / Larry Silverberg, James P. Thrower.
        p.   cm.
    ISBN 0-07-136278-9
    1. Mechanics, Applied—Problems, exercises, etc.   I. Thrower, James P.   II. Title.

TA350.7 .S65   2001
620.1′076—dc21   2001018308

McGraw-Hill

A Division of The McGraw·Hill Companies

1 2 3 4 5 6 7 8 9 0   AGM/AGM   0 7 6 5 4 3 2 1

ISBN 0-07-136278-9

*The sponsoring editor for this book was Scott Grillo, the editing supervisor was David E. Fogarty, and
the production supervisor was Pamela A. Pelton. It was set in Times Roman by Techset Composition
Limited.*

Printed and bound by Quebecor/Martinsburg.

This book is printed on recycled, acid-free paper containing a minimum of 50% recycled,
de-inked fiber.

Contents

MARKS'
MECHANICS
PROBLEM-SOLVING
COMPANION

Introduction

Mechanics deals with relationships between force and displacement. In **Newtonian** mechanics, these relationships are based on **Newton's Laws**. The laws themselves are very simple, yet their implications are far-reaching. Using Newton's Laws, statics problems, dynamics problems, and mechanics of materials problems can be solved. More advanced structural engineering problems, such as stress analysis, machine design, vibration, machine dynamics, and fluid mechanics, build on these principles. These problems are treated in undergraduate courses in engineering.

Marks' Mechanics Problem-Solving Companion treats statics, dynamics and mechanics of materials problems. Each section develops the useful formulas and presents examples, which are followed by unsolved practice problems.

The style here is concise. Reference material and formulas can be found quickly and directly in the theory sections. Valuable points are sometimes made only in the example sections, too, where there is context for them. But the conciseness will sometimes impose on the reader the need for careful repetition to develop a full understanding of the concepts. So, let us begin...

CHAPTER I

Preliminaries

The process of solving a problem involves developing an **understanding** that is mathematical in nature. A problem is converted into a set of mathematical equations, the resulting equations are solved, and the solution is studied.

In mechanics, this process of understanding is divided into the following five steps:

1. A coordinate system is selected for each object of concern.
2. The forces acting on each object are diagrammed.
3. Newton's Laws, or formulas derived from Newton's Laws, are applied to each object.
4. The equations are solved.
5. An understanding of the solution of the equations is sought.

Chapters 1–5 treat situations in which the objects are in **static equilibrium,** that is, the objects are not moving. Chapters 6–9 focus on **dynamics** problems; the objects are now in motion. Chapters 10–16 focus again on statics problems, this time with an emphasis placed on looking **inside** the objects.

Chapter 1 reviews the vector concepts that will be needed throughout the book. Vectors are pairs or triplets of numbers used to express such quantities as position, velocity, acceleration, force, and moment. Through the use of vectors, problems will be solved in a systematic fashion. Chapters 2–5 cover the field of statics. Chapter 2 provides practice at finding forces and moments in typical statics problems.

Chapter 3 describes how to set up the equations that govern static equilibrium. Chapter 4 considers an assortment of applied problems involving trusses, rods, beams, shafts, friction, cables, and ropes. Chapter 5 covers the computation of mass centers, polar moments, area moments, and mass products. These are quantities that will be needed in the remaining chapters of the book.

Chapters 6–9 cover the field of dynamics. Chapter 6 introduces Newton's Laws, which govern the dynamics of mechanical systems. Chapter 7 shows how to use different types of coordinate systems to solve dynamics problems. Energy and momentum methods for solving dynamics problems are developed in Chapter 8. Rigid bodies are treated in Chapter 9.

Chapters 10–16 cover the field of mechanics of materials. Chapter 10 develops the basic concepts that are used in mechanics of materials. The important role of linearity in mechanics problems motivates the definitions of stress and strain. Axial deformation, torsional deformation, and bending deformation problems are developed in Chapters 11, 12, and 13, respectively. In Chapter 14, the relationship between stress and strain is examined in more detail. Problems involving combinations of axial, torsional, and bending deformation are treated in Chapter 15. Chapter 16 reviews applications like pressure vessels and composite beams.

1.1 UNITS OF MEASURE

The physical quantities treated in this book are measured using either the **U.S. Customary** measurement system or the **International System of Units** (Système Internationale, SI). The SI system is also called the **metric** system. Table 1.1-1 and the inside jacket of the book show the two measurement systems and the conversions between them. For simplicity, the **base** units are listed first. In the U.S. system, the base units are taken to be length, time, and force. In the SI system, the base units are taken to be length, time, and mass. Standards committees have adopted precise methods of measuring the base units. The units of measure associated with quantities other than the base units are expressed in terms of the base units. For example, the units associated with a moment are taken to be lb·ft in the U.S. system and N·m in the SI system. Occasionally, these quantities are referred to by specific names, or they need to be converted to the U.S. or SI systems. For example, in the SI system, the special name for a unit of power is **watt**, which is the same as N·m/s. Another unit for power is the horsepower. One unit of horse-power is taken to be equal to 745.7 watts. The units that are given specific names and the prefixes to the units are listed in Tables 1.1-1 and 1.1-2 and are also found in the inside jacket of the book

Table 1.1-1   Conversion between U.S. and SI units

Category	Starting unit		U.S.		SI
Length	1 foot (ft) or (′)	=	base	or	0.3048 m
	1 inch (in) or (″)	=	$\frac{1}{12}$ ft	or	0.0254 m
	1 mile (mi)	=	5280 ft	or	1609 m
	1 meter (m)	=	3.281 ft	or	base
Time	1 second (s)	=	base	or	base
Mass	1 slug (slug)	=	1 lb·s^2/ft	or	14.59 kg
	1 kilogram (kg)	=	0.06852 slug	or	base
Force	1 pound (lb)	=	base	or	4.448 N
	1 newton (N)	=	0.2248 lb	or	1 kg·m/s^2
Pressure	1 kilopound per square inch (kpsi) or (ksi)	=	144 000 lb/ft^2	or	6.895 MPa
	1 megapascal (MPA)	=	0.1450 ksi	or	10^6 kg/m·s^2
Power	1 watt (W)	=	0.001341 hp	or	1 kg·m^2/s^3
	1 horsepower (hp)	=	550.0 ft·lb/s	or	745.7 W
Volume	1 gallon (gal)	=	0.1337 ft^3	or	3.785 L
	1 liter (L)	=	0.2642 gal	or	0.001 m^3
Energy	1 joule (J)	=	0.7376 ft·lb	or	1 kg·m^2/s^2
	1 foot pound (ft·lb)	=	1 ft·lb	or	1.356 J
Temperature	Degree Fahrenheit (°F)	=	°C $\times \frac{9}{5}$ + 32		
	Degree Celsius (°C)	=	(°F − 32) $\times \frac{5}{9}$		
	Kelvin (K)	=	°C + 273.15		

Table 1.1-2 Prefixes

Prefix	Value
micro (μ)	10^{-6}
milli (m)	10^{-3}
centi (c)	10^{-2}
deci (d)	10^{-1}
deka (da)	10
hecto (h)	10^{2}
kilo (k)	10^{3}
mega (M)	10^{6}

1.2 SIGNIFICANT DIGITS

The calculations performed throughout this book use numbers that are not exact. Numbers are given with a certain level of accuracy, and the calculated answers will have a certain level of accuracy, too. One way to indicate accuracy is to count the **number of significant digits**. Expressing numbers in the usual form can be ambiguous, though. For example, 51 200 could have 3, 4, or 5 significant digits. To avoid this ambiguity, numbers can be written in scientific notation. In scientific notation, the number 51 200 is written as 512×10^2, indicating that 51 200 has 3 significant digits, or as 51.200×10^3, indicating that 51 200 has 5 significant digits. The number of digits in front of the exponent unambiguously indicates the number of significant digits.

The errors that result from calculations depend on the calculations themselves. In order to gain a perspective on the nature of the resulting errors, several basic types of calculations are examined. Let's examine the errors produced when numbers are added, subtracted, multiplied, divided, and raised to a power.

Adding and Subtracting Numbers

Consider the equation

$$a = x + y, \tag{1.2-1}$$

where the error in x is e_x, the error in y is e_y, and the error in a, resulting from the addition, is e_a. For example, the number 10.5 could have an error of about 0.05 in either direction, written 10.5 ± 0.05, and the number 25.6 could have an error of about 0.04, written 25.6 ± 0.04. Equation (1.2-1) does not consider the errors in the numbers x and y. The calculation that's actually performed is

$$a \pm e_a = (x \pm e_x) + (y \pm e_y). \qquad (1.2\text{-}2)$$

Subtracting Eq. (1.2-2) from Eq. (1.2-1), and retaining only the positive terms, yields

$$\boxed{e_a = e_x + e_y,} \qquad (1.2\text{-}3)$$

In this example, $x = 10.5$, $e_x = 0.05$, $y = 25.6$, and $e_y = 0.04$, so $a = 31.1$ and $e_a = 0.09$. The error equation, Eq. (1.2-3), is the same equation that would be used for subtracting two numbers. Indeed, **the error resulting from adding or subtracting two numbers is the sum of the errors found in the two numbers.**

Multiplying and Dividing Numbers

Consider the equation

$$b = xy, \qquad (1.2\text{-}4)$$

where the error in b, resulting from the multiplication, is e_b. Equation (1.2-4) does not consider the errors in the numbers. The calculation that's actually performed is

$$b \pm e_b = (x \pm e_x)(y \pm e_y). \qquad (1.2\text{-}5)$$

Subtracting Eq. (1.2-4) from Eq. (1.2-5), dividing by b, neglecting the second-order term $e_x e_y$, which is negligibly small, and retaining only the positive terms yields

$$\boxed{\frac{e_b}{|b|} = \frac{e_x}{|x|} + \frac{e_y}{|y|},} \qquad (1.2\text{-}6)$$

In this example, the normalized errors in x and y are $0.05/10.5 = 0.00476$ and $0.04/25.6 = 0.00156$, which yields $b = 268.8$, and the normalized error in b is $0.00476 + 0.00156 = 0.00632$. The error in b

is $e_b = 0.00632(268.8) = 1.699$. The numbers b and e_b are written together as 268.8 ± 1.699. The error equation for dividing two numbers is the same as the error equation for multiplying two numbers (see Problem 1.2-4). **The normalized error resulting from multiplying or dividing two numbers is the sum of the normalized errors found in the two numbers.**

Powers

Consider the equation

$$c = x^n, \tag{1.2-7}$$

where the error in c, resulting from the power, is e_c. Equation (1.2-7) does not consider the error in the number x. The calculation that's actually performed is

$$c \pm e_c = (x \pm e_x)^n. \tag{1.2-8}$$

Subtracting Eq. (1.2-7) from Eq. (1.2-8), dividing by c, neglecting second- and higher-order terms in the error e_x, and retaining only the positive terms yields

$$\boxed{\frac{e_c}{|c|} = n\left(\frac{e_x}{|x|}\right).} \tag{1.2-9}$$

In this example, letting $n = 3$, we get $c = 1157.625$ and $e_c = 3(1157.625)(0.00476) = 16.53$. Thus, c and e_c are written together as 1157 ± 1653. **The normalized error resulting from raising x to the nth power is n times the normalized error in x.** Notice that the multiplication case and the power case are identical to each other when $x = y$ and $n = 2$.

Examples

| 1.2-1 | Determine the result of subtracting 250 ± 10.3 and 236 ± 10.8.

Solution
$(236 \pm 10.8) - (250 \pm 10.3) = -18 \pm 21.1$.

| 1.2-2 | Determine the result of multiplying 50 ± 4.3 and 428 ± 2.8. How does the error resulting from the multiplication change if the error in the second number is decreased by a factor of 10?

Solution

$$(50 \pm 4.3)(428 \pm 2.8) = 50(428)[1 \pm (4.3/50 + 2.8/428)]$$
$$= 21\,400 \pm 1980.$$

If the second number's error is decreased by a factor of 10,

$$(50 \pm 4.3)(428 \pm 0.28) = 50(428)[1 \pm (4.3/50 + 0.28/428)]$$
$$= 21\,400 \pm 1854.$$

Notice that the error resulting from the multiplication did not change appreciably when the error in the second number was decreased by a factor of 10. The reason is that the resulting error was dominated by the error in the first term.

1.2-3 Determine the result of raising 1.2 ± 0.02 to the fifth power.

Solution

$$(1.2 \pm 0.02)^5 = 1.2^5[1 \pm 5(0.02/1.2)] = 2.49 \pm 0.21.$$

Problems

1.2-1 Determine the result of adding 25 ± 1 and 2.5 ± 0.01.

1.2-2 Determine the result of multiplying 25 ± 1 and 2.5 ± 0.01. How does the error resulting from the multiplication change if the error in the first number is decreased by a factor of 10? How does the error resulting from the multiplication change if the error in the second number is decreased by a factor of 10?

1.2-3 Determine the result of raising 2.5 ± 0.01 to the third power.

1.2-4 Show that the error equation associated with the division $d = x/y$ is the same as the error equation associated with multiplication, that is

$$\frac{e_d}{|d|} = \frac{e_x}{|x|} + \frac{e_y}{|y|}.$$

Hint: Approximate $\dfrac{x \pm e_x}{y \pm e_y}$ using a first-order Taylor series expansion to get

$$\frac{x \pm e_x}{y \pm e_y} = \frac{x}{y} \frac{1 \pm e_x/x}{1 \pm e_y/y} = \frac{x}{y}\left(1 \pm \frac{e_x}{x} \mp \frac{e_y}{y}\right).$$

1.3 SCALAR OPERATIONS WITH VECTORS

Theory

In a planar coordinate system, the position of a point refers to two quantities, for example x and y in a rectangular coordinate system and r and θ in a polar coordinate system. Likewise, two quantities are required in order to specify the velocity of, the acceleration of, or the force at a point. In a three-dimensional coordinate system, three quantities are needed in order to specify these quantities. The important thing to remember is that these are quantities that appear in pairs and in triplets. In the spirit of dealing with these quantities in an **efficient** and **systematic** fashion, the concept of vectors is introduced.

The three positions that locate a point are collected together into the **position vector**, written $\mathbf{r} = (x \quad y \quad z)$. The position vectors of two points can be denoted by $\mathbf{r}_1 = (x_1 \quad y_1 \quad z_1)$ and $\mathbf{r}_2 = (x_2 \quad y_2 \quad z_2)$.

The position vector of a second point relative to the first is the difference between the two vectors, written

$$\boxed{\mathbf{r}_{2/1} = \mathbf{r}_2 - \mathbf{r}_1 = (x_2 - x_1 \quad y_2 - y_1 \quad z_2 - z_1).}$$

Notice that when two vectors are subtracted (or added), each of their components are subtracted (or added). Similarly, when a vector is multiplied by a scalar (a number), each of the vector's components are multiplied by the scalar. Denoting the scalar by a,

$$\boxed{a\mathbf{r}_1 = (ax_1 \quad ay_1 \quad az_1).}$$

The geometric interpretations of **vector addition** and **multiplication of a vector by a scalar** are illustrated in Figs. 1.3a and b. As shown, two vectors are added by placing them end-to-end, and a vector is multiplied

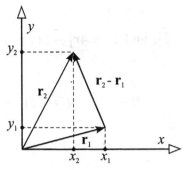

Figure 1.3-1a

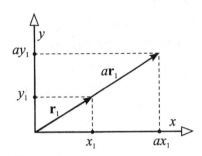

Figure 1.3-1b

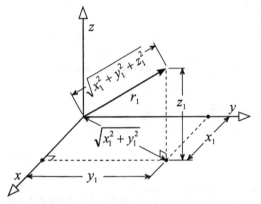

Figure 1.3-2

by a scalar by stretching ($a > 1$ or $a < -1$) or by compressing ($-1 < a < 1$) the vector.

Another vector concept is the **length of a vector**. Using the Pythagorean Theorem, the length of a vector $\mathbf{r}$ is denoted by r (indicated by italics) or by $|\mathbf{r}|$, and is given by

$$|\mathbf{r}| = \sqrt{x^2 + y^2 + z^2},$$

where x, y, and z are rectangular coordinates (see Fig. 1.3-2). A **unit vector** then refers to a vector having unit length. The unit vectors $\mathbf{i} = (1 \quad 0 \quad 0), \mathbf{j} = (0 \quad 1 \quad 0)$, and $\mathbf{k} = (0 \quad 0 \quad 1)$ are called **standard unit vectors**. It is customary in this and in other books to express vectors in terms of unit vectors. Rather than write $\mathbf{r}$ as $\mathbf{r} = (x \quad y \quad z)$, we write $\mathbf{r}$ as $\mathbf{r} = x\mathbf{i} + y\mathbf{j} + z\mathbf{k}$, which is the same as $\mathbf{r} = (x \quad y \quad z)$ since $(x \quad y \quad z) = x(1 \quad 0 \quad 0) + y(0 \quad 1 \quad 0) + z(0 \quad 0 \quad 1) = x\mathbf{i} + y\mathbf{j} + z\mathbf{k}$.

Examples

1.3-1 Given $r_1 = 3i + 2j$ and $r_2 = 7i + 2j$, find $r_3 = 4r_1 - 3r_2$ and illustrate this graphically.

Solution

Referring to the figure, $r_3 = 4r_1 - 3r_2 = 4(3i + 2j) - 3(7i + 2j) = -9i + 2j$.

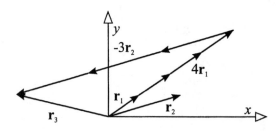

Example 1.3-1

1.3-2 Given $F = 3i + 4j$, find the magnitude of **F** and the angle that **F** makes relative to **i**.

Examples 1.3-2 and 1.3-3

Solution

By the Pythagorean Theorem, the magnitude of **F** is $F = [3^2 + 4^2]^{1/2} = 5$.

Referring to the figure, the angle θ that **F** makes relative to **i** is found from $\tan \theta = \frac{4}{3}$ or $\sin \theta = \frac{4}{5}$ or $\cos \theta = \frac{3}{5}$, so $\theta = 53.13°$.

1.3-3 Find **F**, given that the magnitude of **F** is $F = 5$, that **F** is in the $x-y$ plane, and that the angle that **F** makes relative to **i** is $\theta = 30°$.

Solution

Referring to the figure, $F = 5(\cos 30° \, i + \sin 30° \, j) = [5(\sqrt{3}/2)]i + [5(1/2)]j = 3.61i + 2.50j$.

1.3-4 Find the magnitude of $F = -3i + 4j + 12k$.

Solution

Referring to Fig. 1.3-2, the Pythagorean Theorem is used twice to get the magnitude of **F**, given by $F = \sqrt{(-3)^2 + 4^2 + 12^2} = 13$.

1.3-5 Find the unit vector **n** directed from $r_1 = 2i - 8j$ to $r_2 = 5i - 4j - 12k$.

Solution

The vector $r_{2/1} = r_2 - r_1 = (5 - 2)i + [-4 - (-8)]j + (-12 - 0)k = 3i + 4j - 12k$ is in the direction of n ($r_{2/1}$ is also referred to as r_2 relative to r_1). Now, simply divide by the magnitude and get the unit vector

$$n = \frac{r_{2/1}}{r_{2/1}} = \frac{3i + 4j - 12k}{\sqrt{3^2 + 4^2 + (-12)^2}}$$

$$= \frac{3}{13}i + \frac{4}{13}j - \frac{12}{13}k = 0.23i + 0.31j - 0.92k.$$

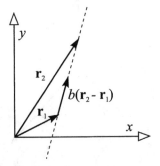

1.3-6 Let $r_1 = 2i + j$ and $r_2 = 3i + 6j$ be two vectors that extend to a given line, as shown. Show for any value of b that $r = r_1 + b(r_2 - r_1)$ is a vector that extends to the same line. Also, show that r is between r_1 and r_2 when b is between 0 and 1.

Example 1.3-6

Solution

It is shown that r extends to the same line by placing the vectors r_1 and $b(r_2 - r_1)$ end-to-end. Notice when $b = 1$ that $r = r_1 + (r_2 - r_1) = r_2$. The resulting vector, $r = r_1 + b(r_2 - r_1)$, is equal to r_1 when $b = 0$ and equal to r_2 when $b = 1$. In other words, points on the line are of the form $r = 2i + j + b[(3i + 6j) - (2i + j)] = (2 + b)i + (1 + 5b)j$.

Problems

1.3-1 Given $r_1 = -7i + 4j$ and $r_2 = 3i + j$, find $r_3 = r_1 - 2r_2$.

1.3-2 Given the vectors shown in the figure overleaf, find and draw $F_3 = F_1 + 3F_2$ and $F_4 = 3F_1 - 2F_2$.

1.3-3 Given $F_1 = 2i + 3j + 4k$ and $F_2 = 2i + 3k$, find $F_3 = F_1 + F_2$.

1.3-4 Given the vectors shown in the figure overleaf, find and draw $F_3 = F_1 + F_2$ and $F_4 = F_1 - F_2$.

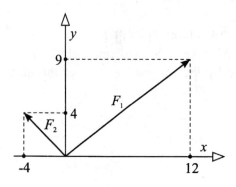

Problem 1.3-2

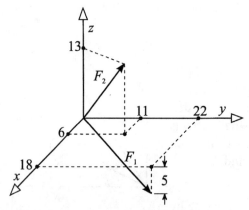

Problem 1.3-4

1.3-5 Find the magnitude of $\mathbf{F} = -3\mathbf{i} + 6\mathbf{j}$ and the angle that $\mathbf{F}$ makes relative to $\mathbf{i}$.

1.3-6 Find the magnitude of $\mathbf{r} = 2\mathbf{i} - 3\mathbf{j} + 4\mathbf{k}$ and the angle that $\mathbf{r}$ makes relative to $\mathbf{k}$.

1.3-7 Find the magnitude of $\mathbf{F} = -8\mathbf{i} + 5\mathbf{j} + \mathbf{k}$.

1.3-8 Find the unit vector in the direction of $\mathbf{r} = -5\mathbf{i} + 12\mathbf{j}$.

1.3-9 Find the unit vector in the direction of $\mathbf{F} = -4\mathbf{i} - 3\mathbf{j} + 12\mathbf{k}$.

1.3-10 Let $\mathbf{r}_1, \mathbf{r}_2$, and $\mathbf{r}_3$ be three vectors that extend to a given plane. Draw and show that $\mathbf{r} = \mathbf{r}_1 + b(\mathbf{r}_2 - \mathbf{r}_1) + c(\mathbf{r}_3 - \mathbf{r}_1)$ is a vector that extends to the same plane. Also show that $\mathbf{r}$ is between $\mathbf{r}_1, \mathbf{r}_2$, and $\mathbf{r}_3$ when b and c are between 0 and 1.

1.3-11 Using Example 1.3-6, show that the **bisector** of two vectors $\mathbf{r}_1$ and $\mathbf{r}_2$ is $\mathbf{r} = \frac{1}{2}(\mathbf{r}_1 + \mathbf{r}_2)$.

1.4 DOT PRODUCTS AND CROSS PRODUCTS

Theory

Now that vector addition, multiplication of a vector by a scalar, and the length of a vector have been defined, two ways in which a vector can be multiplied by a vector are defined. Once again, letting $r_1 = x_1 i + y_1 j + z_1 k$ and $r_2 = x_2 i + y_2 j + z_2 k$, we define the **dot product** and the **cross product** as follows:

Dot Product

$$r_1 \cdot r_2 = x_1 x_2 + y_1 y_2 + z_1 z_2;$$

(1.4-1)

Cross Product

$$r_1 \times r_2 = (y_1 z_2 - z_1 y_2)i + (z_1 x_2 - x_1 z_2)j + (x_1 y_2 - y_1 x_2)k.$$

(1.4-2)

The dot product is also called the **scalar product**, because the result is a scalar (number). The cross product is also called the **vector product**, because the result is a vector.

These products are now examined more closely. Replacing r_1 and r_2 above with the standard unit vectors i, j, and k, the very elegant **unit vector products** are obtained:

$$
\begin{aligned}
i \cdot i &= 1, & i \cdot j &= 0, & i \cdot k &= 0, \\
j \cdot i &= 0, & j \cdot j &= 1, & j \cdot k &= 0, \\
k \cdot i &= 0, & k \cdot j &= 0, & k \cdot k &= 1;
\end{aligned}
$$

(1.4-3)

$$
\begin{aligned}
i \times i &= 0, & i \times j &= k, & i \times k &= -j, \\
j \times i &= -k, & j \times j &= 0, & j \times k &= i, \\
k \times i &= j, & k \times j &= -i, & k \times k &= -j.
\end{aligned}
$$

(1.4-4)

Two other important features of the dot product and the cross product are satisfied for any r_1 and r_2:

$$r_1 \cdot r_2 = r_2 \cdot r_1, \qquad r_1 \times r_2 = -r_2 \times r_1.$$

(1.4-5a,b)

Equation (1.4-3) reveals that **the dot product of a unit vector with itself is equal to one and that the dot product of a unit vector with one of** the other unit vectors is equal to zero. Equation (1.4-4) reveals that the cross product of a unit vector with itself is equal to the zero vector and that a cross product of a unit vector with one of the other unit vectors is equal to the remaining unit vector or the negative of the remaining unit vector depending on the order of the multiplication. As shown in Figs. 1.4-1a and 1.4-1b, the counter-clockwise order yields the remaining unit vector and the clockwise order yields the negative of the remaining unit vector.

As shown in Examples 1.4-1 and 1.4-5, it will be useful to memorize Eqs. (1.4-3) and (1.4-4) and to use them when performing dot product and cross product operations. In fact, it's not necessary to remember the original definitions of the dot product and the cross product, given in Eqs. (1.4-1) and (1.4-2), when performing dot product and cross product operations, provided the reader remembers Eqs. (1.4-3) and (1.4-4).

The dot product of two vectors and the magnitude of the cross product of two vectors are now interpreted graphically. As Examples 1.4-2 and 1.4-6 will show,

$$\mathbf{r}_1 \cdot \mathbf{r}_2 = |\mathbf{r}_1||\mathbf{r}_2| \cos \theta \qquad |\mathbf{r}_1 \times \mathbf{r}_2| = |\mathbf{r}_1||\mathbf{r}_2||\sin \theta|, \qquad (1.4\text{-}6\mathrm{a,b})$$

where θ denotes the angle between $\mathbf{r}_1$ and $\mathbf{r}_2$. The first formula reveals that the dot product of two nonzero vectors is zero when the vectors are perpendicular. The second formula shows that the cross product of two vectors is zero when the two vectors are parallel.

Another operation on vectors that will need to be performed in Chapters 6–9 is time differentiation. The dot product of two vector functions of time and the cross product of two vector functions of time

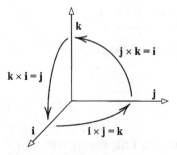

Figure 1.4-1a

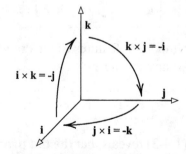

Figure 1.4-1b

will need to be differentiated in time. The product rules for differentiation are

$$\frac{d}{dt}(\mathbf{r}_1 \cdot \mathbf{r}_2) = \frac{d\mathbf{r}_1}{dt} \cdot \mathbf{r}_2 + \mathbf{r}_1 \cdot \frac{d\mathbf{r}_2}{dt},$$ (1.4-7a)

$$\frac{d}{dt}(\mathbf{r}_1 \times \mathbf{r}_2) = \frac{d\mathbf{r}_1}{dt} \times \mathbf{r}_2 + \mathbf{r}_1 \times \frac{d\mathbf{r}_2}{dt}.$$ (1.4-7b)

Notice that these product rules of differentiation for vectors are the same as the product rule of differentiation for scalars.

Understanding of these vector concepts will be useful throughout the study of mechanics.

Examples

1.4-1 Find the dot product of $\mathbf{F}_1 = 4\mathbf{i} + 3\mathbf{j}$ and $\mathbf{F}_2 = \mathbf{i} + 7\mathbf{j}$.

Solution

$$\begin{aligned}
\mathbf{F}_1 \cdot \mathbf{F}_2 &= (4\mathbf{i} + 3\mathbf{j}) \cdot (\mathbf{i} + 7\mathbf{j}) = (4\mathbf{i}) \cdot (\mathbf{i}) + (4\mathbf{i}) \cdot (7\mathbf{j}) + (3\mathbf{j}) \cdot (\mathbf{i}) + (3\mathbf{j}) \cdot (7\mathbf{j}) \\
&= 4(1)(\mathbf{i} \cdot \mathbf{i}) + 4(7)(\mathbf{i} \cdot \mathbf{j}) + 3(1)(\mathbf{j} \cdot \mathbf{i}) + 3(7)(\mathbf{j} \cdot \mathbf{j}) \\
&= 4(1) + 28(0) + 3(0) + 21(1) = 25.
\end{aligned}$$

Notice that the dot product was found by **expansion**. The usual associative and distributive rules of arithmetic were used. Indeed, the dot product obeys all of the rules of algebra, namely, the associative rule of addition, the commutative rule of addition, the associative rule of multiplication, the commutative rule of multiplication, and the distributive rule.

1.4-2 Show, as stated in the theory section, that

$$\mathbf{r}_1 \cdot \mathbf{r}_2 = |\mathbf{r}_1||\mathbf{r}_2| \cos \theta,$$

where $\mathbf{r}_1$ and $\mathbf{r}_2$ are vectors in the x–y plane.

Solution

Write the vectors $\mathbf{r}_1$ and $\mathbf{r}_2$ as $\mathbf{r}_1 = x_1\mathbf{i} + y_1\mathbf{j}$, $\mathbf{r}_2 = x_2\mathbf{i} + y_2\mathbf{j}$, let the

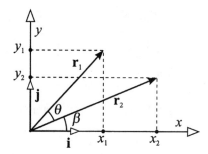

Example 1.4-2

angle between $\mathbf{r}_1$ and $\mathbf{r}_2$ be denoted by θ, and let the angle between $\mathbf{r}_2$ and $\mathbf{i}$ be denoted by β. Thus,

$$\mathbf{r}_1 \cdot \mathbf{r}_2 = x_1 x_2 + y_1 y_2 = r_1 \cos(\theta + \beta) r_2 \cos \beta + r_1 \sin(\theta + \beta) r_2 \sin \beta$$
$$= r_1 r_2 [\cos(\theta + \beta) \cos \beta + \sin(\theta + \beta) \sin \beta] = r_1 r_2 \cos(\theta + \beta - \beta)$$
$$= r_1 r_2 \cos \theta = |\mathbf{r}_1||\mathbf{r}_2| \cos \theta.$$

1.4-3 Find the angle θ between $\mathbf{F}_1 = 4\mathbf{i} + 3\mathbf{k}$ and $\mathbf{F}_2 = \mathbf{i} + 7\mathbf{k}$ using the dot product.

Solution

$$\cos \theta = \frac{\mathbf{F}_1 \cdot \mathbf{F}_2}{F_1 F_2} = \frac{4(1) + 3(7)}{(\sqrt{3^2 + 4^2})(\sqrt{1^2 + 7^2})} = \frac{25}{5\sqrt{50}} = \tfrac{1}{2}\sqrt{2},$$

so $\theta = 45°$.

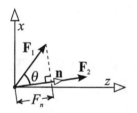

Example 1.4-4

1.4-4 Find the **component** of $\mathbf{F}_1 = 4\mathbf{i} + 3\mathbf{k}$ in the direction of $\mathbf{F}_2 = \mathbf{i} + 7\mathbf{k}$.

Solution
As shown, the sought-after component is $F_n = \mathbf{F}_1 \cdot \mathbf{n} = \mathbf{F}_1 \cdot (\mathbf{F}_2/F_2) = (4\mathbf{i} + 3\mathbf{k}) \cdot (\mathbf{i} + 7\mathbf{k})/\sqrt{1^2 + 7^2} = [4(1) + 3(7)]/\sqrt{50} = 5/\sqrt{2} = 3.54$.
The calculation that's performed in this example illustrates the usefulness of the dot product operation. Dot products are used to easily calculate components of vectors in given directions.

1.4-5 Find the cross product of $\mathbf{F}_1 = 4\mathbf{i} + 3\mathbf{j}$ and $\mathbf{F}_2 = \mathbf{i} + 7\mathbf{j}$.

Solution

$$\mathbf{F}_1 \times \mathbf{F}_2 = (4\mathbf{i} + 3\mathbf{j}) \times (\mathbf{i} + 7\mathbf{j}) = (4\mathbf{i}) \times (\mathbf{i}) + (4\mathbf{i}) \times (7\mathbf{j}) + (3\mathbf{j}) \times (\mathbf{i})$$
$$+ (3\mathbf{j}) \times (7\mathbf{j})$$
$$= 4(1)(\mathbf{i} \times \mathbf{i}) + 4(7)(\mathbf{i} \times \mathbf{j}) + 3(1)(\mathbf{j} \times \mathbf{i}) + 3(7)(\mathbf{j} \times \mathbf{j})$$
$$= 4(0) + 28(\mathbf{k}) + 3(-\mathbf{k}) + 21(0) = 25\mathbf{k}.$$

Notice that the cross product was found by **expanding** the terms. In doing the expansion, the cross product obeys all of the rules of algebra, except for one. The cross product obeys the associative rule of addition, the commutative rule of addition, the associative rule of multiplication, and the distributive rule. The cross product does **not** obey the commutative rule of multiplication. Instead of the commutative rule of multiplication, the cross product obeys the rule $\mathbf{a} \times \mathbf{b} = -\mathbf{b} \times \mathbf{a}$, in which $\mathbf{a}$ and $\mathbf{b}$ are any two vectors (the commutative rule of multiplication, which the cross product does **not** obey, has no minus sign in this expression). Indeed, switching the order of the cross product switches the sign of the cross product.

1.4-6 Show, as stated in the theory section, that

$$|\mathbf{r}_1 \times \mathbf{r}_2| = |\mathbf{r}_1||\mathbf{r}_2||\sin\theta|,$$

where $\mathbf{r}_1$ and $\mathbf{r}_2$ are vectors in the x–y plane.

Solution
Write the vectors $\mathbf{r}_1$ and $\mathbf{r}_2$ as $\mathbf{r}_1 = x_1\mathbf{i} + y_1\mathbf{j}$, $\mathbf{r}_2 = x_2\mathbf{i} + y_2\mathbf{j}$, and let $|\mathbf{r}_1| = r_1$ and $|\mathbf{r}_2| = r_2$. As shown, the angle between $\mathbf{r}_1$ and $\mathbf{r}_2$ is denoted by θ, and the angle between $\mathbf{r}_2$ and $\mathbf{i}$ is denoted by β. Thus,

$$\mathbf{r}_1 \times \mathbf{r}_2 = 0\mathbf{i} + 0\mathbf{j} + (x_1y_2 - x_2y_1)\mathbf{k} = (x_1y_2 - x_2y_1)\mathbf{k}.$$

Notice that $\mathbf{r}_1 \times \mathbf{r}_2$ is perpendicular to both $\mathbf{r}_1$ and $\mathbf{r}_2$. **The cross product of any two vectors is perpendicular to each of these two vectors.** From the figure,

$$
\begin{aligned}
|x_1y_2 - x_2y_1| &= |r_1\cos(\theta + \beta)\, r_2\sin\beta - r_2\cos\beta\, r_1\sin(\theta + \beta)| \\
&= r_1r_2|\cos(\theta + \beta)\sin\beta - \cos\beta\sin(\theta + \beta)| \\
&= r_1r_2|-\sin(\theta + \beta - \beta)| = |\mathbf{r}_1||\mathbf{r}_2||\sin\theta|.
\end{aligned}
$$

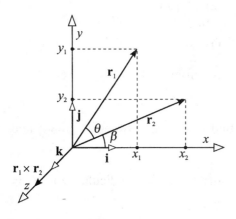

Example 1.4-6

1.4-7 Find the angle θ between $\mathbf{F}_1 = 4\mathbf{i} + 3\mathbf{j}$ and $\mathbf{F}_2 = \mathbf{i} + 7\mathbf{j}$ using the cross product.

Solution

$$|\sin \theta| = \frac{|\mathbf{F}_1 \times \mathbf{F}_2|}{F_1 F_2} = \frac{|4(7)\mathbf{k} - 3(1)\mathbf{k}|}{(\sqrt{4^2 + 3^2})(\sqrt{1^2 + 7^2})} = \frac{|25\mathbf{k}|}{5\sqrt{50}} = \tfrac{1}{2}\sqrt{2}, \text{ so } \theta = 45°.$$

1.4-8 Find a unit vector perpendicular to both $\mathbf{F}_1 = \mathbf{i} + \mathbf{j}$ and $\mathbf{F}_2 = \mathbf{j} + \mathbf{k}$.

Solution
The vector $\mathbf{F}_1 \times \mathbf{F}_2$ is perpendicular to $\mathbf{F}_1$ and $\mathbf{F}_2$, so

$$\mathbf{F}_1 \times \mathbf{F}_2 = (\mathbf{i} + \mathbf{j}) \times (\mathbf{j} + \mathbf{k}) = \mathbf{i} \times \mathbf{j} + \mathbf{i} \times \mathbf{k} + \mathbf{j} \times \mathbf{j} + \mathbf{j} \times \mathbf{k}$$
$$= \mathbf{k} - \mathbf{j} + 0 + \mathbf{i} = \mathbf{i} - \mathbf{j} + \mathbf{k}.$$

The unit vector is $(\mathbf{i} - \mathbf{j} + \mathbf{k})/\sqrt{1^2 + (-1)^2 + 1^2} = (\mathbf{i} - \mathbf{j} + \mathbf{k})/\sqrt{3} = 0.58(\mathbf{i} - \mathbf{j} + \mathbf{k})$.

Problems

1.4-1 Using the dot product, find the angle θ between the vectors shown.

1.4-2 Find the dot product of $\mathbf{r}_1 = 10\mathbf{i} - 4\mathbf{j} - 8\mathbf{k}$ and $\mathbf{r}_2 = -9\mathbf{i} - 2\mathbf{j} + 7\mathbf{k}$.

1.4-3 Using the dot product, find the angle between $\mathbf{F}_1 = -20\mathbf{i} + 12\mathbf{j} + 6\mathbf{k}$ and $\mathbf{F}_2 = 10\mathbf{i} - 6\mathbf{j} - 3\mathbf{k}$.

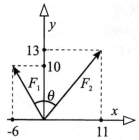

1.4-4 Find the component of $\mathbf{r}_1 = 2\mathbf{i} + 6\mathbf{j}$ in the direction of $\mathbf{r}_2 = 7\mathbf{i} + \mathbf{j}$. Also, find the component of $\mathbf{r}_2$ in the direction of $\mathbf{r}_1$.

Problem 1.4-1

1.4-5 Find and sketch the cross product $\mathbf{F}_1 \times \mathbf{F}_2$, given $\mathbf{F}_1 = -5\mathbf{i} + 3\mathbf{k}$ and $\mathbf{F}_2 = \mathbf{i} - 4\mathbf{k}$.

1.4-6 Using the cross product, find the angle between $\mathbf{F}_1 = 34\mathbf{j}$ and $\mathbf{F}_2 = 13\mathbf{i} - 56\mathbf{k}$.

1.4-7 Find both unit vectors that are perpendicular to $\mathbf{F}_1 = 4\mathbf{i} + 6\mathbf{j} + 3\mathbf{k}$ and $\mathbf{F}_2 = 10\mathbf{i} - 2\mathbf{j} + 7\mathbf{k}$.

CHAPTER 2

Forces and Moments

2.1 RESULTANT FORCES

Theory

A **resultant** force is the result of summing a collection of forces expressed as a single force. For example, assume that an object is being pulled in different directions by several cables attached to it. Each cable force is represented as a vector having components in different directions. In order to determine the resultant cable force, each of the force vectors must be divided into components and the components of the forces summed in each direction separately. The summed components form the resultant cable force vector.

In the US measurement system, the unit of force is the **pound**, abbreviated as lb. In the SI measurement system, the unit of force is the **newton**, abbreviated as N. The units of measure used in this book are listed in Table 1.1-1.

Examples

2.1-1 Three cables are attached to an object as shown. The forces that the cables exert on the object are $F_1 = 10\,\text{N}$, $F_2 = 20\,\text{N}$, and $F_3 = 50\,\text{N}$. The angles that the cable forces make relative to a

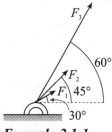

Example 2.1-1

horizontal line are 30°, 45°, and 60°, respectively. Determine the resultant cable force.

Solution

Set up the coordinates so that the horizontal components of the forces are positive to the right and the vertical components of the forces are positive upward. The horizontal components of the cable forces are then $F_{1x} = 10 \cos 30° = 8.67\,\text{N}$, $F_{2x} = 20 \cos 45° = 14.14\,\text{N}$, and $F_{3x} = 50 \cos 60° = 25\,\text{N}$. The vertical components of the cable forces are $F_{1y} = 10 \sin 30° = 5\,\text{N}$, $F_{2y} = 20 \sin 45° = 14.14\,\text{N}$, and $F_{3y} = 50 \sin 60° = 43.30\,\text{N}$. The resultant cable forces in the horizontal and vertical directions are then

$$F_x = 10 \cos 30° + 20 \cos 45° + 50 \cos 60° = 8.67 + 14.14 + 25 = 47.81\,\text{N}$$
$$F_y = 10 \sin 30° + 20 \sin 45° + 50 \sin 60° = 5 + 14.14 + 43.30 = 62.44\,\text{N}.$$

2.1-2 | Solve the previous problem again, this time employing vectors.

Solution

Begin by representing each of the cable forces as a vector. The cable force vectors are

$$\mathbf{F}_1 = 10 \cos 30°\,\mathbf{i} + 10 \sin 30°\,\mathbf{j} = 8.67\mathbf{i} + 5\mathbf{j}\,\text{N},$$
$$\mathbf{F}_2 = 20 \cos 45°\,\mathbf{i} + 20 \sin 45°\,\mathbf{j} = 14.14\mathbf{i} + 14.14\mathbf{j}\,\text{N},$$
$$\mathbf{F}_3 = 50 \cos 60°\,\mathbf{i} + 50 \sin 60°\,\mathbf{j} = 25\mathbf{i} + 43.30\mathbf{j}\,\text{N}.$$

The resultant cable force vector is

$$\mathbf{F} = \mathbf{F}_1 + \mathbf{F}_2 + \mathbf{F}_3$$
$$= (8.67\mathbf{i} + 5.\mathbf{j}) + (14.14\mathbf{i} + 14.14\mathbf{j}) + (25.\mathbf{i} + 43.30\mathbf{j}) = 47.81\mathbf{i} + 62.44\mathbf{j}\,\text{N}.$$

2.1-3 | Kent and Scott want to lift a large table and move it across the room. Kent pulls on one side with $F_1 = 23\,\text{lb}$ and Scott lifts the other side with $F_2 = 27\,\text{lb}$, as shown. Find the resultant force exerted on the table by Kent and Scott.

Solution

The two forces are expressed as vectors as

$$\mathbf{F}_1 = 23[(-3/5)\mathbf{i} + (4/5)\mathbf{j}] = -13.8\mathbf{i} + 18.4\mathbf{j}\,\text{lb},$$
$$\mathbf{F}_2 = 27(\cos 50°\,\mathbf{i} + \sin 50°\,\mathbf{j}) = 17.36\mathbf{i} + 20.68\mathbf{j}\,\text{lb}.$$

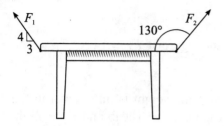

Example 2.1-3

The resultant force is then

$$\mathbf{F} = \mathbf{F}_1 + \mathbf{F}_2 = (-13.8\mathbf{i} + 18.4\mathbf{j}) + (17.36\mathbf{i} + 20.68\mathbf{j}) = 3.56\mathbf{i} + 39.08\mathbf{j}\ \text{lb}.$$

2.1-4 Determine the resultant of the three forces shown if $F_1 = F_2 = 2\,\text{N}$ and $F_3 = 3\,\text{N}$.

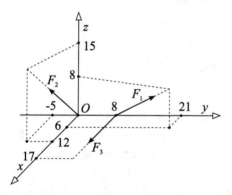

Examples 2.1-4 and 2.2-3

Solution
The three force vectors are

$$\mathbf{F}_1 = \frac{2(6\mathbf{i} + 13\mathbf{j} + 8\mathbf{k})}{\sqrt{6^2 + 13^2 + 8^2}} = 0.73\mathbf{i} + 1.58\mathbf{j} + 0.98\mathbf{k}\ \text{N},$$

$$\mathbf{F}_2 = \frac{2(12\mathbf{i} - 5\mathbf{j} + 15\mathbf{k})}{\sqrt{12^2 + 5^2 + 15^2}} = 1.21\mathbf{i} - 0.50\mathbf{j} + 1.51\mathbf{k}\ \text{N},$$

$$\mathbf{F}_3 = 3(17\mathbf{i})/17 = 3\mathbf{i}\ \text{N}.$$

Thus, the resultant force vector is

$$\mathbf{F} = \mathbf{F}_1 + \mathbf{F}_2 + \mathbf{F}_3 = (0.73\mathbf{i} + 1.58\mathbf{j} + 0.98\mathbf{k})$$
$$+ (1.21\mathbf{i} - 0.50\mathbf{j} + 1.51\mathbf{k}) + (3\mathbf{i})$$
$$= 4.94\mathbf{i} + 1.08\mathbf{j} + 2.49\mathbf{k}\ \text{N}.$$

Problems

2.1-1 Find the resultant force in the figure if $F_1 = 14\,\text{N}$, $F_2 = 9\,\text{N}$, and $\theta = 50°$.

2.1-2 An eye-bolt supports a weight; the weight is **not** shown in the figure. The tension force along each rope is $F_1 = 100\,\text{lb}$ and $F_2 = 71\,\text{lb}$. Find the resultant rope force acting on the bolt.

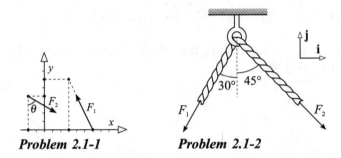

Problem 2.1-1   ***Problem 2.1-2***

2.1-3 A boat pulls two water skiers as shown. The tension force in the first rope is $F_1 = 30\,\text{lb}$ and that in the second rope is $F_2 = 25\,\text{lb}$. Find the resultant rope force exerted on the boat by the two water skiers.

2.1-4 Find the resultant force shown if $F_1 = 24\,\text{N}$ and $F_2 = 16\,\text{N}$.

2.1-5 Find the resultant of the three forces acting on the eyelet shown if $F_1 = 300\,\text{N}$, $F_2 = 500\,\text{N}$, and $F_3 = 450\,\text{N}$.

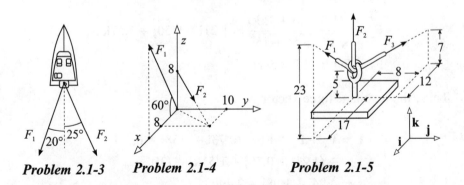

Problem 2.1-3   ***Problem 2.1-4***   ***Problem 2.1-5***

2.2 RESULTANT MOMENTS

Theory

A force can cause an object to translate as well as to rotate about a point of rotation. The force multiplied by the perpendicular distance between the line of force and the point of rotation is called the **moment of the force**, or simply the **moment**. Referring to Fig. 2.2-1, a force vector $\mathbf{F}$ acts at the point A, and the position vector from the point O to the point A is denoted by $\mathbf{r}$. The magnitude of the force $\mathbf{F}$ is denoted by F, and the perpendicular distance between the line of force and the point O is denoted by $r_\perp$.

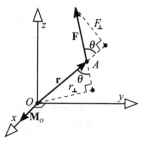

Figure 2.2-1

The **moment of the force about the point O**, denoted by $\mathbf{M}_O$, is a vector. The magnitude of the moment, denoted by M_O, is defined as $M_O = r_\perp F$. The direction of the moment is taken to be perpendicular to $\mathbf{r}$ and $\mathbf{F}$. The direction of the moment is also called the **axis of the moment**.

In Fig. 2.2-1, the angle between the position vector and the force vector is denoted by θ. The magnitude of the position vector $\mathbf{r}$ is denoted by r, and the component of the force that is perpendicular to $\mathbf{r}$ is denoted by $F_\perp = F \sin \theta$. Notice that

$$M_O = r_\perp F = r \sin \theta \, F = rF \sin \theta = rF_\perp.$$

The magnitude of the moment is calculated by any of the following three expressions:

$$
\begin{array}{ll}
M_O = r_\perp F, & \text{(2.2-1a)} \\[2mm]
M_O = rF \sin \theta, & \text{(2.2-1b)} \\[2mm]
M_O = rF_\perp, & \text{(2.2-1c)}
\end{array}
$$

where $r_\perp = r \sin \theta$ and $F_\perp = F \sin \theta$.

Recall that both the vector $\mathbf{r} \times \mathbf{F}$ and the axis of the moment are perpendicular to $\mathbf{r}$ and $\mathbf{F}$. Also, recall that $|\mathbf{r} \times \mathbf{F}| = |rF \sin \theta|$. Thus, the magnitude of $\mathbf{r} \times \mathbf{F}$ is the same as the magnitude of the moment [see Eq. (2.2-1b)], and the direction of $\mathbf{r} \times \mathbf{F}$ is the same as the direction of the moment. The moment of the force $\mathbf{F}$ about the point O is precisely the vector

$$\boxed{\mathbf{M}_O = \mathbf{r} \times \mathbf{F}.} \qquad (2.2\text{-}2)$$

The magnitude of the moment about the point O is given by any of Eqs. (2.2-1), and the moment vector about O is given by Eq. (2.2-2). Just as force vectors are added to produce a resultant force vector, moment vectors are added to produce a resultant moment vector.

Examples

2.2-1 Use Eqs. (2.2-1) to determine the magnitude of the moment about the point O produced by the force $\mathbf{F}$ shown. Assume that $r = 5\,\text{m}$ and $F = 40\,\text{N}$.

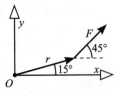

Example 2.2-1

Solution

The angles between $\mathbf{r}$ and the horizontal line and between $\mathbf{F}$ and the horizontal line are $15°$ and $45°$, respectively. So the angle from $\mathbf{r}$ to $\mathbf{F}$ is $45° - 15° = 30°$. The magnitude of the moment is now calculated in three ways:

$$M_O = r_\perp F = (5 \sin 30°)(40) = 100\,\text{N·m.}$$
$$M_O = rF \sin \theta = 5(40) \sin 30° = 100\,\text{N·m.}$$
$$M_O = rF_\perp = 5(40 \sin 30°) = 100\,\text{N·m.}$$

2.2-2 Use Eq. (2.2-2) to determine the moment about the point O produced by the force $\mathbf{F}$ in the previous problem.

Solution

First calculate the position vector $\mathbf{r}$ and the force vector $\mathbf{F}$, to get

$$\mathbf{r} = 5 \cos 15° \,\mathbf{i} + 5 \sin 15° \,\mathbf{j} = 4.830\mathbf{i} + 1.294\mathbf{j,}$$
$$\mathbf{F} = 40 \cos 45° \,\mathbf{i} + 40 \sin 45° \,\mathbf{j} = 28.284\mathbf{i} + 28.284\mathbf{j}$$

The moment vector about the point O is then

$$
\begin{aligned}
\mathbf{M}_O &= \mathbf{r} \times \mathbf{F} = (4.830\mathbf{i} + 1.294\mathbf{j}) \times (28.284\mathbf{i} + 28.284\mathbf{j}) \\
&= 28.284(4.830\mathbf{i} + 1.294\mathbf{j}) \times (\mathbf{i} + \mathbf{j}) = 28.284(4.830\mathbf{k} - 1.294\mathbf{k}) \\
&= 100\mathbf{k}\,\text{N·m}.
\end{aligned}
$$

The magnitude of the moment about O is 100 N·m and the direction is $\mathbf{k}$.

2.2-3 Determine the moment about the point O produced by the three forces shown in Example 2.1-4.

Solution
The moment produced by each of the forces is

$$
\begin{aligned}
\mathbf{M}_{O1} &= 8\mathbf{j} \times (0.73\mathbf{i} + 1.58\mathbf{j} + 0.98\mathbf{k}) = -5.84\mathbf{k} + 7.84\mathbf{i}\,\text{N·m}, \\
\mathbf{M}_{O2} &= \mathbf{0} \times (1.21\mathbf{i} - 0.50\mathbf{j} + 1.51\mathbf{k}) = \mathbf{0}\,\text{N·m}, \\
\mathbf{M}_{O3} &= 8\mathbf{j} \times (3\mathbf{i}) = -8.00\mathbf{k}\,\text{N·m}.
\end{aligned}
$$

The resultant moment is then

$$
\begin{aligned}
\mathbf{M}_O &= \mathbf{M}_{O1} + \mathbf{M}_{O2} + \mathbf{M}_{O3} = -5.84\mathbf{k} + 7.84\mathbf{i} + \mathbf{0} - 8.00\mathbf{k} \\
&= -5.84\mathbf{i} - 0.16\mathbf{k}\,\text{N·m}.
\end{aligned}
$$

Problems

2.2-1 Letting $F_1 = 300$ lb, $F_2 = 400$ lb, and $\theta = 40°$, use Eq. (2.2-1a) to find the magnitude of the resultant moment about the point O produced by the forces shown.

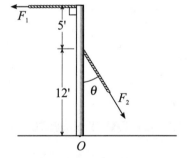

2.2.-2 Letting $F_1 = 300$ lb, $F_2 = 400$ lb, and $\theta = 40°$, use Eq. (2.2-1c) to find the magnitude of the resultant moment about the point O produced by the forces shown.

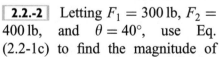

Problems 2.2-1 to 2.2-3

2.2-3 Find the resultant moment vector about the point O produced by the forces pulling on the mast. The forces are $F_1 = 300$ lb, $F_2 = 400$ lb, and $\theta = 40°$. Use Eq. (2.2-2).

2.2-4 Find the resultant moment about the point O produced by the three forces shown. Let $F_1 = 35\,\text{N}$, $F_2 = 40\,\text{N}$, and $F_3 = 65\,\text{N}$.

2.2-5 Find the resultant moment about the point A produced by the three forces shown. Let $F_1 = 35\,\text{N}$, $F_2 = 40\,\text{N}$, and $f_3 = 65\,\text{N}$.

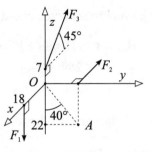

Problems 2.2-4 and 2.2-5

2.2-6 Find the resultant moment about the point O produced by the forces and moments shown. Let $F_1 = 10\,\text{N}$, $F_2 = 15\,\text{N}$, $M_1 = 100\,\text{N·m}$, and $M_2 = 75\,\text{N·m}$.

2.2-7 Find the resultant moment about the point A produced by the forces and moments shown. Let $F_1 = 10\,\text{N}$, $F_2 = 15\,\text{N}$, $M_1 = 100\,\text{N·m}$, and $M_2 = 75\,\text{N·m}$.

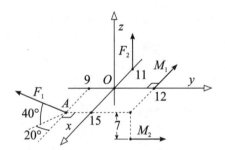

Problems 2.2-6 and 2.2-7

2.3 RESULTANT FORCES AND MOMENTS

Theory

Assume that n forces act on an object. The goal is to replace the individual forces with a resultant force and a resultant moment about a point O. The relationship between the resultant moment about the point O and the associated resultant moment about another point A is also shown.

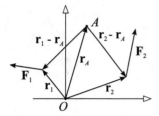

Figure 2.3-1

Let the forces be denoted by $\mathbf{F}_1, \mathbf{F}_2, \ldots, \mathbf{F}_n$ and let the positions at which the forces act be denoted by $\mathbf{r}_1, \mathbf{r}_2, \ldots, \mathbf{r}_n$. The resultant force and the resultant moment about the point O are

$$\mathbf{F} = \mathbf{F}_1 + \mathbf{F}_2 + \ldots + \mathbf{F}_n, \tag{2.3-1}$$
$$\mathbf{M}_O = \mathbf{r}_1 \times \mathbf{F}_1 + \mathbf{r}_2 \times \mathbf{F}_2 + \ldots + \mathbf{r}_n \times \mathbf{F}_n. \tag{2.3-2}$$

Using the resultant force $\mathbf{F}$ and the resultant moment $\mathbf{M}_O$ about the point O, the resultant moment $\mathbf{M}_A$ about a different point A can be calculated. From Fig. 2.3-1, observe that

$$\mathbf{M}_A = (\mathbf{r}_1 - \mathbf{r}_A) \times \mathbf{F}_1 + (\mathbf{r}_2 - \mathbf{r}_A) \times \mathbf{F}_2 + \ldots + (\mathbf{r}_n - \mathbf{r}_A) \times \mathbf{F}_n$$
$$= \mathbf{r}_1 \times \mathbf{F}_1 + \mathbf{r}_2 \times \mathbf{F}_2 + \ldots + \mathbf{r}_n \times \mathbf{F}_n - \mathbf{r}_A \times (\mathbf{F}_1 + \mathbf{F}_2 + \ldots + \mathbf{F}_n),$$

so

$$\boxed{\mathbf{M}_A = \mathbf{M}_O - \mathbf{r}_A \times \mathbf{F}.} \tag{2.3-3}$$

Equation (2.3-3) can be used to determine the resultant moment $\mathbf{M}_A$ when the resultant force $\mathbf{F}$ and the resultant moment $\mathbf{M}_O$ are given.

Examples

2.3-1 Determine the resultant force and the resultant moment about the point O produced by the forces shown in the figure overleaf, using Eqs. (2.3-1) and (2.3-2). Let $F_1 = 10\,\text{N}$, $F_2 = 20\,\text{N}$, and $F_3 = 40\,\text{N}$. Then, compute the resultant moment about the point A in two ways, using Eq. (2.3-2) and then using Eq. (2.3-3).

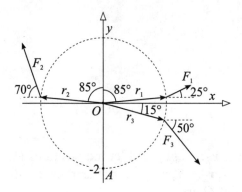

Example 2.3-1

Solution

The forces and positions shown are

$$\mathbf{F}_1 = 10\cos 25° \,\mathbf{i} + 10\sin 25° \,\mathbf{j} = 9.063\mathbf{i} + 4.226\mathbf{j} \,\text{N},$$
$$\mathbf{r}_1 = 2\cos 5° \,\mathbf{i} + 2\sin 5° \,\mathbf{j} = 1.992\mathbf{i} + 0.174\mathbf{j} \,\text{m},$$
$$\mathbf{F}_2 = 20\cos 110° \,\mathbf{i} + 20\sin 110° \,\mathbf{j} = -6.840\mathbf{i} + 18.794\mathbf{j} \,\text{N},$$
$$\mathbf{r}_2 = 2\cos 175° \,\mathbf{i} + 2\sin 175° \,\mathbf{j} = -1.992\mathbf{i} + 0.174\mathbf{j} \,\text{m},$$
$$\mathbf{F}_3 = 40\cos(-50°)\,\mathbf{i} + 40\sin(-50°)\,\mathbf{j} = 25.71\mathbf{i} - 30.64\mathbf{j} \,\text{N}$$
$$\mathbf{r}_3 = 2\cos(-15°)\,\mathbf{i} + 2\sin(-15°)\,\mathbf{j} = 1.932\mathbf{i} - 0.518\mathbf{j} \,\text{m}.$$

Using Eqs. (2.3-1) and (2.3-2), the resultant force and the resultant moment about the point O become

$$\mathbf{F} = \mathbf{F}_1 + \mathbf{F}_2 + \mathbf{F}_3 = (9.063\mathbf{i} + 4.226\mathbf{j}) + (-6.840\mathbf{i} + 18.794\mathbf{j})$$
$$+ (25.71\mathbf{i} - 30.64\mathbf{j})$$
$$= 27.93\mathbf{i} - 7.58\mathbf{j} \,\text{N}.$$
$$\mathbf{M}_O = \mathbf{r}_1 \times \mathbf{F}_1 + \mathbf{r}_2 \times \mathbf{F}_2 + \mathbf{r}_3 \times \mathbf{F}_3$$
$$= (1.992\mathbf{i} + 0.174\mathbf{j}) \times (9.063\mathbf{i} + 4.226\mathbf{j})$$
$$+ (-1.992\mathbf{i} + 0.174\mathbf{j}) \times (-6.840\mathbf{i} + 18.794\mathbf{j})$$
$$+ (1.932\mathbf{i} - 0.518\mathbf{j}) \times (25.71\mathbf{i} - 30.64\mathbf{j})$$
$$= [1.992(4.226) - 0.174(9.063) - 1.992(18.794)$$
$$- 0.174(-6.840) + 1.932(-30.64) + 0.518(25.71)]\mathbf{k}$$
$$= -75.28\mathbf{k} \,\text{N·m}.$$

The resultant moment about the point A is now found in two ways, using Eq. (2.3-2) and then using Eq. (2.3-4). Using Eq. (2.3-2),

$$
\begin{aligned}
\mathbf{M}_A &= \mathbf{r}_{1/A} \times \mathbf{F}_1 + \mathbf{r}_{2/A} \times \mathbf{F}_2 + \mathbf{r}_{3/A} \times \mathbf{F}_3 \\
&= (1.992\mathbf{i} + 0.174\mathbf{j}) \times (9.063\mathbf{i} + 4.226\mathbf{j}) \\
&\quad + (-1.992\mathbf{i} + 0.174\mathbf{j}) \times (-6.840\mathbf{i} + 18.794\mathbf{j}) \\
&\quad + (1.932\mathbf{i} - 0.518\mathbf{j}) \times (25.71\mathbf{i} - 30.64\mathbf{j}) \\
&= [1.743(4.226) - 2.174(9.063) - 1.992(18.794) \\
&\quad - 2.174(-6.840) + 1.932(-30.64) - 1.482(25.71)]\mathbf{k} \\
&= -131\mathbf{k}\,\text{N·m.}
\end{aligned}
$$

Using Eq. (2.3-4),

$$
\begin{aligned}
\mathbf{M}_A &= \mathbf{M}_O - \mathbf{r}_A \times \mathbf{F} = -75.28\mathbf{k} - (-2\mathbf{j}) \times (27.93\mathbf{i} - 7.58\mathbf{j}) \\
&= [-75.28 - 2(27.93)]\mathbf{k} = -131\mathbf{k}\,\text{N·m.}
\end{aligned}
$$

Problems

2.3-1 Letting $F_1 = 12$ lb, $F_2 = 9$ lb, and $F_3 = 8$ lb, as shown, find the resultant force, the resultant moment about the point O, and the resultant moment about the point A. Distance is given in feet.

2.3-2 Letting $F_1 = 5$ lb, $F_2 = 7$ lb, $F_3 = 14$ lb, and $F_4 = 10$ lb, as shown, find the resultant force, the resultant moment about the point O, and the resultant moment about the point A. Distance is given in feet.

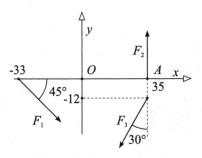

Problem 2.3-1

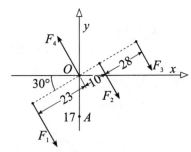

Problem 2.3-2

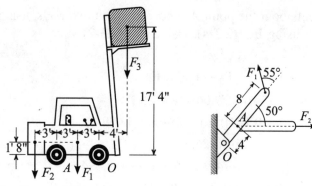

Problem 2.3-3   **Problem 2.3-4**

2.3-3 A fork lift is subjected to the three weight forces shown. Find the resultant weight force and the resultant moment about the point O if $F_1 = 500$ lb, $F_2 = 750$ lb, and $F_3 = 200$ lb.

2.3-4 Find the resultant external force, the resultant external moment about the point O, and the resultant external moment about the point A exerted on the object by the external forces shown. Assume that $F_1 = 22$ N and $F_2 = 40$ N. Distance is given in meters.

2.3-5 Find the resultant force, the resultant moment about the point O, and the resultant moment about the point A. Assume that $F_1 = 5$ N, $F_2 = 75$ N, and $F_3 = 10.5$ N. Distance is given in meters.

2.3-6 Find the resultant force, the resultant moment about the point O, and the resultant moment about the point A. Let $F_1 = F_2 = F_3 = 100$ lb. Distance is given in feet.

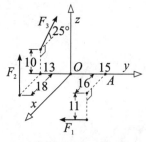

Problem 2.3-5

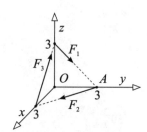

Problem 2.3-6

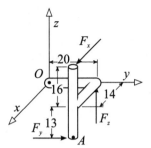

Problem 2.3-7

2.3-7 Find the resultant external force, the resultant external moment about the point O, and the resultant external moment about the point A exerted on the object by the external forces shown. Assume that $F_x = 25\,\mathrm{N}$, $F_y = 30\,\mathrm{N}$, and $F_z = 35\,\mathrm{N}$. Distance is given in meters.

2.3-8 Find the resultant force, the resultant moment about the point O, and the resultant moment about the point A exerted on the object by the three forces shown. Assume that $F_1 = 125\,\mathrm{lb}$, $F_2 = 175\,\mathrm{lb}$, $F_3 = 200\,\mathrm{lb}$, $\theta_1 = 75°$, and $\theta_2 = 20°$.

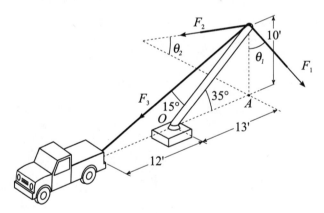

Problem 2.3-8

CHAPTER 3

Free Body Diagrams

3.1 STATIC EQUILIBRIUM

Theory

The field of **Newtonian mechanics** is governed by Newton's First, Second, and Third Laws and by Newton's Law of Gravitation. The field of statics is based on Newton's First and Third Laws, and Newton's Law of Gravitation. The **static equilibrium equations** are governed by Newton's First and Third Laws. The Second Law is considered in Chapters 6–9, and governs the equations associated with dynamic systems.

Newton's First Law states:

A particle moves with constant speed and direction when the resultant force is zero.

Newton's Second Law states:

*A particle's acceleration is linearly proportional to the resultant force. The proportionality constant is called the particle's **mass**.*

Newton's Third Law states:

Two particles interact with each other through forces that are equal in magnitude, opposite in direction, and colinear.

Newton's Law of Gravitation states:

Two particles are attracted to each other through forces that are in linear proportion to their masses and in inverse proportion to the square of the distance between them.

These laws describe basic relationships between forces acting on particles and their subsequent motion. (See Figs. 3.1-1 and 3.1-2.)

In statics problems, the constant speed referred to in Newton's First Law is often taken to be zero. Newton's First Law is expressed mathematically as

$$\mathbf{F} = \mathbf{0},$$

(3.1-1)

where $\mathbf{F}$ denotes the resultant force acting on the particle.

Newton's Law of Gravitation is expressed mathematically as (see Fig. 3.1-2)

$$\mathbf{f}_{12} = G\frac{m_1 m_2}{r_{12}^2}\mathbf{n}_{12},$$

(3.1-2)

where $\mathbf{f}_{12}$ denotes the gravitational force on particle 1 due to particle 2, m_1 is the mass of particle 1, m_2 is the mass of particle 2, $\mathbf{n}_{12}$ is the

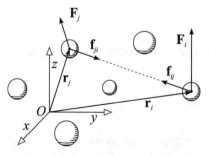

Figure 3.1-1

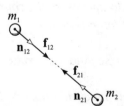

Figure 3.1-2

unit vector directed from particle 1 to particle 2, and $G = 6.67 \times 10^{-11} \mathrm{m}^3/(\mathrm{s}^2 \cdot \mathrm{kg})$ is the gravitational constant. If particle 1 is the Earth, and particle 2 is a much smaller particle *on* the Earth, then $m_1 = M_E = 5.97 \times 10^{24}$ kg is the mass of the Earth, $m_2 = m$ is the mass of the particle *on* the Earth, and $r_{12} = R_E = 6370$ km is the radius of the Earth. In this case, Eq. (3.1-2) reduces to

$$\boxed{\mathbf{f} = mg\mathbf{n},} \tag{3.1-3a}$$

where

$$g = \frac{GM_E}{R_E^2}. \tag{3.1-3b}$$

A particle on or near the Earth is subjected to a force in linear proportion to its mass and directed toward the center of the Earth. From Eq. (3.1-3b), the proportionality constant is approximately $g = 9.81 \mathrm{m/s}^2 = 32.2 \mathrm{ft/s}^2$. Also, referring to the inside jacket of the book, notice that the unit of measure for mass in the US system is **slug**, and the unit of measure for mass in the SI system is **kilogram**, abbreviated kg.

Now consider a **system** of particles. Letting particles be distinguished from one another by subscripts, we pick out the ith particle and the jth particle and examine the interaction forces between them using Newton's Third Law. Denoting the force that the jth particle exerts on the ith particle by $\mathbf{f}_{ij}$ and denoting the force that the ith particle exerts on the jth particle by $\mathbf{f}_{ji}$, Newton's Third Law is expressed mathematically as (see Fig. 3.1-1)

$$\boxed{\mathbf{f}_{ij} = -\mathbf{f}_{ji}, \qquad \mathbf{r}_i \times \mathbf{f}_{ij} = -\mathbf{r}_j \times \mathbf{f}_{ji},} \tag{3.1-4a,b}$$

where $\mathbf{r}_i$ is the position of the ith particle and $\mathbf{r}_j$ is the position of the jth particle. Equation (3.1-4a) is the mathematical statement expressing that interacting forces are equal in magnitude and opposite in direction. Equation (3.1-4b) is the mathematical statement expressing that the associated interacting moments are also equal in magnitude and opposite in direction. Although Newton's Third Law does not explicitly state that interacting moments are equal in magnitude and opposite in direction, this follows directly from the fact that the forces are equal in magnitude, opposite in direction, and colinear (see Example 3.1-6).

Let's now show a simple but far-reaching result that follows from Newton's First and Third Laws. Consider the **system** of n particles shown in Fig. 3.1-1. From Newton's First Law, the resultant force acting on the ith particle is given by

$$\mathbf{F}_i + \mathbf{f}_{i1} + \mathbf{f}_{i2} + \ldots + \mathbf{f}_{in} = \mathbf{0}$$

for $i = 1, 2, \ldots, n$, where $\mathbf{F}_i$ denotes the resultant **external** force acting on the ith particle. For convenience, let's now use the summation notation. The equation above is rewritten as

$$\mathbf{F}_i + \sum_{j=1}^{n} \mathbf{f}_{ij} = \mathbf{0}. \qquad (3.1\text{-}5)$$

Next, we sum Eqn. (3.1-5) for each i and get

$$\sum_{i=1}^{n}\left(\mathbf{F}_i + \sum_{j=1}^{n} \mathbf{f}_{ij}\right) = \left(\sum_{i=1}^{n}\mathbf{F}_i\right) + \left(\sum_{i=1}^{n}\sum_{j=1}^{n}\mathbf{f}_{ij}\right) = \mathbf{0}. \qquad (3.1\text{-}6)$$

The first term on the right-hand side is the sum of the external forces acting on the entire system, that is $\mathbf{F} = \sum_{i=1}^{n}\mathbf{F}_i$, and the second term is the sum of the internal interacting forces in the entire system, which, by Eq. (3.1-4a), is equal to zero. Equation (3.1-6) therefore reduces to

$$\boxed{\mathbf{F} = \mathbf{0}.} \qquad (3.1\text{-}7)$$

Equation (3.1-7) mathematically states that **the resultant external force acting on a system of particles that is in static equilibrium is zero.** This equation is the same as Eq. (3.1-1) for a single particle. Since in Eq. (3.1-7) the internal forces canceled out, Eq. (3.1-7) indicates that the behavior of macroscopic bodies can be analyzed without regard to the internal forces acting inside the bodies.

The resultant moment acting on the ith particle is now examined. Referring to Fig. 3.1-1, the resultant moment about the point O acting on the ith particle is

$$\mathbf{r}_i \times \left(\mathbf{F}_i + \sum_{j=1}^{n} \mathbf{f}_{ij} \right) = \mathbf{0}. \tag{3.1-8}$$

Summing the resultant moments in Eq. (3.1-8) yields

$$\sum_{i=1}^{n} \left[\mathbf{r}_i \times \left(\mathbf{F}_i + \sum_{j=1}^{n} \mathbf{f}_{ij} \right) \right] = \left(\sum_{i=1}^{n} \mathbf{r}_i \times \mathbf{F}_i \right) + \left(\sum_{i=1}^{n} \sum_{j=1}^{n} \mathbf{r}_i \times \mathbf{f}_{ij} \right) = \mathbf{0}.$$
$$\tag{3.1-9}$$

The first term on the right-hand side represents the sum of the external moments acting on the system, that is $\mathbf{M}_O = \sum_{i=1}^{n} \mathbf{r}_i \times \mathbf{F}_i$, and the second term is the sum of the internal interacting moments in the entire system, which, by Eq. (3.1-4b), is equal to zero. Equation (3.1-9) therefore reduces to

$$\boxed{\mathbf{M}_O = \mathbf{0}.} \tag{3.1-10}$$

Equation (3.1-10) states that **the sum of the external moments acting on a system of particles that is in static equilibrium is equal to zero**. Again, the internal moments cancel out, and hence don't need to be considered in solving statics problems.

In summary, the state of static equilibrium of a **body** is governed by Eqs. (3.1-7) and (3.1-10), rewritten as

$$\boxed{\mathbf{F} = \mathbf{0} \quad \text{and} \quad \mathbf{M}_O = \mathbf{0}} \tag{3.1-11a,b}$$

Equations (3.1-11a,b) mathematically state that **the resultant external force acting on a body in static equilibrium is zero and that the resultant external moment acting on a body in static equilibrium is zero**. Equations (3.1-11a,b) are the equations that are used to solve statics problems.

THE TWO-FORCE SYSTEM AND THE THREE-FORCE SYSTEM

When a body is subjected to only two external forces, in order for the body to be in static equilibrium, based on Newton's Laws, the two forces

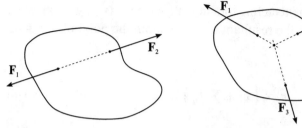

Figure 3.1-3   **Figure 3.1-4**

must line up (be collinear) and be of equal magnitude and opposite in direction. (Can the reader see why this is true?) This situation is shown in Fig. 3.1-3.

When a body is subjected to only three external forces, in order for the body to be in static equilibrium, based on Newton's Laws, the lines of force must intersect at a single point. Notice that if this were not the case, the resultant moment about the intersection point of two of the lines of force would not be equal to zero. (Can the reader see why this is true?) This situation is shown in Fig. 3.1-4.

Examples

3.1-1 A body is subjected to the three forces shown. Find the angles θ_1 and θ_2 that keep the body in static equilibrium. Assume that $F_1 = 15.00$ lb, $F_2 = 8.66$ lb, and $F_3 = 17.32$ lb.

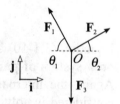

Example 3.1-1

Solution

The body is in static equilibrium if the resultant force is zero and the resultant moment about the point O (or about any other point) is zero. The resultant moment about point O is clearly zero in this example, since $r_1 = r_2 = r_3 = 0$. The resultant force is

$$0 = \mathbf{F} = \mathbf{F}_1 + \mathbf{F}_2 + \mathbf{F}_3 = 15(-\cos\theta_1\,\mathbf{i} + \sin\theta_1\,\mathbf{j})$$
$$+ 8.66(\cos\theta_2\,\mathbf{i} + \sin\theta_2\,\mathbf{j}) - 17.32\mathbf{j}.$$

Therefore,

$$0 = -15\cos\theta_1 + 8.66\cos\theta_2,$$
$$0 = 15\sin\theta_1 + 8.66\sin\theta_2 - 17.32,$$

which are two trigonometric equations expressed in terms of the two unknown angles. Usually trigonometric equations cannot be solved analytically (by hand), in which case one resorts to a numerical method (solved by computer). But in this case, the problem can be solved analytically using an appropriate substitution. Letting $x = \cos\theta_1$ and $y = \cos\theta_2$, and using $\sin\theta = (1 - \cos^2\theta)^{1/2}$, these two equations can be rewritten as

$$0 = -15x + 8.66y, \qquad 17.32 = 15(1 - x^2)^{1/2} + 8.66(1 - y^2)^{1/2}.$$

Substituting for x from the first equation into the second yields

$$17.32 = 15(1 - 0.33y^2)^{1/2} + 8.66(1 - y^2)^{1/2}.$$

This equation is manipulated to get a quadratic equation in y^2. Solving this quadratic equation yields y and thus x, which then yield the angles $\theta_1 = 60.00°$ and $\theta_2 = 30.00°$.

3.1-2 A body is subjected to the three forces shown. Let the angles be given by $\theta_1 = \theta_2 = 30°$ and let $F_1 = 100\,\text{N}$. Determine the location d of the force F_3, and the forces F_2 and F_3 that keep the body in static equilibrium.

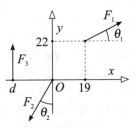

Example 3.1-2

Solution

The resultant force and the resultant moment about the point O are zero. Thus,

$$\sum F_x = 0 = 100\cos 30° - F_2\sin 30°,$$
$$\sum F_y = 0 = 100\sin 30° - F_2\cos 30° + F_3,$$
$$\sum M = 0 = -22(100)\cos 30° + 19(100)\sin 30° - F_3 d.$$

The first equation is solved to determine F_2, then F_2 is substituted into the second equation to yield F_3, and then F_3 is substituted into the third equation to yield d. Thus $F_2 = 173.21\,\text{N}$, $F_3 = 100.00\,\text{N}$, $d = -9.55\,\text{m}$. Notice that the third force is located to the right of the y axis.

3.1-3 Determine the force F and the angle θ that keeps the pulley system shown in the figure overleaf in static equilibrium.

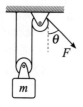

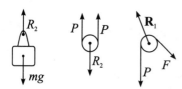

Example 3.1-3

Solution

In order to solve this problem, the forces acting on the right pulley, the left pulley, and the block will need to be diagrammed. The diagram of forces acting on an object is called a **free body diagram**. First consider the free body diagram of the right pulley. As shown, the right pulley is subjected to the applied force F, a rope force P, and a reaction force $\mathbf{R}_1$ at the center pin. The vector $\mathbf{R}_1$, is bold because we do not immediately know its direction. The resultant moment about the center pin is zero only if the rope force P is equal to the applied force F. Next, consider the free body diagram of the left pulley. The resultant moment acting about the center pin of the left pulley is equal to zero only if the two rope forces acting on the left pulley are each equal to P. The resultant force acting on the left pulley is equal to zero only if the reaction force is equal to $2P$; that is, $R_2 = 2P$. Next, consider the free body diagram for the block. The block is subjected to the reaction force R_2 and the weight force mg. The resultant force on the block is equal to zero only if the reaction force is equal to the weight force mg; thus, $R_2 = mg = 2P = 2F$. Thus, $F = \frac{1}{2}mg$, and the angle θ is free to assume any value. Pulleys like this can be used to reduce the force that is required to lift an object.

3.1-4 A human force $F = 120\,\text{lb}$ and a vertical reaction $R_{1y} = 1500\,\text{lb}$ at a truck wheel's center pin are

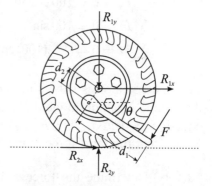

Example 3.1-4

applied, as shown. Determine the horizontal component of the reaction at the wheel center and the reactions at the point of contact with the road. Assume that the wheel radius is 2 ft, and that $d_1 = 3$ ft, $d_2 = 0.25$ ft, and $\theta = 30°$.

Solution
As shown in the free body diagram of the wheel, the wheel is subject to horizontal and vertical reaction components at the center pin, horizontal and vertical reaction components at the road contact point, and an applied force. The wheel is in static equilibrium provided the resultant force is zero and the resultant moment is zero; that is,

$$\sum F_x = 0 = R_{1x} + R_{2x} - 120 \sin 30°,$$
$$\sum F_y = 0 = -1500 + R_{2y} - 120 \cos 30°,$$
$$\sum M = 0 = 2R_{2x} - 3(120).$$

Thus, the reactions are

$$R_{1x} = -120 \text{ lb}, \qquad R_{2y} = 1604 \text{ lb}, \qquad R_{2x} = 180 \text{ lb}.$$

Notice that d_2 does not affect the answers, because it did not enter into the calculations.

3.1-5 Three ropes support an "I" beam, as shown. Determine the tension forces in each of the ropes. Assume that the "I" beam weighs 175 N.

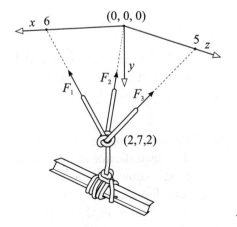

Example 3.1-5

Solution

The knot in the middle of the ropes is subjected to the three rope forces and the gravitational force. Thus,

$$0 = \mathbf{F}_1 + \mathbf{F}_2 + \mathbf{F}_3 + \mathbf{F}_4,$$

in which

$$\mathbf{F}_1 = \frac{F_1(4\mathbf{i} - 7\mathbf{j} - 2\mathbf{k})}{\sqrt{4^2 + (-7)^2 + (-2)^2}} = \frac{F_1(4\mathbf{i} - 7\mathbf{j} - 2\mathbf{k})}{8.307}\,\text{N},$$

$$\mathbf{F}_2 = \frac{F_2(-2\mathbf{i} - 7\mathbf{j} - 2\mathbf{k})}{\sqrt{(-2)^2 + (-7)^2 + (-2)^2}} = \frac{F_2(-2\mathbf{i} - 7\mathbf{j} - 2\mathbf{k})}{7.550}\,\text{N},$$

$$\mathbf{F}_3 = \frac{F_3(-2\mathbf{i} - 7\mathbf{j} + 3\mathbf{k})}{\sqrt{(-2)^2 + (-7)^2 + 3^2}} = \frac{F_3(-2\mathbf{i} - 7\mathbf{j} + 3\mathbf{k})}{7.874}\,\text{N},$$

$$\mathbf{F}_4 = 175\mathbf{j}\,\text{N}.$$

Thus,

$$0 = 0.482F_1 - 0.265F_2 - 0.254F_3,$$
$$0 = -0.843F_1 - 0.927F_2 - 0.889F_3 + 175,$$
$$0 = -0.241F_1 - 0.265F_2 + 0.381F_3.$$

Solving this set of three linear algebraic equations yields

$$F_1 = 69.2\,\text{N}, \qquad F_2 = 50.3\,\text{N}, \qquad F_3 = 78.8\,\text{N}.$$

3.1-6 Show that the forces between two bodies produce equal and opposite moments provided the two forces are equal and opposite and colinear. Recall that this result was used in Eq. (3.1-4b).

Solution

Denote the two forces by $\mathbf{f}_1$ and $\mathbf{f}_2$, and the positions of the forces by $\mathbf{r}_1$ and $\mathbf{r}_2$, respectively. We know that $\mathbf{f}_2 = -\mathbf{f}_1$, since the forces are equal and opposite. Since the forces line up, we also know that the position vector $\mathbf{r}_2$ lies on the line of action of the forces, so $\mathbf{r}_2$ is of the form $\mathbf{r}_2 = \mathbf{r}_1 + a\mathbf{f}_1$ for some a. (See Example 1.3-6.) Therefore,

$\mathbf{r}_2 \times \mathbf{f}_2 = (\mathbf{r}_1 + a\mathbf{f}_1) \times (-\mathbf{f}_1) = -\mathbf{r}_1 \times \mathbf{f}_1 - a\mathbf{f}_1 \times \mathbf{f}_1 = -\mathbf{r}_1 \times \mathbf{f}_1$. Thus, the moments produced by these forces are equal and opposite.

Problems

3.1-1 A body is subjected to four forces, as shown. Let $F_1 = 62\,\text{N}$, $\theta_1 = 50°$, and $F_2 = 35\,\text{N}$. Find the forces F_3 and F_4 and the angle θ_2 that keeps the body in static equilibrium.

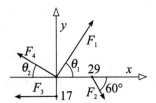

Problem 3.1-1

3.1-2 A 20 lb traffic light is suspended from two poles, as shown. Determine the force F exerted in the cable if $\theta = 5°$.

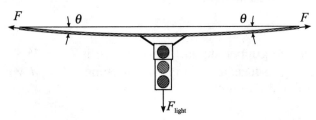

Problem 3.1-2

3.1-3 A block is supported by a rope, as shown. Letting $F = 80\,\text{N}$ and $m = 10\,\text{kg}$, determine the bend angle θ in the rope.

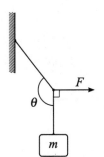

Problem 3.1-3

3.1-4 A block is supported by a rope, as shown. Letting $F_1 = F_2 = 80\,\text{N}$ and $m = 20\,\text{kg}$, determine the bend angles θ_1 and θ_2 in the rope.

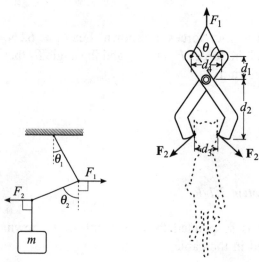

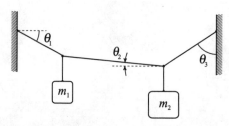

Problem 3.1-4   **Problem 3.1-5**

3.1-5 The tension in the gripper support cable shown is $F_1 = 45\,\text{N}$. Determine the resulting squeezing force F_2, letting $d_1 = d_3 = d_4 = 0.2\,\text{m}$, $d_2 = 0.6\,\text{m}$, and $\theta = 60°$.

3.1-6 The angles of the rope were determined to be $\theta_1 = 30°$, $\theta_2 = 5°$, and $\theta_3 = 45°$, as shown. Determine the tension in the rope to the left and right of each weight. Then determine m_2. Let $m_1 = 35\,\text{kg}$ and assume that m_2 is unknown.

Problem 3.1-6

3.1-7 Scott and Kent are carrying Scott's old computer up the stairs. Kent is pulling a rope at an angle $\theta = 60°$. Find the force exerted on the platform by Scott. Assume that Scott does not exert a moment on the platform and that the system is in static equilibrium. How much greater is the magnitude of the force exerted on the platform by Scott than by Kent? Let $a = 2$ ft and $W = 100$ lb.

3.2 REACTIONS

Theory

Among the external forces that act on a body, some are specified in advance and others need to be determined. The external forces acting on a body that need to be determined can be regarded as **reactions**. The usual situation consists of a body subjected to known external forces. The body is also connected in some manner to other bodies. Reaction forces and reaction moments develop between the bodies, depending on the connections between the bodies. These reactions are typically unknown.

Examples

3.2-1 The system shown is subjected to two applied forces and a weight force. Reaction forces act at the pin connection and the roller connection. Determine the reaction forces if $F_1 = 20\,\text{N}$, $W = 30\,\text{N}$, $F_2 = 16\,\text{N}$, and $\theta = 23°$.

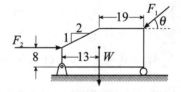

Solution
The pin connection shown is a connection that can carry forces but no moments. The roller connection shown can only carry a force perpendicular to the surface. Thus, the free body diagram of the body consists of the forces shown in the figure,

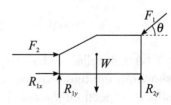

Example 3.2-1

in addition to two reaction forces at the pin connection, R_{1x} and R_{1y}, and a reaction force R_{2y} at the roller connection. The resultant forces in the horizontal and vertical directions are zero and the resultant moment about the pin connection is zero, so

$$\sum F_x = 0 = -(20)\cos 23° + 16 + R_{1x},$$
$$\sum F_y = 0 = -(20)\sin 23° - 30 + R_{1y} + R_{2y},$$
$$\sum M = 0 = (8 + 13/2)(20)\cos 23° - (13 + 19)(20)\sin 23°$$
$$- 8(16) - 13(30) + (13 + 19)R_{2y}.$$

The reactions are

$$R_{1x} = 2.41\,\text{N}, \qquad R_{1y} = 22.15\,\text{N}, \qquad R_{2y} = 15.66\,\text{N}.$$

3.2-2 A bent bar is subjected to the applied forces shown. Neglecting the weight of the bar, determine the wall reactions. Let $F_1 = 200\,\text{lb}$, $F_2 = 50\,\text{lb}$, $d_1 = 15\,\text{ft}$, $d_2 = 12\,\text{ft}$, and $\theta = 30°$.

Solution
The bent bar is fixed to the wall. The wall connection can carry reaction forces, R_x and R_y, and a reaction moment M. The resultant of the forces and the moment about the fixed connection are zero, so

$$\sum F_x = 0 = R_x + 200\sin 30°,$$
$$\sum F_y = 0 = R_y + 200\cos 30° - 50,$$
$$\sum M = 0 = 15(200)\cos 30° - (15 + 12)50 + M.$$

Example 3.2-2

Thus, the reactions are given by

$$R_x = -100.0\,\text{lb}, \qquad R_y = -123.2\,\text{lb}, \qquad M = -1248\,\text{lb·ft}.$$

3.2-3 A roller–slider mechanism is subjected to two applied forces as shown in the figure overleaf. Determine the reactions R_{1x} and M in the slider connection, the roller pin reactions R_{2x} and R_{2y}, and the inclined surface reaction N. Let $F_1 = 15\,\text{N}$, $F_2 = 25\,\text{N}$, $d_1 = 0.5\,\text{m}$, $d_2 = 0.2\,\text{m}$, and $\theta = 30°$. Neglect the weight of the roller and the slider

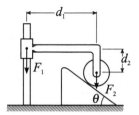

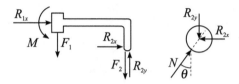

Example 3.2-3

Solution

The roller–slider mechanism is composed of two parts: a slider and a roller. The forces acting on each are diagrammed. The forces acting on the slider consist of the applied forces F_1 and F_2, and the reactions at the slider connection and the pin connection. The horizontal reaction force R_{1x} and a reaction moment M act at the slider connection, and the two reaction forces R_{2x} and R_{2y} act at the pin connection. The forces acting on the roller consist of the pin reaction forces R_{2x} and R_{2y} and the contact force with the inclined surface, N. The contact force N is directed normal to the inclined surface. Looking at the slider part, the resultant forces and the resultant moment about the slider connection are zero, so

$$\sum F_x = 0 = R_{1x} + R_{2x},$$
$$\sum F_y = 0 = -15 - 25 + R_{2y},$$
$$\sum M = 0 = 0.5(R_{2y} - 25) + 0.2R_{2x} + M.$$

Looking at the roller part, the resultant forces are zero, so

$$\sum F_x = 0 = N \sin 30° - R_{2x},$$
$$\sum F_y = 0 = N \cos 30° - R_{2y}.$$

These five linear algebraic equations are expressed in terms of the five unknowns: R_{1x}, R_{2x}, R_{2y}, M, and N. The reactions are given by

$$R_{1x} = -23.09\,\text{N}, \quad R_{2x} = 23.09\,\text{N}, \quad R_{2y} = 40.00\,\text{N},$$
$$M = -12.12\,\text{N·m}, \quad N = 46.19\,\text{N}.$$

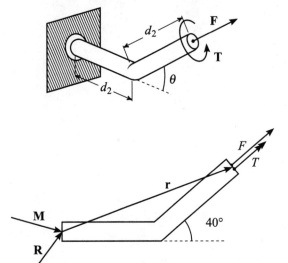

Example 3.2-4

3.2-4 A bent beam is subjected to a tip force and tip moment (an applied torque T) as shown. Determine the reactions at the base of the beam. Let $F = 125\,\text{N}$, the torque $T = 2350\,\text{N·m}$, $d_2 = 5\,\text{m}$, and $\theta = 40°$.

Solution

The beam is fixed to the wall at its base, so the wall connection can carry all forces and all moments. This problem will be solved using vectors, with the force reaction and the moment reaction at the base being represented by the vectors **R** and **M**, respectively. The position vector of the tip relative to the base is denoted by **r**. The resultant force is zero and the resultant moment acting about the base point is zero, so

$$\sum \mathbf{F} = 0 = \mathbf{R} + \mathbf{F},$$
$$\sum \mathbf{M} = 0 = \mathbf{M} + \mathbf{T} + \mathbf{r} \times \mathbf{F},$$

in which

$$\mathbf{F} = 125(\cos 40°\,\mathbf{i} + \sin 40°\mathbf{j}),$$
$$\mathbf{T} = 2350(\cos 40°\,\mathbf{i} + \sin 40°\mathbf{j}),$$
$$\mathbf{r} = (5 + 5\cos 40°\,\mathbf{i} + 5\sin 40°\mathbf{j}).$$

Substituting these expressions into the three equations yields

$$\mathbf{R} = -\mathbf{F} = -125(\cos 40°\mathbf{i} + \sin 40°\mathbf{j}) = -95.75\mathbf{i} - 80.35\mathbf{j},$$
$$\mathbf{M} = -\mathbf{T} - \mathbf{r} \times \mathbf{F} = -2350(\cos 40°\mathbf{i} + \sin 40°\mathbf{j})$$
$$- [(5 + 5\cos 40°)\mathbf{i} + 5\sin 40°\mathbf{j}] \times 125(\cos 40°\mathbf{i} + \sin 40°\mathbf{j})$$
$$= -1800.2\mathbf{i} - 1510.6\mathbf{j} - 401.70\mathbf{k}.$$

3.2-5 A small ring part is subjected to two applied forces, as shown. Determine the reactions at the base. Let $F_1 = 0.75$ lb, $F_2 = 0.45$ lb, $d = 0.075$ ft, $r = 0.055$ ft, $\theta_1 = 50°$, and $\theta_2 = 210°$.

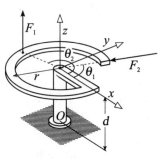

Example 3.2-5

Solution
The base is fixed, and hence can carry reaction forces $\mathbf{R}$ and reaction moments $\mathbf{M}$ in all three directions. The resultant force acting on the ring part is zero and the resultant moment about the base point acting on the ring part is zero, so

$$\sum \mathbf{F} = 0 = \mathbf{R} + 0.75\mathbf{k} - 0.45(\cos 50°\mathbf{i} + \sin 50°\mathbf{j}),$$
$$\sum \mathbf{M} = 0 = \mathbf{M} + [0.055(\cos 210°\mathbf{i} + \sin 210°\mathbf{j}) + 0.075\mathbf{k}] \times 0.75\mathbf{k}$$
$$+ [0.055(\cos 50°\mathbf{i} + \sin 50°\mathbf{j}) + 0.075\mathbf{k}]$$
$$\times [-0.45(\cos 50°\mathbf{i} + \sin 50°\mathbf{j})]$$

Thus, the force and moment reactions are

$$\mathbf{R} = 0.29\mathbf{i} + 0.345\mathbf{j} - 0.75\mathbf{k} \text{ lb}, \qquad \mathbf{M} = 0.0192\mathbf{i} + 0.0345\mathbf{j} \text{ lb·ft.}$$

Problems

3.2-1 Three applied forces act on the system shown in the figure overleaf. Determine the reactions at the pin connection and the roller connection. Let $F_1 = 2000$ lb, $F_2 = 3000$ lb, $F_3 = 1600$ lb, and $\theta = 45°$.

3.2-2 A force F acts on a uniform pendulum as shown in the figure overleaf. Find the reaction forces at the pin connection and the angle θ, letting $F = 100$ N, $d = 1.6$ m, and $W = 300$ N. Notice that this is a three-force system.

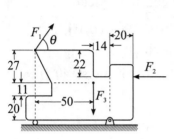

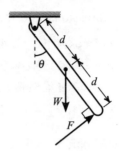

Problem 3.2-1

Problem 3.2-2

3.2-3 A uniform ring part of radius r is subject to a force F as shown. Determine the reactions at the pin connection and F_2. Let $F = 150\,\text{N}$, $r = 2.5\,\text{m}$, $\theta_1 = 15°$, and $\theta_2 = 60°$.

3.2-4 Determine the reactions at the base of the bent bar shown. Let $F_1 = 120\,\text{lb}$, $F_2 = 75\,\text{lb}$, $T = 100\,\text{lb·ft}$, $d_1 = 0.8\,\text{m}$, $d_2 = 0.6\,\text{m}$, and $\theta = 35°$.

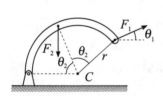

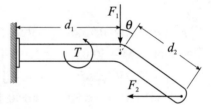

Problem 3.2-3

Problem 3.2-4

3.2-5 A bent bar is subject to a tip force and a tip moment as shown. Determine the base reactions if $\mathbf{F} = 70\text{k}\,\text{N}$, $T = 150\,\text{N·m}$, and $\theta = 30°$.

3.2-6 A hinged door is subject to an applied automatic door force at the top as shown. A pin is removed from the bottom hinge, making it useless. Reactions build up at the hinge and the door-stop on the floor. Determine the reactions if $F = 75\,\text{lb}$, $W = 60\,\text{lb}$, $h = 7\,\text{ft}$, $d_1 = 4\,\text{ft}$, $d_2 = 3\,\text{ft}$, $d_3 = 5\,\text{ft}$, and $d_4 = 1\,\text{ft}$.

3.2-7 An applied force acts on a crank as shown. Determine the reactions at the point O to keep the crank in static equilibrium. Let $F = 85\,\text{N}$, $d_1 = d_2 = 1.2\,\text{m}$, and $d_3 = 0.8\,\text{m}$.

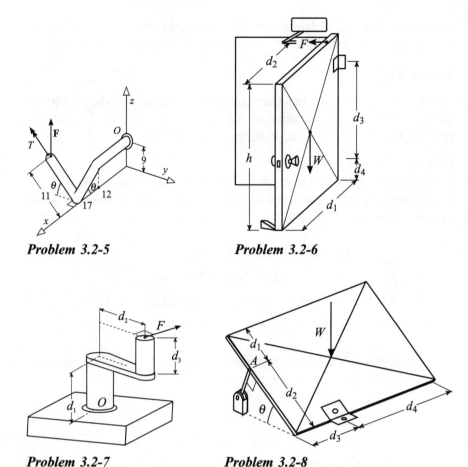

Problem 3.2-5

Problem 3.2-6

Problem 3.2-7

Problem 3.2-8

3.2-8 A basement door is propped up at the angle shown. Determine the reactions in the hinges and the prop. Let $W = 100\,\text{N}$, $d_1 = 0.4\,\text{m}$, $d_2 = 0.6\,\text{m}$, $d_3 = 0.3\,\text{m}$, $d_4 = 0.8\,\text{m}$, and $\theta = 20°$. Assume that the prop joint A can carry forces but not moments.

3.3 MULTI-BODIES

Theory

This section considers statics problems involving two or more bodies. These **multi-bodies** are analyzed in the same way that single bodies are analyzed. Free body diagrams are determined for each body. In the case of multi-bodies, when determining the free body diagrams, it is important **not** to carry over forces from one free body diagram to another.

Examples

3.3-1 Determine the reactions in the two-member system shown. Assume that the two members are identical, and let $F = 250\,N$, $W = 120\,N$, $d_1 = 0.25\,m$, $d_2 = 0.1\,m$, and $\theta = 50°$.

Solution
A free body diagram is drawn for each member as shown. Looking at the left member, the resultant force is zero, and the resultant moment about the point O is zero, so

$$\sum F_x = 0 = R_{1x} + 250\cos 50° + R_{2x},$$
$$\sum F_y = 0 = R_{1y} - 120 - 250\sin 50° + R_{2y},$$
$$\sum M = 0 = -(0.25/2)\sin 50°(120) - (0.25 - 0.1)250$$
$$+ 0.25\sin 50°\ R_{2y} - 0.25\cos 50°R_{2x}.$$

Looking at the right member, the resultant force is zero and the resultant moment about the point A is zero, so

$$\sum F_x = 0 = -R_{2x} + R_{3x},$$
$$\sum F_y = 0 = -R_{2y} - 120 + R_{3y},$$
$$\sum M = 0 = -(0.25/2)(120) + 0.25R_{3y}.$$

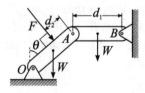

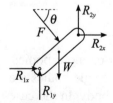
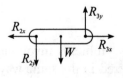

Example 3.3-1

These six linear algebraic equations are expressed in terms of the six unknowns R_{1x}, R_{1y}, R_{2x}, R_{2y}, R_{3x}, and R_{3y}. The reactions are given by $R_{1x} = 216\,\text{N}$, $R_{1y} = 372\,\text{N}$, $R_{2x} = -376\,\text{N}$, $R_{2y} = -60\,\text{N}$, $R_{3x} = -376\,\text{N}$, and $R_{3y} = 60\,\text{N}$.

3.3-2 Two cylinders are stacked on top of each other as shown. Determine the normal reactions at the contact points. Let $W = 100\,\text{N}$ and $r = 1\,\text{m}$.

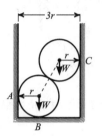

Solution
By forming a right triangle connecting the centers of the two cylinders, it is first determined that $\theta = 60°$. The free body diagram for each cylinder is then drawn as shown. The resultant forces acting on each cylinder are zero. For the top cylinder,

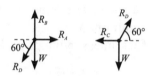

$$\sum F_x = 0 = -R_C + R_D \cos 60°,$$
$$\sum F_y = 0 = -W + R_D \sin 60°.$$

Example 3.3-2

For the bottom cylinder:

$$\sum F_x = 0 = -R_D \cos 60° + R_A,$$
$$\sum F_y = 0 = R_B - W - R_D \sin 60°.$$

These are four linear algebraic equations expressed in terms of the four unknowns R_A, R_B, R_C, and R_D. The reactions are found to be

$$R_A = 115.5\,\text{N}, \qquad R_B = 200\,N, \qquad R_C = 57.7\,\text{N}, \qquad R_D = 115.5\,\text{N}.$$

Problems

3.3-1 The system shown in the figure overleaf was observed to hang at the angles $\theta_1 = 60°$ and $\theta_2 = 70°$. Determine the weight of the center bar, if the weights of the left and right bars are each 15 lb.

3.3-2 Determine the reactions at the point A in the system shown in the figure overleaf. Let $D = 1.8\,\text{m}$ and $W = 25\,\text{N}$.

3.3-3 The tension in the serpentine belt shown in the figure overleaf is $T = 25\,\text{lb}$. Determine the reactions in each of the pin connections. Let

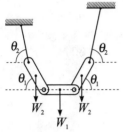

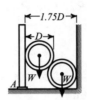

Problem 3.3-1

Problem 3.3-2

the radii of the disks be $r_A = r_B = 0.2\,\mathrm{m}$ and $r_C = r_D = 0.1\,\mathrm{m}$, and let $d_1 = 0.8\,\mathrm{m}$, $d_2 = 0.4\,\mathrm{m}$, and $d_3 = d_4 = 0.3\,\mathrm{m}$.

3.3-4 Determine the angle θ that keeps the system shown in static equilibrium. Let $R = 1\,\mathrm{m}$ and $r = 0.2\,\mathrm{m}$, and let the weight of each frictionless cylinder be $W = 100\,\mathrm{N}$.

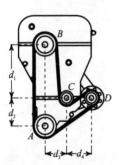

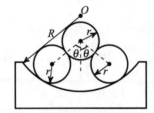

Problem 3.3-3

Problem 3.3-4

CHAPTER 4

Applications

This chapter considers statics problems involving trusses, bars, shafts, beams, compound structures, and cables. Also considered are frictional forces between surfaces: wedges, screws, belts, disks, and bearings.

4.1 TRUSSES

Theory

A **truss** is a system composed of slender members that are pinned together. At the pinned joints, the members are free to rotate, from which it follows that the joints carry forces but do not carry moments. Furthermore, all of the external forces acting on the system are assumed to act at the joints. Since no moments act at the joints, the free body diagram of each slender member in a truss system is a two-force system, with one force at each end of the member. Recall that two-force systems have the property that the forces are equal and colinear. With these facts, a simple procedure for analyzing trusses is developed.

When analyzing trusses, **planar** trusses are distinguished from **three-dimensional** trusses. Planar trusses are trusses whose members lie in a plane and whose external forces act on the truss at the joints in the plane of the truss. If the members of the truss do not lie in a plane or if the external forces do not lie in the plane of the truss, then the truss is a three-dimensional truss. If the forces do not act at the joints, then the system is not regarded as a truss (although it can be converted into an equivalent truss, which is considered later).

In **stable** trusses, there is a simple relationship between the number of joints and the number of members. First consider planar trusses. In planar trusses, summing the forces at one joint produces two equations. Letting n denote the number of joints, it follows by summing the forces at the joints that we get $2n$ equations (n for x and n for y). On the other hand, one unknown reaction acts on each truss member. Letting m denote the number of members, and letting p denote the number of unknown external forces (including wall reactions), the total number of unknowns is $m + p$. There are $2n$ equations and $m + p$ unknowns. Thus, in a stable planar truss, the following condition holds:

$$2n = m + p. \qquad (4.1\text{-}1)$$

The relationship between the number of joints and the number of members in a three-dimensional truss is developed in a similar way. In three-dimensional trusses, though, summing forces at each joint produces three equations instead of two. Thus, in a stable three-dimensional truss, the condition that holds is

$$3n = m + p. \qquad (4.1\text{-}2)$$

It was stated earlier that if external forces act on members that are **not** joints, then the system is not regarded as a truss. The weight force is a good example of a force that commonly acts on a member **not** at its joints. This situation is handled by replacing the external force acting on the member with two equivalent forces acting on each end (at each joint) of the member. This produces an equivalent system that **is** a truss. The equivalent forces acting on each end of the member are calculated to produce the same force and the same moment as the originally applied external force. Examples below illustrate this situation.

Examples

4.1-1 Consider the two-member truss system shown. Determine the reactions at the joints A, B, and C, letting $F = 100\,\text{lb}$ and $\theta = 60°$.

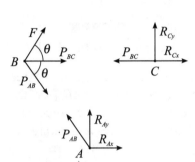

Example 4.1-1

Solution

Notice that this is a stable planar truss in which $n = 3$, $m = 2$, and $p = 4$ (since $2n = m + p$). At the joint A,

$$\sum F_x = 0 = -P_{AB} \cos 60° + R_{Ax},$$
$$\sum F_y = 0 = P_{AB} \sin 60° + R_{Ay}.$$

At the joint B,

$$\sum F_x = 0 = P_{AB} \cos 60° + F \cos 60° + P_{BC},$$
$$\sum F_y = 0 = -P_{AB} \sin 60° + F \sin 60°.$$

At the joint C,

$$\sum F_x = 0 = R_{Cx} - P_{BC},$$
$$\sum F_y = 0 = R_{Cy}.$$

These are six equations expressed in terms of the six unknowns P_{AB}, P_{BC}, R_{Ax}, R_{Ay}, R_{Cx}, and R_{Cy}. The solution is $P_{AB} = 100\,\text{lb}$, $P_{BC} = -173.2\,\text{lb}$, $R_{Ax} = 50\,\text{lb}$, $R_{Ay} = -86.6\,\text{lb}$, $R_{Cx} = -173.2\,\text{lb}$, and $R_{Cy} = 0\,\text{lb}$. Notice that the bar AB is in **tension** and that the bar BC is in **compression**. Also, notice that the reaction at A acts along the line AB and that the reaction at C acts along the line BC.

4.1-2 Consider the system shown. Determine the reactions at B and C. Let $F = 100\,\text{N}$, $d_1 = 3\,\text{m}$, $d_2 = 4\,\text{m}$, $a = 1.5\,\text{m}$, and $b = 2.5\,\text{m}$.

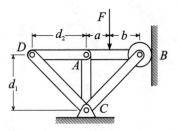

Solution

Notice that this is not a truss system because the force F does not act at a joint. Nevertheless, an equivalent truss

Examples 4.1-2 and 4.1-3

can be developed from this truss by replacing the force F with a force at each end of the member AB. Looking at the member AB [see the diagrams in (a)], the equivalent forces produce the same force and moment about the point A as the force F. Summing forces and taking the moment about the point A yields

$$F = F_A + F_B,$$
$$aF = (a + b)F_B.$$

Thus,

$$\boxed{F_A = F\frac{b}{a+b}, \qquad F_B = F\frac{a}{a+b}.} \qquad (4.1\text{-}3)$$

(a)

$$F_A = \frac{b}{a+b}F \qquad F_B = \frac{a}{a+b}F$$

(b)

Diagrams for Example 4.1-2

In this example, we get $F_A = 62.5\,\text{N}$ and $F_B = 37.5\,\text{N}$. Now, we have an equivalent planar truss in which $n = 4$, $m = 5$, and $p = 3$. The equations at the four joints [see the diagrams in (b)]

$$\sum F_x = 0 = -P_{AB} - \tfrac{4}{5}P_{BC} + R_{Bx},$$

$$\sum F_y = 0 = -F_B - \tfrac{3}{5}P_{BC},$$

$$\sum F_x = 0 = -P_{AD} + P_{AB},$$

$$\sum F_y = 0 = -F_A - P_{AC},$$

$$\sum F_x = 0 = P_{AD} + \tfrac{4}{5}P_{DC},$$

$$\sum F_y = 0 = -\tfrac{3}{5}P_{DC},$$

$$\sum F_x = 0 = -\tfrac{4}{5}P_{DC} + \tfrac{4}{5}P_{BC} + R_{Cx},$$

$$\sum F_y = 0 = \tfrac{3}{5}P_{DC} + P_{AC} + \tfrac{3}{5}P_{BC} + R_{Cy}.$$

Notice above that fractions represent trigonometric functions, for example $\cos\theta = \text{adjacent/diagonal}$. These are eight equations expressed in terms of the eight unknowns P_{AB}, P_{BC}, P_{AC}, P_{DC}, P_{AD}, R_{Cx}, R_{Cy}, and R_{Bx}. The solution is $P_{AB} = 0\,\text{N}$, $P_{BC} = -62.5\,\text{N}$, $P_{AC} = -62.5\,\text{N}$, $P_{DC} = 0\,\text{N}$, $P_{AD} = 0\,\text{N}$, $R_{Cx} = 50\,\text{N}$, $R_{Cy} = 100\,\text{N}$, and $R_{Bx} = -50\,\text{N}$. Notice that the internal forces to the left of A are zero and that the vertical reaction at C supports the load.

4.1-3 Consider the system from Example 4.1-2. Whereas in that example, the weight of each member was neglected, now consider the weight of each member in the problem statement. Thus, determine the reactions at B and C, letting the weight of each member be equal to $W = 25\,\text{N}$, and let $F = 0\,\text{N}$, $d_1 = 3\,\text{m}$, $d_2 = 4\,\text{m}$, $a = 1.5\,\text{m}$, and $b = 2.5\,\text{m}$.

Solution
Notice that the weight force does not act at a joint. An equivalent truss can be developed from this truss by replacing the weight force with a point force at each end of member AB. Notice that the previous example showed how **point forces are moved to each end** of a member and this example shows how **distributed forces are moved to each end** of a member. Looking at the member AB (see the diagram) the equivalent

(a)

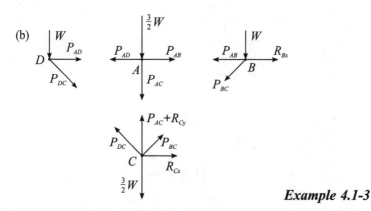

(b)

Example 4.1-3

forces produce the same force and moment about the point A as the distributed force. Since the force is distributed, it is expressed as a force per unit length $f(x)$ over the length of member AB. Summing forces (by integration) and taking the moment about point A yields

$$\int_0^c f(x)\,dx = F_A + F_B,$$

$$\int_0^c xf(x)\,dx = (a+b)F_B.$$

Thus,

$$F_A = \frac{1}{c}\left(c\int_0^c f\,dx - \int_0^c xf\,dx\right), \qquad F_B = \frac{1}{c}\int_0^c xf\,dx. \qquad (4.1\text{-}4)$$

When the force per unit length is uniform, Eq. (4.14) reduces to

$$F_A = F_B = \tfrac{1}{2}cf. \qquad (4.1\text{-}5)$$

In this example, the force per unit length is uniform and equal to $f = W/c$. Thus, $F_A = F_B = \frac{1}{2}W = 12.5\,\text{N}$. Referring to the diagram (b), the eight equations at the four joints are

$$0 = -P_{AB} - \tfrac{4}{5}P_{BC} + R_{Bx},$$

$$0 = -W - \tfrac{3}{5}P_{BC},$$

$$0 = -P_{AD} + P_{AB},$$

$$0 = -\tfrac{3}{2}W - P_{AC},$$

$$0 = P_{AD} + \tfrac{4}{5}P_{DC},$$

$$0 = -W - \tfrac{3}{5}P_{DC},$$

$$0 = -\tfrac{4}{5}P_{DC} + \tfrac{4}{5}P_{BC} + R_{Cx},$$

$$0 = -\tfrac{3}{2}W + \tfrac{3}{5}P_{DC} + P_{AC} + \tfrac{3}{5}P_{BC} + R_{Cy}.$$

These equations are expressed in terms of the eight unknowns P_{AB}, P_{BC}, P_{AC}, P_{DC}, P_{AD}, R_{Cx}, R_{Cy}, and R_{Bx}. The solution is $P_{AB} = 33.3\,\text{N}$, $P_{BC} = -41.7\,\text{N}$, $P_{AC} = -37.5\,\text{N}$, $P_{DC} = -41.7\,\text{N}$, $P_{AD} = 33.3\,\text{N}$, $R_{Cx} = 0\,\text{N}$, $R_{Cy} = 125\,\text{N}$, and $R_{Bx} = 0\,\text{N}$. Notice that the x components of the wall reactions are zero and that the vertical reaction at C supports the total weight.

4.1-4 Determine the floor reactions in the truss system shown in the figure overleaf. Let $F = 1000\,\text{lb}$, $a = 3\,\text{ft}$, and $b = 4\,\text{ft}$.

Solution
The system shown is a three-dimensional truss in which $n = 4$, $m = 3$, and $p = 9$. The ball joints at the floor carry forces but not moments. The forces in the members acting on the joint A are given by

$$\mathbf{P}_{AB} = P_{AB}(\tfrac{3}{5}\mathbf{i} - \tfrac{4}{5}\mathbf{k}),$$

$$\mathbf{P}_{AC} = P_{AC}(3\mathbf{i} + 3\mathbf{j} - 4\mathbf{k})/\sqrt{34},$$

$$\mathbf{P}_{AD} = P_{AD}(\tfrac{3}{5}\mathbf{j} - \tfrac{4}{5}\mathbf{k}),$$

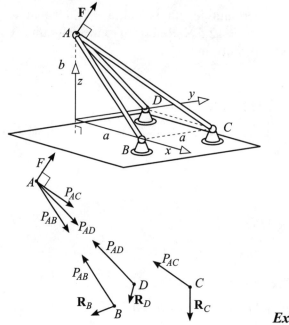

Example 4.1-4

and the applied force is given by

$$\mathbf{F} = F(2\mathbf{i} + 2\mathbf{j} + 3\mathbf{k})/\sqrt{17}.$$

Thus, the equations at the joints are written in vector form as

$$0 = \mathbf{P}_{AB} + \mathbf{P}_{AC} + \mathbf{P}_{AD} + \mathbf{F},$$
$$0 = -\mathbf{P}_{AB} + \mathbf{R}_{B},$$
$$0 = -\mathbf{P}_{AC} + \mathbf{R}_{C},$$
$$0 = -\mathbf{P}_{AD} + \mathbf{R}_{D}.$$

These are 12 equations expressed in terms of the 12 unknowns P_{AB}, P_{AC}, P_{AD}, $\mathbf{R}_{B}$, $\mathbf{R}_{C}$, and $\mathbf{R}_{D}$. The solution is $P_{AB} = 1718$ lb, $P_{AC} = -2946$ lb, $P_{AD} = 1718$ lb, $\mathbf{R}_{B} = \mathbf{P}_{AB} = 1031\mathbf{i} - 1374\mathbf{k}$ lb, $\mathbf{R}_{C} = \mathbf{P}_{AC} = -1516\mathbf{i} - 1516\mathbf{j} + 2021\mathbf{k}$ lb, and $\mathbf{R}_{D} = \mathbf{P}_{AD} = 1031\mathbf{j} - 1374\mathbf{k}$ lb. Notice that the members AB and AD act in tension and that the member AC acts in compression.

4.1-5 Determine whether either of the systems (a) or (b) shown is a truss.

(a)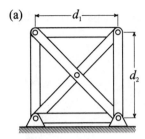

(b)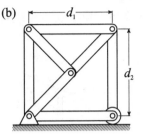

Example 4.1-5

Solution

The system in part (a) of the figure cannot be a truss, because the cross members are pinned in the middle, not at the ends. However, it might be possible to convert this system into a truss by moving the middle pin reactions to the ends of these members. Once this is done, one obtains an equivalent system consisting of $n = 4$ joints (the middle one has been removed), $m = 6$ members, and $p = 6$ reactions (including the two middle reactions). Notice that $2n$ and $m + p$ are not equal, so this system is still not a truss. In fact, this system is **indeterminate** because the reactions cannot be determined without considering the deformations in the members. (Indeterminate structures are treated in Chapters 10 through 15.) Note that by removing one of the cross members, one obtains $n = 4$, $m = 5$, and $p = 4$, which is still not a truss; there is still one too many reactions. One of the pin connections needs to be replaced with a roller

Now consider the system in (b). Notice that $n = 5$, $m = 7$, and $p = 3$. Thus, $2n = m + p$, and so this system is a truss.

Problems

4.1-1 Determine whether the system shown is a truss. Determine n, m, and p in Eq. (4.1-1).

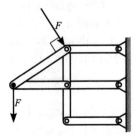

Problem 4.1-1

4.1-2 Determine whether the system shown is a truss. Determine $n, m,$ and p in Eq. (4.1-2).

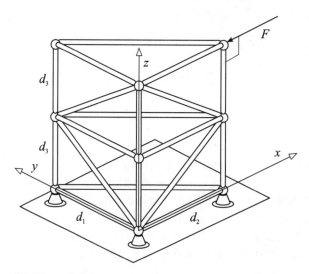

Problem 4.1-2

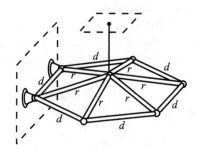

Problem 4.1-3

4.1-3 Determine whether the system shown is a truss. Determine $n, m,$ and p in Eq. (4.1-2).

4.1-4 Consider the system shown. First verify that this system is a truss. Then determine the wall reactions. Let $F = 250\,\text{N}$ and $d = 0.5\,\text{m}$. *Hint:* First consider the joint equations at the point A, solving for the internal forces in the members AG and AB. Then move to the right to the joints B and G.

4.1-5 Consider the system shown. First verify that this system is a truss. Then determine the floor reactions. Assume that the members AC

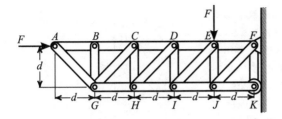

Problem 4.1-4

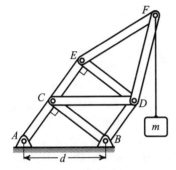

Problem 4.1-5   *Problems 4.1-6 and 4.1-7*

and *CE* are 3 m long, that the members *CB* and *ED* are 4 m long, that the member *CD* is 5 m long, and that the members *DF* and *EF* are each 5.2 m long. Also, neglect the weight of the members and let $m = 500\,\text{kg}$ and $d = 5m$.

4.1-6 Consider the system shown. First verify that this system is a truss. Assume that the applied forces act at the centers of the members and that the horizontal members are of length *d*. Determine the reactions in the members. Neglect the weight of the members. Let $F = 100\,\text{lb}$ and $d = 2.5\,\text{ft}$.

4.1-7 Solve Problem 4.1-6 again, this time letting the weight per unit length of each member be $w = 5\,\text{lb/ft}$ and removing the applied forces $(F = 0\,\text{lb})$.

4.2 BARS, SHAFTS, AND BEAMS

Theory

A **bar** refers to a long member that is subjected to external forces along its long axis. Unlike a truss member, a bar is **not** required to be pinned at each one of its ends and is **not** required to be subjected to forces only at its ends. The external forces to which a bar is subjected are called **longitudinal** forces. The external longitudinal forces F induce internal longitudinal forces P. These internal forces cause longitudinal deformations. The longitudinal deformations are frequently important, because they can reach prohibitively high levels, causing failure.

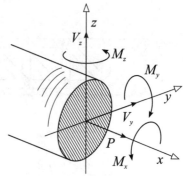

Figure 4.2-1

A **shaft** refers to a long member that is subjected to external twisting moments about its long axis. These external moments are called **twisting** moments or **torsional** moments. The external torsional moments T induce internal torsional moments M (M_x in Fig. 4.2-1), which are responsible for twisting the shaft. Failure occurs when the twisting deformations reach excessive levels.

A **beam** refers to a long member that is subjected to external **lateral** forces and external lateral moments. The external lateral forces F and external lateral moments T cause internal lateral forces V called **shear** forces and internal lateral moments M (M_y and M_z in Fig. 4.2-1) called **bending** moments. The shear forces and the bending moments induce lateral deformations, also called bending. Excessive bending deformations can also result in failure.

This section looks at the internal forces and the internal moments created by the external forces and the external moments described above. Deformation is examined in more detail in Chapters 10–16.

Examples

4.2-1 Determine the internal longitudinal forces throughout the length of the bar shown. Let $F_1 = 240\,\text{N}$, $F_2 = 140\,\text{N}$, $d_1 = 1.2\,\text{m}$, and $d_2 = 2\,\text{m}$.

Solution
Static equilibrium dictates that the resultant force acting on the entire bar be zero, so the reaction at the wall balances the two applied forces; that is, $R_x = 240 + 140 = 380\,\text{N}$. Next, **cut** the bar between the wall and the left force, and examine the left piece. As shown, since the resultant force is zero, the internal force is equal to $P = 380\,\text{N}$. Notice that the right piece could be examined to arrive at the same conclusion, namely that $P = 380\,\text{N}$. Finally, **cut** the bar between the left force and the right force,

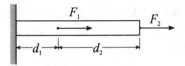

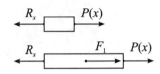

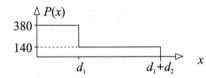

Example 4.2-1

and examine the left piece. As shown, since the resultant force is zero, the internal force is equal to $P = 140\,\text{N}$. Notice again that the right piece could have been examined to conclude the same thing, namely that $P = 140\,\text{N}$.

The internal force in the bar is a constant 380 N between the wall and the left applied force, and then it jumps down to a new constant value of 140 N between the left and right applied forces.

4.2-2 A uniform bar of weight W stands on its end as shown. Determine the internal longitudinal force at the point A. Let $W = 1500\,\text{lb}$, $d_1 = 2\,\text{ft}$, and $d_2 = 4\,\text{ft}$.

Solution
The resultant force acting on the entire bar is zero, so the reaction at the floor balances the weight, that is $R_y = 1500\,\text{lb}$. Next, **cut** the bar at the point A and examine the top piece. As shown, since the resultant force is zero, the internal force at the point A balances with the weight of the top

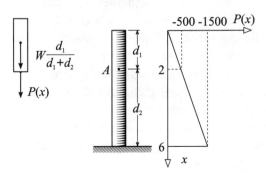

Example 4.2-2

piece; that is, $P = -Wd_1/(d_1 + d_2) = -1500(2/6) = -1500 \, \text{lb}$. In this example, it is not difficult to show that the internal compressive force at the top of the bar is zero and increases linearly with the distance from the top until it is equal to the total weight at the bottom of the bar.

4.2-3 Determine the internal torsional moment throughout the length of the shaft shown in (a). Let $T_1 = 100 \, \text{N·m}$, $T_2 = 150 \, \text{N·m}$, $d_1 = 1.2 \, \text{m}$, and $d_2 = 2 \, \text{m}$.

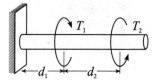

Example 4.2-3a

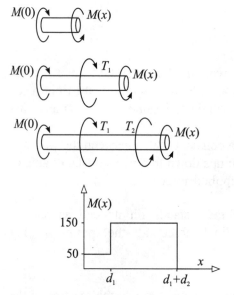

Example 4.2-3b

Solution

The resultant moment acting on the entire shaft is zero, so the reaction at the wall balances the two applied moments; that is, $M(0) = -100 + 150 = 50 \, \text{N·m}$. Next, **cut** the shaft between the wall and the left applied moment and examine the left piece. As shown in (b), since the resultant moment is zero, the internal moment is equal to $M = 50 \, \text{N·m}$. Next, **cut** the shaft between the left applied moment and the right applied moment and examine the left piece. As shown in (b), since the resultant moment is zero, the internal moment is equal to $M = 150 \, \text{N·m}$. The internal moment to the right of the right applied moment is equal to $M = 0 \, \text{N·m}$. Thus, the internal moment in the bar is a constant 50 N·m between the wall and the left applied moment, then it jumps up to a new constant value of 150 N·m between the left and right applied moments, and it is zero to the right of the right applied moment.

4.2-4 A beam is subjected to an applied bending moment and an applied tip force as shown. Determine the internal shear forces and the internal bending moments throughout the beam. Let $T = 100$ N·m, $F = 250$ N, $d_1 = 2$ m, and $d_2 = 3$ m.

Solution
The free body diagram of the entire beam reveals three unknowns: two unknown reactions at the pin connection and an unknown reaction at the roller. Referring to the figures, the resultant moment about the roller is zero and the resultant force is zero, so $R_{Ox} = 0$ N, $R_{Oy} = -20$ N, and $R_{Ay} = 270$ N. Next, cut the beam between the pin connection and the applied moment, and look at the left piece. The length of the left piece is x. The resultant moment about the point O
is zero and the resultant force is zero, so $V(x) = 20$ N and $M(x) = -20x$ N·m. Next, cut the beam to the right of the applied moment and look at the right piece. The resultant moment about the point A is zero and the resultant force is zero, so $V(x) = 20$ N and

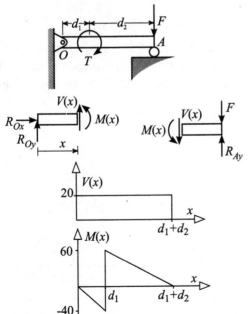

Example 4.2-4

$M(x) = 20(5 - x)$ N·m. Notice that the internal shear force is constant throughout the beam and that the internal bending moment is zero at the pin connection and varies linearly with x until it reaches the applied moment. At the applied moment, the internal moment jumps 100 N·m, from -40 N·m to $+60$ N·m. After the applied moment, the internal moment decreases linearly with x until it is zero at the roller.

Problems

4.2-1 A block of mass m rests on top of a uniform pylon of mass M. Determine the internal force at the point A. Let $m = 210$ kg, $M = 230$ kg, $d_1 = 2$ m, and $d_2 = 5$ m.

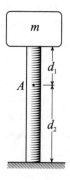

Problem 4.2-1

4.2-2 The chimney duct shown, of weight W, has uniform mass density (per unit volume) and uniform thickness. Determine the internal force as a function of the distance x measured up from the base. Let $W = 75$ lb, $w = 2$ ft, and $h = 3$ ft.

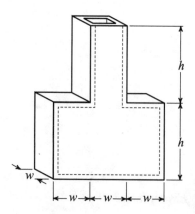

Problem 4.2-2

4.2-3 A shaft of length L is subjected to three applied moments as shown. Determine the internal moment at point A. Let $T_1 = 25 \, \text{N·m}$, $T_2 = 30 \, \text{N·m}$, $T_3 = 50 \, \text{N·m}$, $d_1 = 1 \, \text{m}$, $d_2 = d_3 = 2 \, \text{m}$, and $L = 5.5 \, \text{m}$.

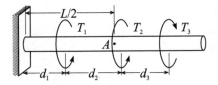

Problem 4.2-3

4.2-4 A beam is subjected to two applied forces as shown. Determine the internal shear forces and the internal bending moments at the point B, just to the left of the point A and just to the right of the point A. Let $F_1 = 70 \, \text{lb}$, $F_2 = 35 \, \text{lb}$, $d_1 = 1 \, \text{m}$, $d_2 = d_3 = 2.5 \, \text{m}$.

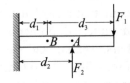

Problem 4.2-4

4.2-5 A beam of length L is subjected to a distributed applied force as shown. The force per unit length increases linearly until it reaches the constant level of f. Determine the internal shear force and the internal bending moment at the points A and B. Let $f = 55 \, \text{lb/ft}$, $L = 5 \, \text{ft}$, $d_1 = 2.5 \, \text{ft}$, and $d_2 = d_3 = 1.5 \, \text{ft}$.

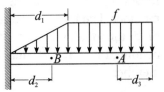

Problem 4.2-5

4.2-6 A beam of length L is subjected to a distributed applied force as shown. Determine the internal shear force and the internal bending moment at the point A. Let $f = 100 \, \text{N/m}$, $L = 10 \, \text{m}$, and $d = 6 \, \text{m}$.

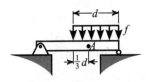

Problem 4.2-6

4.2-7 A beam of length L is subjected to an applied force and an applied moment as shown. Determine the internal shear force and the internal bending moment at the point B and just to the left and right of the point A. Let $F = 100 \, \text{N}$, $T = 800 \, \text{N·m}$, $L = 20 \, \text{m}$, $d_1 = 8 \, \text{m}$, $d_2 = 6 \, \text{m}$, and $d_3 = 4 \, \text{m}$.

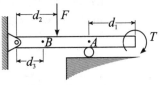

Problem 4.2-7

4.3 COMPOUND STRUCTURES

Theory

Compound structures are multi-bodies composed of members that are attached together. In this section, each of the bodies that we consider is a slender member. The internal forces and the internal moments in the members are combinations of the internal forces and the internal moments that were found in bars, shafts, and beams, as shown in Fig. 4.3-1. The internal forces at a point in a compound member consist of a

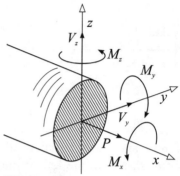

Figure 4.3-1

longitudinal force P along the axis of the member, and two shear forces V_y and V_z in the directions perpendicular to the axis of the member. The internal moments at a point in a compound member consist of a torsional moment M_x about the axis of the member, and two bending moments M_y and M_z about axes perpendicular to the axis of the member.

Examples

4.3-1 Determine the reactions at the points $A, C,$ and D, and the internal forces and internal moments in the section CD of the system shown. Let $AD = 2\,\text{m}$, $DB = BC = 3\,\text{m}$, $DE = 2\,\text{m}$, $\theta = 45°$, and $f = 120\,\text{N/m}$.

Solution

The system shown consists of two thin members. The free body diagram of each member is also shown. Looking at the first body, the reactions $R_{Cx}, R_{Cy},$ and R_D are determined first:

$$0 = R_{Cx} + R_D \sin 45°$$

$$0 = -2(120) - R_D \cos 45° + R_{Cy},$$

$$0 = 2(120)(1) + 3\sqrt{2}R_{Cy}.$$

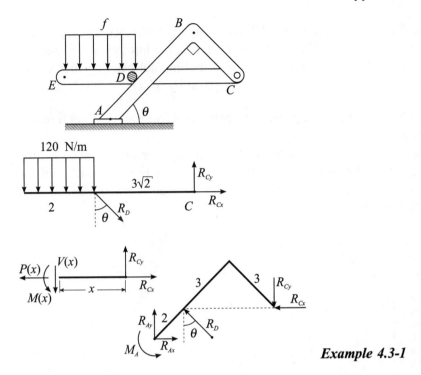

Example 4.3-1

The reactions at C and D are $R_{Cx} = 183.4\,\text{N}$, $R_{Cy} = 56.6\,\text{N}$, and $R_D = -259.4\,\text{N}$. Looking at the second body, the reactions R_{Ax}, R_{Ay}, and M_A are determined next:

$$0 = R_{Ax} - R_D \sin 45° - R_{Cx},$$
$$0 = R_{Ay} + R_D \cos 45° - R_{Cy},$$
$$0 = M_A + 2R_D + \sqrt{2}R_{Cx} - 4\sqrt{2}R_{Cy}.$$

The reactions at A are $R_{Ax} = 0\,\text{N}$, $R_{Ay} = 240\,\text{N}$, and $M_A = 99.4\,\text{N·m}$. Notice that, as expected, these reactions balance the distributed load and are independent of the internal forces in the system. Next, cut the section CD and look at the right piece. Letting x denote the distance from the point C, we have

$$0 = -P(x) + R_{Cx},$$
$$0 = -V(x) + R_{Cy},$$
$$0 = M(x) + xR_{Cy}.$$

Thus, the internal longitudinal force throughout the section CD is the constant $P(x) = 183.4$ N, the internal shear force throughout the section CD is the constant $V(x) = 56.6$ N, and the internal bending moment in the section CD varies linearly, and is given by $M(x) = 56.6x$ N·m.

4.3-2 A system is subjected to a distributed force and a point moment as shown.

Determine the reactions at the pinned connection O and the end of the cord D. Also determine the internal forces and internal moments in the section OA. Let $OA = 2$ m, $AB = BC = 3$ m, $f = 1000$ N/m, and $T = 250$ N·m.

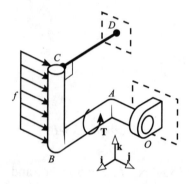

Example 4.3-2

Solution

The system shown consists of one thin member pinned about the y axis. The cord CD prevents the system rotating about the y axis. The pin connection can carry forces in all three directions and moments about all three axes except for the y axis. From the free body diagram shown in (a),

$$0 = R_{Ox} - R_C,$$

$$0 = 1000(3) + R_{Oy},$$

$$0 = R_{Oz},$$

$$0 = M_{Ox} - 1000\left(\frac{3}{2}\right)(3),$$

$$0 = 250 - 3R_C,$$

$$0 = M_{Oz} + 1000(3)(3).$$

The reactions are $R_{Ox} = 83.3$ N, $R_{Oy} = -3000$ N, $R_{Oz} = 0$ N, $M_{Ox} = 4500$ N·m, and $M_{Oz} = -9000$ N·m. Next cut member OA and

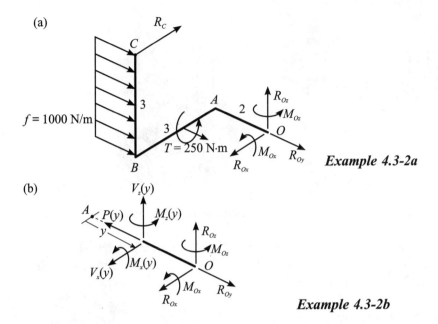

(a)

Example 4.3-2a

(b)

Example 4.3-2b

look at the piece to the right. Letting y denote the distance between the point A and the cut [see (b)],

$$0 = -P(y) + R_{Oy},$$
$$0 = V_x(y) + R_{Ox},$$
$$0 = V_z(y) + R_{Oz},$$
$$0 = M_x(y) + M_{Ox} + 2R_{Oz} + yV_z(y),$$
$$0 = M_y(y),$$
$$0 = M_z(y) + M_{Oz} - 2R_{Ox} - yV_x(y).$$

The internal reactions are $P(y) = -3000\,\text{N}$, $V_x(y) = -83.3\,\text{N}$, $V_z(y) = 0\,\text{N}$, $M_x(y) = -4500\,\text{N·m}$, $M_y(y) = 0\,\text{N·m}$, and $M_z(y) = 83.3(2 - y) + 9000\,\text{N·m}$.

Problems

4.3-1 A hinged shelf is subjected to a distributed load as shown in the figure overleaf. Determine the reactions at the points A and B. Let $CD = 2\,\text{ft}$, $CB = 6\,\text{ft}$, $d_1 = 3\,\text{ft}$, $d_2 = 1\,\text{ft}$, $f = 150\,\text{lb/ft}$, and $\theta = 60°$.

4.3-2 A bent member is subjected to a distributed force and a point moment as shown in the figure overleaf. Determine the reactions at the

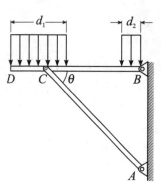

Problem 4.3-1

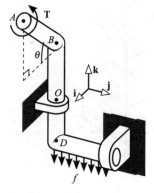

Problem 4.3-2

points O and E. Then determine the internal forces and internal moments in the section DE. Let $AB = 4\,\text{m}$, $BO = 5\,\text{m}$, $OD = 3\,\text{m}$, $DE = 4\,\text{m}$, $f = 850\ \text{N/m}$, and $\theta = 50°$.

4.3-3 A doubly hinged bar is subjected to a point moment as shown. Determine the force reactions at the points A and B in the plane of rotation. Let $AD = BC = 5\,\text{m}$, $CD = 2\,\text{m}$, $T = 75\ \text{N·m}$, and $\theta = 30°$.

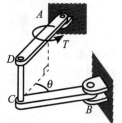

Problem 4.3-3

4.4 FRICTION

Theory

The friction force is a reaction force between two bodies that depends on the properties of the two materials in contact. Precise prediction of the friction force can be difficult because of the dependence on material properties. This section considers the friction force that is exerted between two solid bodies. The friction force acts to resist relative motion between the bodies, as shown in Fig. 4.4-1. For illustrative purposes, one body is a block and the other body is the ground. A force F acts on the body and, in reaction, a force acts between the body and the ground. This reaction is broken down into

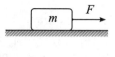

Figure 4.4-1

vertical and horizontal components. The vertical component, which is normal (perpendicular) to the surface, is the **normal reaction** N, and the horizontal component is the **friction force** F_f. The normal reaction N balances with the gravitational force mg acting on the block, so $N = mg$.

The friction force resists relative motion between the block and the ground. The friction force balances with the applied force, and the block stays in static equilibrium, provided the applied force is sufficiently small. Thus, if the body is in static equilibrium, $F_f = F$. As the applied force continues to be increased, a point is reached at which the block begins to slide. According to the **dry friction** model, the block begins to slide when the friction force reaches a level that is proportional to the normal force between the bodies; that is,

$$F_f = \mu_s N,$$

(4.4-1)

just before sliding occurs. The proportionality constant μ_s is called the **static friction coefficient**. The following situations assume a dry friction model, as described above.

WEDGES

A force pushes on the wedge shown in Fig. 4.4-2. Referring to the free body diagram, just before slipping occurs, the friction force and the normal force are related by Eq. (4.4-1), in which case

$$0 = 2N \sin \theta + 2\mu_s N \cos \theta - F,$$

so the normal force acting on the wedge is

$$N = \frac{F}{2(\sin \theta + \mu_s \cos \theta)}.$$

(4.4-2)

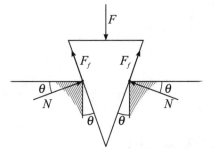

Figure 4.4-2

SQUARE THREADED SCREWS

Square threaded screws are screws that accommodate relatively large forces. As shown in Fig. 4.4-3, a force F pushes on a square threaded screw and a moment M is also applied. The spacing p between revolutions is called the **pitch** of the thread. The angle θ of the thread is called the **lead angle**. The mean radius r of the thread is related to the pitch and the lead angle by $p = 2\pi r \tan \theta$. To help visualize the forces acting on the thread, imagine the thread being **unwound**. The free body diagram of the unwound thread is drawn. As shown, the applied moment produces a force P on the thread that is related to the applied moment by $M = rP$.

First determine the applied moment M needed to rotate the screw upward. The free body diagram and the conditions described earlier yield the four equations

$$0 = -F + N \cos \theta - F_f \sin \theta,$$
$$0 = P - N \sin \theta - F_f \cos \theta,$$
$$M = rP, \tag{4.4-3}$$
$$F_f = \mu_s N,$$

in terms of the unknowns M, N, P, and F_f. The applied moment just before the screw begins to move upward is obtained:

$$M = rF \frac{\sin \theta + \mu_s \cos \theta}{\cos \theta - \mu_s \sin \theta}. \tag{4.4-4}$$

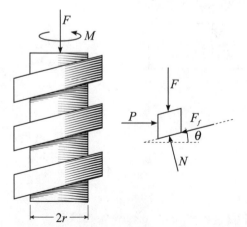

Figure 4.4-3

Notice that Eq. (4.4-4) makes sense when M is positive, that is when $\mu_s < \cot\theta$.

Next, determine the applied moment M needed to rotate the screw downward. The free body diagram remains the same, except that the friction force F_f, the applied moment M, and the associated force P switch direction. The applied moment just before the screw begins to move downward is

$$M = rF\frac{-\sin\theta + \mu_s\cos\theta}{\cos\theta + \mu_s\sin\theta}. \qquad (4.4\text{-}5)$$

Notice that Eq. (4.4-5) makes sense when M is positive, that is when $\mu_s > \tan\theta$.

Finally, the free body diagram in the absence of an applied moment is examined. The applied force F acts on the screw alone. Referring to Fig. 4.4-4, the system is in static equilibrium provided

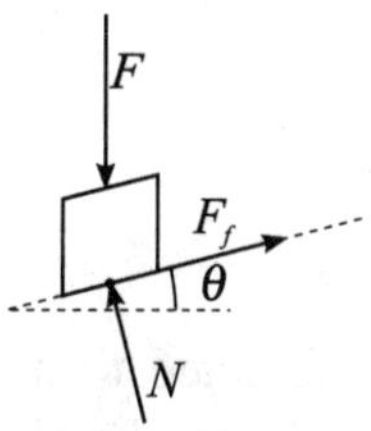

Figure 4.4-4

$$0 = -F + N\cos\theta + F_f\sin\theta,$$
$$0 = -N\sin\theta + F_f\cos\theta.$$

These two equations yield

$$N = F\cos^2\theta, \qquad F_f = F\cos\theta\sin\theta. \qquad (4.4\text{-}6)$$

Notice from Eq. (4.4-6) that $F_f = N\tan\theta$. When μ_s is greater than $\tan\theta$, the thread reaches static equilibrium before slipping occurs. A screw of this type is referred to as **self-locking**. Thus, a screw is self-locking provided

$$\mu_s > \tan\theta. \qquad (4.4\text{-}7)$$

FLAT BELTS

A flat belt passes over the curved surface shown in Fig. 4.4-5. The belt is in contact with the surface over the angle β. Assume that the belt is

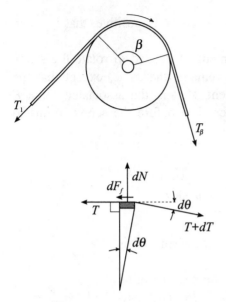

Figure 4.4-5

subjected to a tension force T_1 at the left. Determine the tension force $T(\beta)$ on the right just before the belt slides to the right.

A free body diagram of a small (differential) piece of the belt is shown. The resultant force is zero, so

$$dF_f = dT, \qquad (4.4\text{-}8a)$$

$$dN = T \, d\theta. \qquad (4.4\text{-}8b)$$

According to the dry friction model, the condition $dF_f = \mu_s \, dN$ holds just before slipping to the right, so, from Eq. (4.4-8), the following differential equation is obtained:

$$\frac{dT}{d\theta} = \mu_s T. \qquad (4.4\text{-}9)$$

The solution of Eq. (4.4-9) is

$$\boxed{T(\beta) = T_1 e^{\mu_s \beta}.} \qquad (4.4\text{-}10)$$

In Eq. (4.4-10), the angle β can be much larger than 2π, indicating multiple turns.

DISKS AND BEARINGS

A disk is pressed against a surface with force F as shown in Fig. 4.4-6. The interest lies in computing the moment acting on the disk just before it begins to rotate. Assuming that the force F is balanced with an evenly distributed force per unit area f over the cross section,

$$dF = \mu_s \, dN = \mu_s f \, dA = \frac{\mu_s F}{\pi(R_2^2 - R_1^2)} \, dA,$$

$$0 = M - \int r \, dF.$$

The applied moment just before the disk begins to slip is

$$M = \int_{R_1}^{R_2} \int_0^{2\pi} r \frac{\mu_s F}{\pi(R_2^2 - R_1^2)} r \, d\theta \, dr$$

$$= \tfrac{2}{3} \mu_s F \frac{R_2^3 - R_1^3}{R_2^2 - R_1^2}.$$

(4.4-11)

When $R_1 = 0$ and $R_2 = R$, Eq. (4.4-11) simplifies, so

$$M = \tfrac{2}{3} \mu_s F R.$$

(4.4-12)

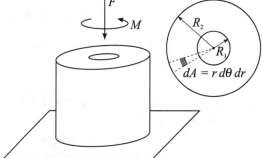

Figure 4.4-6

Examples

4.4-1 A block rests on an inclined surface as shown. Determine a formula for the largest angle for which the block does not slip. The coefficient of friction between the block and the surface is μ_s.

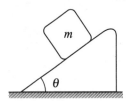
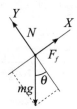

Example 4.4-1

Solution

The resultant force acting on the block is zero, so

$$0 = N - mg \cos \theta,$$
$$0 = F_f - mg \sin \theta.$$

From these equations,

$$F_f/N = \tan \theta.$$

It is clear from this equation that when θ is small, $F_f/N < \mu_s$. On the other hand, at the angle $\theta = \tan^{-1} \mu_s$, one finds that $F_f = \mu_s N$ just before the block begins to slip.

Problems

4.4-1 Determine an expression for the critical force F above which the block shown just begins to slide upward.

Problem 4.4-1

4.4-2 Determine the angle θ for which the blocks shown begin to slide. Let $m_1/m_2 = 5$ and $\mu_s = 0.1$.

4.4-3 Determine the critical force F above which the lower block shown just begins to slide to the right. Assume that the floor is smooth, that the rollers are frictionless, and that the friction coefficient between the blocks is μ_s.

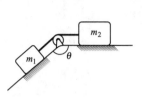

Problem 4.4-2

Problem 4.4-3

4.4-4 Find the smallest angle as a function of μ_s for which the uniform triangle shown can remain in static equilibrium. The weight force acts at the point D. Let $AB = BC$, and $FB = EB = \frac{1}{3}AB$.

Problem 4.4-4

4.4-5 Determine the smallest static coefficient of friction in Problem 3.1-5.

4.4-6 Determine the moment M required to initiate upward motion of the block W shown. Let $W = 1500\,\text{lb}$, $r = 0.2\,\text{inches}$, $p = 0.125\,\text{inches}$, and $\theta = 25°$.

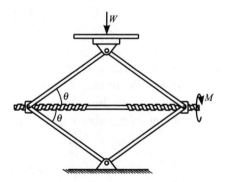

Problem 4.4-6

4.4-7 If a person pulls on the rope shown with $F = 50\,\text{lb}$, how many turns of rope around the capstan will resist the motion of the rope resulting from the ship force $P = 5000\,\text{lb}$. Let $\mu_s = 0.1$.

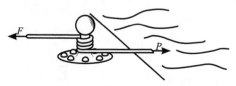

Problem 4.4-7

4.4-8 A collar bearing supports an axial load $F = 2500\,\text{N}$ as shown. Determine the applied moment M required to begin the turning of the collar. Let $\mu_s = 0.4$, $R_1 = 0.03\,\text{m}$, and $R_2 = 0.06\,\text{m}$.

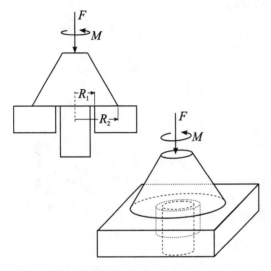

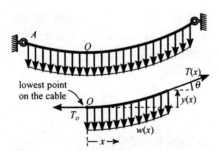

Problem 4.4-8

4.5 CABLES

Theory

Cables are elastic objects that internally carry tension forces but do **not** carry compression forces, shear forces, or bending moments.

First consider cables that are subjected to distributed forces. As shown in Fig. 4.5-1, since the resultant force acting on the cable section is zero, it follows that the horizontal component of the internal tension $T(x)$ is constant; that is,

$$T_0 = T \cos \theta \qquad (4.5\text{-}1)$$

Figure 4.5-1

is constant, where T and θ are functions of x. If the cable has uniform mass density, and the sag in the cable is sufficiently small, the differential length along the cable, ds, given by $ds^2 = dx^2 + dy^2$, can be taken to be approximately the same as dx, and the weight per unit length of the cable, denoted by w, can be taken to be uniform. Then, referring to Fig. 4.5-1, summing forces in the y direction (by integration) yields

$$T_0 \tan \theta = \int_0^x w \, dx. \tag{4.5-2}$$

Since $\tan \theta = dy/dx$, Eq. (4.5-2) can be integrated with respect to x to get

$$y(x) = \frac{wx^2}{2T_0}. \tag{4.5-3}$$

Equation (4.5-3) shows that a cable assumes a parabolic shape under a uniformly distributed load. Also, notice that x is measured from the lowest point on the cable, namely, the point O in Fig. 4.5-1.

Next, consider cables that are subjected to n point loads. A free body diagram of the ith joint is shown in Fig. 4.5-2. The resultant forces at the joints are zero, so

$$
\begin{aligned}
T_0 &= T_i \cos \theta_i = T_{i+1} \cos \theta_{i+1}, \\
W_i &= T_{i+1} \sin \theta_{i+1} - T_i \sin \theta_i.
\end{aligned}
\tag{4.5-4}
$$

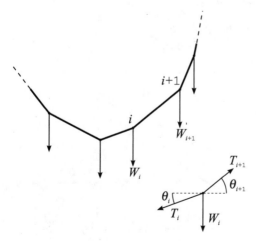

Figure 4.5-2

Equations (4.5-4) are a set of $2n$ equations and $2n$ unknowns. Depending on the problem statement, a combination of weights, angles, and tensions are given, while the remaining quantities are unknown. When the angles are unknown, Eq. (4.5-4) represents a set of nonlinear algebraic equations, which usually need to be solved numerically (by computer). When the angles are known, Eq. (5.4-4) represents a set of linear algebraic equations, which can be solved analytically (by hand).

Examples

4.5-1 The length of the uniformly distributed cable shown is decreased until the reaction at B is $T_B = 600$ lb. Determine the left reaction and the shape of the cable. Let $L = 60$ ft, $W = 1000$ lb, and $h = 40$ ft.

Solution

From the free body diagram of the entire cable, summing the moments about the point A, yields

$$0 = -(60/2)1000 + 40(600)\cos\theta + 60(600)\sin\theta.$$

The angle at the point B is $36.87°$ ($\cos 36.87° = 0.8$ and $\sin 36.87° = 0.6$). Next, denote the unknown location of the lowest point of the cable by x_0. Looking at the free body diagram of the portion of the cable that is to the right of the lowest point (identical to Fig. 4.5-1), and summing forces in the vertical direction, yields

$$0 = -1000\left(\frac{60 - x_0}{60}\right) + 600\sin 36.87°$$

so $x_0 = 38.4$ ft. The shape of the cable is

$$y(x) = \frac{(1000/60)x^2}{2(480)} = \frac{5}{288}x^2.$$

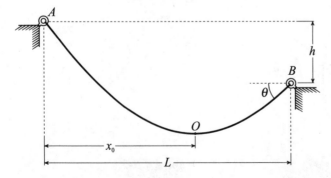

Example 4.5-1

This function is measured from the lowest point on the cable. At $x = L - x_0 = 21.6\,\text{ft}$, we get $y = (5/288)21.6^2 = 8.1\,\text{ft}$. The lowest point of the cable is 8.1 ft below the point B.

4.5-2 Consider a cable supporting the two point loads as shown. Determine the load W_2 that creates the given cable angles. Also, determine the reactions at the points A and B. Let $W_1 = 200\,\text{N}$, $\theta_1 = -30°$, $\theta_2 = 10°$, and $\theta_3 = 40°$.

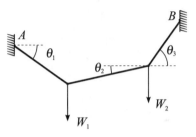

Solution

Example 4.5-2 and Problem 4.5-2

The static equilibrium of the cable is governed by Eqs. (4.5-4), letting $i = 1$

$$T_1 \cos \theta_1 = T_2 \cos \theta_2, \quad W_1 = T_2 \sin \theta_2 - T_1 \sin \theta_1,$$

The letting $i = 2$

$$T_2 \cos \theta_2 = T_3 \cos \theta_3, \quad W_2 = T_3 \sin \theta_3 - T_2 \sin \theta_2.$$

These are four linear algebraic equations expressed in terms of the four unknowns T_1, T_2, T_3, and W_2. The solution is $T_1 = 306.42\,\text{N}$, $T_2 = 269.46$, $T_3 = 346.41\,\text{N}$, and $W_2 = 175.88\,\text{N}$.

Problems

4.5-1 A cable is subjected to a uniformly distributed weight, as shown. Determine the tension T_A on the left end of the cable if the tension T_B on the right end of the cable is given. Let $L = 20\,\text{m}$, $T_B = 500\,\text{N}$, $W = 50\,\text{N}$, and $h = 5\,\text{m}$.

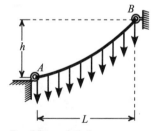

Problem 4.5-1

4.5-2 A cable is subjected to two point loads (see Example 4.5-2). Determine the angle θ_1 and the tension in the cable. Let $W_1 = 200\,\text{N}$, $W_2 = 500\,\text{N}$, $\theta_2 = 10°$, and $\theta_3 = 40°$.

CHAPTER 5

Mass Integrals and Area Integrals

In the remaining chapters of this book, masses, mass centers, polar moments, area moments, and products of inertia of different bodies will need to be calculated. These quantities are called mass integrals and area integrals. This chapter develops a methodical way of computing mass integrals and area integrals.

5.1 CLASSIFICATION OF BODIES

Theory

In order to calculate a body's mass, mass center, polar moments, area moments, and products of inertia, the body must first be classified. The mass of the body can be **lumped** at points, along lines, or over surfaces, depending on the shape of the body, or distributed throughout the volume of the body. Depending on how the mass of the body is lumped, the body is classified as a system of particles (lumped at points), a line body, a surface body, or a general body (not lumped at all). Of course, lumping the mass at points, over lines, or over surfaces, are approximations. These approximations are made in order to simplify the calculations.

Let's begin somewhat abstractly with a general expression for mass integrals and area integrals in the form

$$F = \int_D f(x, y, z) \, dD, \qquad (5.1\text{-}1)$$

where dD denotes a differential quantity associated with a differential element of the body. Depending on the type of body, the differential quantity in Eq. (5.1-1) is length, area, volume, or mass, and $f(x, y, z)$ denotes a **density** function of x, y, and z.

As an example, let F in Eq. (5.1-1) represent mass. Let's now compute the mass of different types of bodies.

POINT MASS

When the body looks like a system of n particles, the mass can be lumped at points (see Fig. 5.1-1), in which case Eq. (5.1-1) is rewritten as

$$M = \sum_{i=1}^{n} m_i, \qquad (5.1\text{-}2)$$

where M is the mass of the body. In Eq. (5.1-1), the integral sign (the symbol for an infinite sum) has been replaced by the finite summation symbol, we let $f = 1$, and we let $m_i = dD$ denote the lumped mass at the point (x_i, y_i, z_i), which is no longer infinitesimal. Of course, it's simpler to calculate M using Eq. (5.1-2) than to use Eq. (5.1-1).

LINE MASS

When the body's cross section is small compared with the length of the body (see Fig. 5.1-2), the mass can be lumped over the length L of the

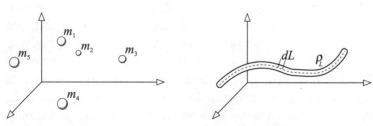

Figure 5.1-1   *Figure 5.1-2*

body. Equation (5.1-1) then becomes

$$M = \int_L \rho_L \, dL, \qquad (5.1\text{-}3)$$

where $dD = dL$ is the differential length of the body (that is, the length of the associated differential element), $f = \rho_L$ is the mass per unit length of the body, and the integration is over the length L.

SURFACE MASS

When the body's cross-sectional thickness is small compared with the surface area of the body (see Fig. 5.1-3), the mass can be lumped over the surface A of the body. Equation (5.1-1) then becomes

$$M = \int_A \rho_A \, dA, \qquad (5.1\text{-}4)$$

where $dD = dA$ is the differential area of the body, $f = \rho_A$ is the mass per unit area of the body, and the integration is over the surface area A. Note that the single integral over A is often shorthand for a double integral over x and y or r and θ.

GENERAL MASS

When the body's cross section is **not** small compared with the body length and when the cross-sectional thickness is **not** small compared with the surface area of the body (see Fig. 5.1-4), the mass **cannot** be lumped over a line or over a surface of the body. Equation (5.1-1), is then written as

$$M = \int_V \rho \, dV, \qquad (5.1\text{-}5)$$

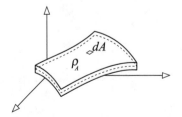

Figure 5.1-3

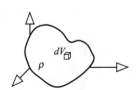

Figure 5.1-4

Table 5.1-1   The Integrand of F in Eq. (5.1-1)

Quantity (F)	Particles (F_i)	Lines (dF)	Surfaces (dF)	Volumes (dF)
Mass (M)	m_i	$\rho_L\,dL$	$\rho_A\,dA$	$\rho\,dV$
Mass center (x_C)	$(x_i/M)m_i$	$(x/M)\rho_L\,dL$	$(x/M)\rho_A\,dA$	$(x/M)\rho\,dV$
Mass center (y_C)	$(y_i/M)m_i$	$(y/M)\rho_L\,dL$	$(y/M)\rho_A\,dA$	$(y/M)\rho\,dV$
Mass center (z_C)	$(z_i/M)m_i$	$(z/M)\rho_L\,dL$	$(z/M)\rho_A\,dA$	$(z/M)\rho\,dV$
Mass center ($\mathbf{r}_C$)	$(\mathbf{r}/M)m_i$	$(\mathbf{r}/M)\rho_L\,dL$	$(\mathbf{r}/M)\rho_A\,dA$	$(\mathbf{r}/M)\rho\,dV$
Polar moment (I_O)	$r_i^2 m_i$	$r^2\rho_L\,dL$	$r^2\rho_A\,dA$	$r^2\rho\,dV$
Area moment (I_x or I_z)	—	—	$y^2\,dA$	—
Mass product (I_{xy})	$-x_i y_i m_i$	$-xy\rho_L\,dL$	$-xy\rho_A\,dA$	$-xy\rho\,dV$
Mass product (I_{yz})	$-y_i z_i m_i$	$-yz\rho_L\,dL$	$-yz\rho_A\,dA$	$-yz\rho\,dV$
Mass product (I_{zx})	$-z_i x_i m_i$	$-zx\rho_L\,dL$	$-zx\rho_A\,dA$	$-zx\rho\,dV$

where $dD = dV$ is the differential volume of the body, $f = \rho$ is the mass per unit volume of the body, and the integration is over the volume V.

The assorted mass integrals and area integrals differ in their integrands $dF = f(x, y, z)\,dD$ in Eq. (5.1-1). These quantities are defined in Table 5.1-1. In this table, ρ_L is mass per unit length, ρ_A is mass per unit area, and ρ is mass per unit volume. The masses per unit length and the masses per unit area are determined from the masses per unit volume through the relationships

$$\rho_L = \rho A, \qquad \rho_A = \rho t, \qquad (5.1\text{-}6a,b)$$

where A denotes cross-sectional area and t cross-sectional thickness. The next sections deal with the mass integrals and area integrals given in this table.

The polar moments and the mass products tabulated above arise in problems involving rotating rigid bodies and the area moments arise in problems involving bending.

5.2 Particles

Theory

The mass can be lumped at points when the body looks like a system of particles as shown in Fig. 5.1-1. The integral sign in Eq. (5.1-1) turns into a finite summation sign. The second column of Table 5.1-1 treats this case.

Examples

5.2-1 Determine the mass of the system of particles shown. Let $m_1 = 7.0$ kg, $m_2 = 4.0$ kg, $m_3 = 9.0$ kg, $r_1 = (0.0 \quad 8.0 \quad 0.0)$ m, $r_2 = (-3.0 \quad 4.0 \quad 2.0)$ m, and $r_3 = (8.0 \quad 0.0 \quad 6.0)$ m.

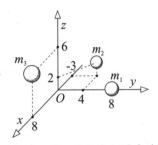

Solution

The mass of the system of particles is calcu-
lated from Eq. (5.1-2). The mass is simply   *Examples 5.2-1 to 5.2-4*
$M = 7.0 + 4.0 + 9.0 = 20$ kg.

5.2-2 Determine the mass center of the system of particles shown. Let $m_1 = 7.0$ kg, $m_2 = 4.0$ kg, $m_3 = 9.0$ kg, $r_1 = (0.0 \quad 8.0 \quad 0.0)$ m, $r_2 = (-3.0 \quad 4.0 \quad 2.0)$ m, and $r_3 = (8.0 \quad 0.0 \quad 6.0)$ m.

Solution

From the second column of Table 5.1-1,

$$x_C = \frac{1}{M}(m_1 x_1 + m_2 x_2 + m_3 x_3) = \frac{1}{20}[7(0) + 4(-3) + 9(8)] = 3.0 \text{ m},$$

$$y_C = \frac{1}{M}(m_1 y_{21} + m_2 y_2 + m_3 y_3) = \frac{1}{20}[7(8) + 4(4) + 9(0)] = 3.6 \text{ m},$$

$$z_C = \frac{1}{M}(m_1 z_1 + m_2 z_2 + m_3 z_3) = \frac{1}{20}[7(0) + 4(2) + 9(6)] = 3.1 \text{ m}.$$

Thus, $r_C = (3.0 \quad 3.6 \quad 3.1)$ m.

5.2-3 Determine the polar moments of the system of particles shown.
Let $m_1 = 7.0$ kg, $m_2 = 4.0$ kg, $m_3 = 9.0$ kg, $r_1 = (0.0 \quad 8.0 \quad 0.0)$ m, $r_2 = (-3.0 \quad 4.0 \quad 2.0)$ m, and $r_3 = (8.0 \quad 0.0 \quad 6.0)$ m.

Solution

One distinguishes between polar moments about the x, y, and z axes. The quantities r_i in the second column of Table 5.1-1 indicate the distances between the ith point and the axes in question. Thus

$$I_{xO} = m_1 r_{x1}^2 + m_2 r_{x2}^2 + m_3 r_{x3}^2 = 7(8^2 + 0^2) + 4(4^2 + 2^2) + 9(0^2 + 6^2)$$
$$= 852 \text{ kg·m}^2,$$

$$I_{yO} = m_1 r_{y1}^2 + m_2 r_{y2}^2 + m_3 r_{y3}^2 = 7(0^2 + 0^2) + 4[(-3)^2 + 2^2] + 9(8^2 + 6^2)$$
$$= 952 \text{ kg·m}^2,$$

$$I_{zO} = m_1 r_{z1}^2 + m_2 r_{z2}^2 + m_3 r_{z3}^2 = 7(0^2 + 8^2) + 4[(-3)^2 + 4^2] + 9(8^2 + 0^2)$$
$$= 1124 \text{ kg·m}^2.$$

5.2-4 Determine the mass products about the point O of the system of particles shown. Let $m_1 = 7.0$ kg, $m_2 = 4.0$ kg, $m_3 = 9.0$ kg, $\mathbf{r}_1 = (0.0 \quad 8.0 \quad 0.0)$ m, $\mathbf{r}_2 = (-3.0 \quad 4.0 \quad 2.0)$ m, and $\mathbf{r}_3 = (8.0 \quad 0.0 \quad 6.0)$ m.

Solution
From the second column of Table 5.1-1,

$$I_{xyO} = -(m_1 x_1 y_1 + m_2 x_2 y_2 + m_3 x_3 y_3) = -[7(0)8 + 4(-3)4 + 9(8)0]$$
$$= -48 \text{ kg·m}^2,$$

$$I_{yzO} = -(m_1 y_1 z_1 + m_2 y_2 z_2 + m_3 y_3 z_3) = -[7(8)0 + 4(4)2 + 9(0)6]$$
$$= -32 \text{ kg·m}^2,$$

$$I_{zxO} = (m_1 z_1 x_1 + m_2 z_2 x_2 + m_3 z_3 x_3) = -[7(0)0 + 4(2)(-3) + 9(6)8]$$
$$= -408 \text{ kg·m}^2.$$

Problems

5.2-1 Three identical point masses lie in a plane as shown. Each point mass has mass m and each chain is d units long. Determine the position vectors of the point masses. Then determine the total mass and the position of the mass center. Let $m = 24$ slug and $d = 1.2$ ft.

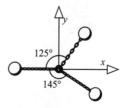

Problems 5.2-1 to 5.2-3

5.2-2 Three identical point masses lie in a plane as shown. Each point mass has mass m and each chain is d units long. Determine the position vectors of the point masses, and the polar moments. Let $m = 24$ slug and $d = 1.2$ ft.

5.2-3 Three identical point masses lie in a plane as shown. Each point mass has mass m and each chain is d units long. Determine the position vectors of the point masses and the mass products. Let $m = 24$ slug and $d = 1.2$ ft.

5.2-4 Each of the four identical point masses shown has mass m, and they are separated from one another by an amount d. Determine the position vectors of the point masses. Then determine expressions for the total mass and the position of the mass center.

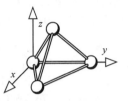

Problems 5.2-4 to 5.2-6

5.2-5 Each of the four identical point masses shown has mass m, and they are separated from one another by an amount d. Determine the position vectors of the point masses and then determine expressions for the polar moments.

5.2-6 Each of the four identical point masses shown has mass m, and they are separated from one another by an amount d. Determine the position vectors of the point masses and then determine expressions for the mass products.

5.3 LINES

Theory

When the body is thin and long, that is, when the body has one long dimension and two short dimensions, the mass can be lumped along a center line. The third column of Table 5.1-1 treats this case. The differential is the differential line dL as shown in Fig. 5.1-2. In order to compute these mass integrals and area integrals, dL needs to be expressed in terms of the variables used in the integration, and the proper limits of integration need to be determined. Two types of problems are considered below: the first is associated with straight bodies, for which rectangular coordinates are used, while the second is associated with round bodies, for which polar coordinates are used.

Examples

5.3-1 Consider the thin bar shown. Assume that its mass density of $\rho = 15$ slug/ft^3 is uniform, that its length is $L = 10$ ft, and that its cross-sectional area is $A = 0.05$ ft^2. Determine the mass of the bar.

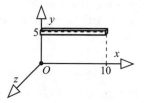

Examples 5.3-1 to 5.3-4

Solution

The mass of the thin bar is calculated from the third column of Table 5.1-1, letting $dL = dx$. The mass is being lumped over the line x. From Eq. (5.1-6a), $\rho_L = \rho A = 15(0.05) = 0.75$ slug/ft, so

$$M = \int_0^{10} 0.75\, dx = 7.5 \text{ slug.}$$

5.3-2 Consider the thin bar shown. Assume that its mass density of $\rho = 15$ slug/ft^3 is uniform, that its length is $L = 10$ ft, and that its cross-sectional area is $A = 0.05$ ft^2. Determine the position of the mass center of the bar.

Solution

The mass center of the thin bar is calculated from the third column of Table 5.1-1. Using the results from Example 5.3-1 yields

$$x_C = \frac{1}{M} \int_0^{10} 0.75x\, dx = \frac{1}{7.5}(0.75)50 = 5 \text{ ft,}$$

$$y_C = \frac{1}{M} \int_0^{10} 0.75y\, dx = \frac{1}{7.5}(0.75)5(10) = 5 \text{ ft,}$$

$$z_C = \frac{1}{M} \int_0^{10} 0.75z\, dx = \frac{1}{7.5}(0.75)0 = 0 \text{ ft.}$$

Thus, the mass center is located at (5 5 0) ft, as the reader might expect after looking at the figure.

5.3-3 Consider the thin bar shown. Assume that its mass density of $\rho = 15$ slug/ft^3 is uniform, that its length is $L = 10$ ft, and that its cross-sectional area is $A = 0.05$ ft^2. Determine the polar moments of the bar.

Solution

The polar moments of the thin bar are calculated using Table 5.1-1:

$$I_{x0} = \int_0^{10} 0.75(5^2)\, dx = (0.75)250 = 187.5 \text{ slug·ft}^2,$$

$$I_{y0} = \int_0^{10} 0.75(x^2)\, dx = (0.75)100/3 = 25 \text{ slug·ft}^2,$$

$$I_{z0} = \int_0^{10} 0.75(5^2 + x^2)\, dx = (0.75)(250 + 100/3) = 212.5 \text{ slug·ft}^2.$$

In the above, the parenthetic terms inside the integrals represent r^2, which is the squared distance between the differential element and the **axis** in question.

5.3-4 Consider the thin bar shown. Assume that its mass density of $\rho = 15$ slug/ft^3 is uniform, that its length is $L = 10$ ft, and that its cross-sectional area is $A = 0.05$ ft^2. Determine the mass products of the bar.

Solution

The mass products of the thin bar are calculated using Table 5.1-1. The mass per unit length $\rho_L = \rho A = 15(0.05) = 0.75$ slug/ft, so

$$I_{xy0} = -\int_0^{10} 0.75x(5)\, dx = -(0.75)5(100/2) = -187.5 \text{ slug·ft}^2,$$

$$I_{yz0} = -\int_0^{10} 0.75(5)0\, dx = 0 \text{ slug·ft}^2,$$

$$I_{zx0} = -\int_0^{10} 0.75(0)x\, dx = 0 \text{ slug·ft}^2.$$

5.3-5 Consider the thin quarter ring shown. Assume that its mass density of $\rho = 10$ kg/m^3 is uniform, that its radius is $R = 2$ m, and that its cross-sectional area is $A = 0.01$ m^2. Determine the mass of the quarter ring.

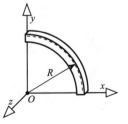

Solution

The mass of the quarter ring is calculated from the third column of Table 5.1-1, letting $dL = R\, d\theta$. The mass is

Examples 5.3-5 to 5.3-8

being lumped over the arc. The ring is thin enough that we can estimate the mass per unit length to be $\rho_L = \rho A = 10(0.01) = 0.1$ kg/m, so

$$M = \int_0^{\pi/2} 0.1(2)\, d\theta = 0.1\pi = 0.314 \text{ kg}.$$

5.3-6 Consider the thin quarter ring shown. Assume that its mass density of $\rho = 10$ kg/m^3 is uniform, that its radius is $R = 2$ m, and that its cross-sectional area is $A = 0.01$ m^2. Determine the mass center of the quarter ring.

Solution
The mass center of the quarter ring is calculated from the third column of Table 5.1-1, letting $dL = R\, d\theta$. Using the results from Example 5.3-5 yields

$$x_C = \frac{1}{M}\int_0^{\pi/2} (R\cos\theta)\rho_L R\, d\theta = \frac{0.1(2)^2}{0.1\pi}\int_0^{\pi/2}\cos\theta\, d\theta = \frac{4}{\pi} = 1.27 \text{ m},$$

$$y_C = \frac{1}{M}\int_0^{\pi/2} (R\sin\theta)\rho_L R\, d\theta = \frac{0.1(2)^2}{0.1\pi}\int_0^{\pi/2}\sin\theta\, d\theta = \frac{4}{\pi} = 1.27 \text{ m},$$

$$z_C = \frac{1}{M}\int_0^{\pi/2} (0)\rho_L R\, d\theta = 0 \text{ m}.$$

5.3-7 Consider the thin quarter ring shown. Assume that its mass density of $\rho = 10$ kg/m^3 is uniform, that its radius is $R = 2$ m, and that its cross-sectional area is $A = 0.01$ m^2. Determine the polar moments of the quarter ring about the point O.

Solution
The polar moments of the quarter ring are calculated from the third column of Table 5.1-1, letting $dL = R\, d\theta$. Using the results from Example 5.3-5 yields

$$I_{xO} = \int_0^{\pi/2} [(R\sin\theta)^2 + 0^2]\rho_L R\, d\theta = \rho_L R^3 \int_0^{\pi/2} \sin^2\theta\, d\theta = 0.628 \text{ kg·m}^2,$$

$$I_{yO} = \int_0^{\pi/2} [(R\cos\theta)^2 + 0^2]\rho_L R\, d\theta = \rho_L R^3 \int_0^{\pi/2} \cos^2\theta\, d\theta = 0.628 \text{ kg·m}^2,$$

$$I_{zO} = \int_0^{\pi/2} [R^2]\rho_L R\, d\theta = \rho_L R^3 \pi/2 = MR^2 = 0.4\pi = 1.256 \text{ kg·m}^2.$$

The terms in the brackets [ ] inside the integrals represent r^2, which is the squared distance between the differential element and the **axis** in question.

5.3-8   Consider the thin quarter ring shown. Assume that its mass density of $\rho = 10$ kg/m^3 is uniform, that its radius is $R = 2$ m, and that its cross-sectional area is $A = 0.01$ m^2. Determine the mass products of the quarter ring about the point O.

Solution
The mass products of the quarter ring are calculated from the third column of Table 5.1-1, letting $dL = R\, d\theta$. The mass per unit length $\rho_L = \rho A = 10(0.01) = 0.1$ kg/m, so

$$I_{xyO} = -\int_0^{\pi/2} (R\cos\theta)(R\sin\theta)\, \rho_L R\, d\theta$$

$$= -\rho_L R^3 \int_0^{\pi/2} \sin\theta \cos\theta\, d\theta = -0.4 \text{ kg·m}^2,$$

$$I_{yzO} = -\int_0^{\pi/2} (R\sin\theta)(0)\rho_L R\, d\theta = 0 \text{ kg·m}^2,$$

$$I_{zxO} = -\int_0^{\pi/2} (0)(R\cos\theta)\rho_L R\, d\theta = 0 \text{ kg·m}^2.$$

Problems

5.3-1   Consider a uniform bar of mass M and length $L = 2d_2$ and offset by $d = d_2$, as shown. Determine expressions for the bar's mass center, polar moments, and mass products.

5.3-2   Consider a thin wedge of mass density ρ and length L, as shown. The cross-sectional area is given by $A = A_0 x/L$, in which A_0 is a

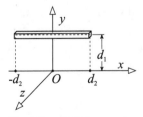

Problem 5.3-1

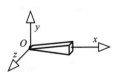

Problem 5.3-2

constant. Determine expressions for the wedge's mass center, polar moments, and mass products.

5.3-3   Consider a thin ring of mass M and radius R, as shown. Determine expressions for the ring's polar moments about the point O, and mass products about O. Notice that $x = R \sin \theta$ and $y = R + R \cos \theta$.

5.3-4   Consider a thin ring element, as shown. The mass of the ring element is M. Determine expressions for the ring element's mass center, polar moments about the point O, and mass products about O.

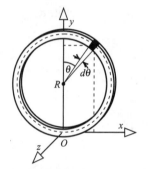

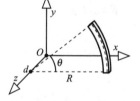

Problem 5.3-3   **Problem 5.3-4**

5.4 Surfaces

Theory

When a body looks like a thin surface, that is, when a body has two long dimensions and one short dimension, the mass can be lumped over a center surface. The fourth column of Table 5.1-1 treats this case. The differential in the fourth column is the differential area dA, as shown in Fig. 5.1-3. In order to compute these mass integrals and area integrals, dA needs to be expressed in terms of the variables used in the integration, and the proper limits of integration need to be determined. Two types of problems are considered below: the first is associated with straight bodies, for which rectangular coordinates are used, while the second is associated with round bodies, for which polar or cylindrical or spherical coordinates are used.

Examples

5.4-1   Consider the thin plate shown. Assume that its mass density of $\rho = 6$ slug/ft^3 is uniform and that its cross-sectional thickness is

$t = 0.0833$ ft (1 inch). Determine the mass of the plate. Also let $a = 3$ ft, $b = 8$ ft, and $c = 4$ ft.

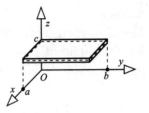

Examples 5.4-1 to 5.4-5

Solution

The mass of the plate is calculated from the fourth column of Table 5.1-1, letting $dA = dx\,dy$. The mass per unit area ρ_A is equal to the product of the mass per unit volume and the cross-sectional thickness, that is $\rho_A = \rho t = 6(0.0833) = 0.5$ slug/ft^2, so

$$M = \int_0^3 \int_0^8 0.5\,dx\,dy = 12 \text{ slug.}$$

5.4-2 Consider the thin plate shown. Assume that its mass density of $\rho = 6$ slug/ft^3 is uniform and that its cross-sectional thickness is $t = 0.0833$ ft (1 inch). Determine the mass center of the plate. Also let $a = 3$ ft, $b = 8$ ft, and $c = 4$ ft.

Solution

The mass center of the plate is calculated from the fourth column of Table 5.1-1. Using the results from Example 5.4-1, we get

$$x_C = \frac{1}{M} \int_0^3 \int_0^8 x 0.5\,dx\,dy = \frac{1}{12} 0.5 \int_0^3 x\,dx \int_0^8 dy = \frac{1}{12} 0.5 \left(\frac{9}{2}\right)(8) = 1.5 \text{ ft,}$$

$$y_C = \frac{1}{M} \int_0^3 \int_0^8 y 0.5\,dx\,dy = \frac{1}{12} 0.5 \int_0^3 dx \int_0^8 y\,dy = \frac{1}{12} 0.5(3)(32) = 4 \text{ ft,}$$

$$z_C = \frac{1}{M} \int_0^3 \int_0^8 z 0.5\,dx\,dy = \frac{1}{12} 0.5(4) \int_0^3 dx \int_0^8 dy = \frac{1}{12} 0.5(4)(3)(8) = 4 \text{ ft.}$$

5.4-3 Consider the thin plate shown. Assume that its mass density of $\rho = 6$ slug/ft^3 is uniform and that its cross-sectional thickness is $t = 0.0833$ ft (1 inch). Determine the polar moments of the plate about the point O. Also let $a = 3$ ft, $b = 8$ ft, and $c = 4$ ft.

Solution

The polar moments of the plate about the point O are calculated from the fourth column of Table 5.1-1, letting $dA = dx\,dy$. The mass per unit area $\rho_A = \rho t = 6(0.0833) = 0.5$ slug/ft^2, so

$$I_{xO} = \int_0^3 \int_0^8 (y^2 + 4^2)0.5\,dx\,dy = 0.5 \int_0^3 dx \int_0^8 (y^2 + 4^2)\,dy$$

$$= 0.5(3)\left[\frac{8^3}{3} + 16(8)\right] = 528 \text{ slug·ft}^2,$$

$$I_{yO} = \int_0^3 \int_0^8 (x^2 + 4^2)0.5\,dx\,dy = 0.5 \int_0^3 (x^2 + 4^2)\,dx \int_0^7 dy$$

$$= 0.5\left[\frac{3^3}{3} + 16(3)\right](8) = 228 \text{ slug·ft}^2,$$

$$I_{zO} = \int_0^3 \int_0^8 (x^2 + y^2)0.5\,dx\,dy = 0.5\left[(8)\int_0^3 x^2\,dx + (3)\int_0^8 y^2\,dy\right]$$

$$= 292 \text{ slug·ft}^2.$$

The parenthetic terms inside the integrals represent r^2, which is the squared distance between the differential element and the **axis** in question.

5.4-4   Consider the thin plate shown. Assume that its mass density of $\rho = 6$ slug/ft^3 is uniform and that its cross-sectional thickness is $t = 0.0833$ ft (1 inch). Determine the area moment of the plate about the x axis. Also let $a = 3$ ft, $b = 8$ ft, and $c = 4$ ft.

Solution

Consider the area moment about the x axis as given in Table 5.1-1. Notice that the area moment only applies to cross-sectional surfaces, which are treated in the fourth column of Table 5.1-1. The squared distance between the x axis and the differential element in the cross section is y^2. The mass per unit area $\rho_A = \rho t = 6(0.0833) = 0.5$ slug/ft^2, so

$$I_x = \int_0^3 \int_0^8 y^2\,dx\,dy = \int_0^3 dx \int_0^8 y^2\,dy = (3)\left(\frac{8^3}{3}\right) = 528 \text{ ft}^4.$$

5.4-5   Consider the thin plate shown. Assume that its mass density of $\rho = 6$ slug/ft^3 is uniform and that its cross-sectional thickness is $t = 0.0833$ ft (1 inch). Determine the mass products of the plate about the point O. Let $a = 3$ ft, $b = 8$ ft, and $c = 4$ ft.

Solution
General expressions for the mass products of the plate about the point O are given in Table 5.1-1. The mass per unit area $\rho_A = \rho t = 6(0.0833) = 0.5$ slug/ft^2, so

$$I_{xyO} = -\int_0^3 \int_0^8 xy\rho_A \, dx \, dy = -0.5 \int_0^3 x \, dx \int_0^8 y \, dy$$

$$= -0.5\left(\frac{9}{2}\right)\left(\frac{64}{2}\right) = -72 \text{ slug·ft}^2,$$

$$I_{yzO} = -\int_0^3 \int_0^8 yz\rho_A \, dx \, dy = -0.5(4) \int_0^3 dx \int_0^8 y \, dy$$

$$= -0.5(4)(3)\left(\frac{64}{2}\right) = -192 \text{ slug·ft}^2,$$

$$I_{zxO} = -\int_0^3 \int_0^8 zx\rho_A \, dx \, dy = -0.5(4) \int_0^3 x \, dx \int_0^8 dy$$

$$= -0.5(4)\left(\frac{9}{2}\right)(8) = -72 \text{ slug·ft}^2.$$

5.4-6   Consider the thin quarter disk shown. Assume that its mass density of $\rho = 10$ kg/m^3 is uniform and that its thickness is $t = 0.1$ m. Determine the mass of the disk. Let $R = 2$ m.

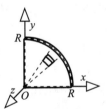

Solution
The mass of the quarter disk is calculated from the fourth column of Table 5.1-1, letting $dA = r \, d\theta \, dr$. The mass per unit area ρ_A is equal to the product of the mass per unit volume and the cross-sectional thickness, that is $\rho_A = \rho t = 10(0.1) = 1$ kg/m^2, so

Examples 5.4-6 to 5.4-10

$$M = \int_0^{\pi/2} \int_0^2 (1)r \, d\theta \, dr = \int_0^{\pi/2} d\theta \int_0^2 r \, dr = \pi = 3.14 \text{ kg}.$$

5.4-7 Consider the thin quarter disk shown. Assume that its mass density of $\rho = 10$ kg/m³ is uniform and that its thickness is $t = 0.1$ m. Determine the mass center of the disk. Let $R = 2$ m.

Solution

The mass center of the quarter disk is calculated from the fourth column of Table 5.1-1, letting $dA = r \, d\theta \, dr$. Using the results from Example 5.4-6 yields

$$x_C = \frac{1}{M} \int_0^{\pi/2} \int_0^2 (r \cos \theta)(1)r \, d\theta \, dr = \frac{1}{M} \int_0^{\pi/2} \cos \theta \, d\theta \int_0^2 r^2 \, dr = \frac{8}{3\pi} = 0.85 \text{ m},$$

$$y_C = \frac{1}{M} \int_0^{\pi/2} \int_0^2 (r \sin \theta)(1)r \, d\theta \, dr = \frac{1}{M} \int_0^{\pi/2} \sin \theta \, d\theta \int_0^2 r^2 \, dr = \frac{8}{3\pi} = 0.85 \text{ m},$$

$$z_C = \frac{1}{M} \int_0^{\pi/2} \int_0^2 (0)(1)r \, d\theta \, dr = 0 \text{ m}.$$

5.4-8 Consider the thin quarter disk shown. Assume that its mass density of $\rho = 10$ kg/m³ is uniform and that its thickness is $t = 0.1$ m. Determine the polar moments of the disk about the point O. Let $R = 2$ m.

Solution

The polar moments of the quarter disk are calculated from the fourth column of Table 5.1-1, letting $dA = r \, d\theta \, dr$. The mass per unit area ρ_A is equal to the product of the mass per unit volume and the cross-sectional thickness, that is $\rho_A = \rho t = 10(0.1) = 1$ kg/m², so

$$I_{xO} = \int_0^{\pi/2} \int_0^2 (r \sin \theta)^2 (1)r \, d\theta \, dr = \int_0^{\pi/2} \sin^2 \theta \, d\theta \int_0^2 r^3 \, dr = \frac{\pi}{4} \frac{2^4}{4}$$

$$= 3.14 \text{ kg·m}^2,$$

$$I_{yO} = \int_0^{\pi/2} \int_0^2 (r \cos \theta)^2 (1)r \, d\theta \, dr = \int_0^{\pi/2} \cos^2 \theta \, d\theta \int_0^2 r^3 \, dr = \frac{\pi}{4} \frac{2^4}{4}$$

$$= 3.14 \text{ kg·m}^2,$$

$$I_{zO} = \int_0^{\pi/2} \int_0^2 r^2 (1)r \, dr = \int_0^{\pi/2} d\theta \int_0^2 r^3 \, dr = \frac{\pi}{2} \frac{2^4}{4} = 6.28 \text{ kg·m}^2.$$

5.4-9 Consider the thin quarter disk shown. Determine the area moment of the disk about the x axis. Let $R = 2$ m.

Solution

The area moment of the quarter disk is calculated from the fourth column of Table 5.1-1, letting $dA = r \, d\theta \, dr$, so

$$I_x = \int_0^{\pi/2} \int_0^2 y^2 r \, d\theta \, dr = \int_0^{\pi/2} \int_0^2 (r \cos \theta)^2 r \, d\theta \, dr = \int_0^{\pi/2} \cos^2 \theta \, d\theta \int_0^2 r^3 \, dr$$

$$= \frac{\pi}{4} \frac{2^4}{4} = 3.14 \text{ m}^4.$$

5.4-10 Consider the thin quarter disk shown. Assume that its mass density of $\rho = 10 \text{ kg/m}^3$ is uniform and that its thickness is $t = 0.1$ m. Determine the mass products of the disk about the point O. Let $R = 2$ m.

Solution

The mass products of the quarter disk are calculated from the fourth column of Table 5.1-1, letting $dA = r \, d\theta \, dr$. The mass per unit area ρ_A is equal to the product of the mass per unit volume and the cross-sectional thickness, that is $\rho_A = \rho t = 10(0.1) = 1 \text{ kg/m}^2$, so

$$I_{xyO} = -\int_0^{\pi/2} \int_0^2 [(r \cos \theta)(r \sin \theta)](1) r \, d\theta \, dr = -\int_0^{\pi/2} \cos \theta \sin \theta \, d\theta \int_0^2 r^3 \, dr$$

$$= -\frac{2^4}{4} = -4.0 \text{ kg·m}^2,$$

$$I_{yzO} = -\int_0^{\pi/2} \int_0^2 [(r \sin \theta)(0)](1) r \, d\theta \, dr = 0 \text{ kg·m}^2,$$

$$I_{zxO} = -\int_0^{\pi/2} \int_0^2 [(0)(r \cos \theta)](1) r \, d\theta \, dr = 0 \text{ kg·m}^2.$$

Problems

5.4-1 Consider the thin triangular plate shown. Assume that its mass density of $\rho = 100 \text{ kg/m}^3$ is uniform and that its thickness is $t = 0.01$ m. Determine the mass of the plate. Let $d_1 = 0.4$ m and $d_2 = 0.3$ m.

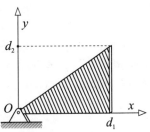

Problems 5.4-1 to 5.4-5

5.4-2 Consider the thin triangular plate shown. Assume that its mass density of $\rho = 100 \text{ kg/m}^3$ is uniform and that its thickness is $t = 0.01$ m. Determine the mass center of the plate. Let $d_1 = 0.4$ m and $d_2 = 0.3$ m.

5.4-3 Consider the thin triangular plate shown. Assume that its mass density of $\rho = 100 \text{ kg/m}^3$ is uniform and that its thickness is $t = 0.01$ m. Determine the polar moments of the plate about the point O. Let $d_1 = 0.4$ m and $d_2 = 0.3$ m.

5.4-4 Consider the thin triangular plate shown. Determine the area moment of the plate about the x axis. Let $d_1 = 0.4$ m and $d_2 = 0.3$ m.

5.4-5 Consider the thin triangular plate shown. Assume that its mass density of $\rho = 100 \text{ kg/m}^3$ is uniform and that its thickness is $t = 0.01$ m. Determine the mass products of the plate about the point O. Let $d_1 = 0.4$ m and $d_2 = 0.3$ m.

5.4-6 Consider the cylindrical shell shown. Assume that its mass density of $\rho = 50 \text{ kg/m}^3$ is uniform and that its thickness is $t = 0.015$ m. Determine the mass of the shell. Let $a = 0.5$ m and $R = 0.6$ m.

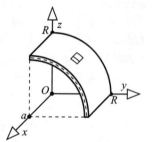

5.4-7 Consider the cylindrical shell shown. Assume that its mass density of $\rho = 50 \text{ kg/m}^3$ is uniform and that its thickness is $t = 0.015$ m. *Problems 5.4-6 to 5.4-9* Determine the mass center of the shell. Let $a = 0.5$ m and $R = 0.6$ m.

5.4-8 Consider the cylindrical shell shown. Assume that its mass density of $\rho = 50 \text{ kg/m}^3$ is uniform and that its thickness is $t = 0.015$ m. Determine the polar moments of the shell about the point O. Let $a = 0.5$ m and $R = 0.6$ m.

5.4-9 Consider the cylindrical shell shown. Assume that its mass density of $\rho = 50 \text{ kg/m}^3$ is uniform and that its thickness is $t = 0.015$ m. Determine the mass products of the shell about the point O. Let $a = 0.5$ m and $R = 0.6$ m.

5.5 VOLUMES

Theory

When the body looks neither like a thin line nor like a thin surface, that is, when the body has neither one long dimension and two short dimensions, nor two long dimensions and one short dimension, the mass cannot be lumped. The mass integrals and the area integrals have to be calculated without these approximations. The fifth column of Table 5.1-1 treats this case. The differential in the fifth column is the differential volume dV. In order to compute these mass integrals and area integrals, dV must to be expressed in terms of the variables used in the integration, and the proper limits of integration need to be determined. Two types of problems are considered below: the first is associated with straight bodies, for which rectangular coordinates are used, while the second is associated with round bodies, for which polar or cylindrical or spherical coordinates are used.

Examples

5.5-1   Consider the solid rectangular box shown. Its mass density is ρ and its dimensions are $a, b,$ and c. Develop expressions for the box's mass and its mass center.

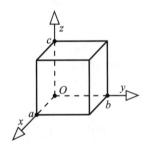

Examples 5.5-1 to 5.5-3

Solution

The mass of the box and its mass center are calculated from the fifth column of Table 5.1-1, letting $dV = dx\,dy\,dz$, so

$$M = \int_0^a \int_0^b \int_0^c \rho\,dx\,dy\,dz = \rho abc,$$

$$x_C = \frac{1}{M} \int_0^a \int_0^b \int_0^c x\rho\,dx\,dy\,dz = \frac{1}{M}\rho \int_0^a x\,dx \int_0^b dy \int_0^c dz = \tfrac{1}{2}a,$$

$$y_C = \frac{1}{M} \int_0^a \int_0^b \int_0^c y\rho\,dx\,dy\,dz = \frac{1}{M}\rho \int_0^a dx \int_0^b y\,dy \int_0^c dz = \tfrac{1}{2}b,$$

$$z_C = \frac{1}{M} \int_0^a \int_0^b \int_0^c z\rho\,dx\,dy\,dz = \frac{1}{M}\rho \int_0^a dx \int_0^b dy \int_0^c z\,dz = \tfrac{1}{2}c.$$

5.5-2 Consider the rectangular box shown. Its mass density is ρ and its dimensions are a, b, and c. Develop expressions for the box's polar moments about the point O.

Solution

The mass moments of the box about the point O are calculated from the fifth column of Table 5.1-1, letting $dV = dx\, dy\, dz$. Using the results from Example 5.5-1 yields

$$I_{xO} = \int_0^a \int_0^b \int_0^c (y^2 + z^2)\rho\, dx\, dy\, dz = \rho\left(ac\frac{b^3}{3} + ab\frac{c^3}{3}\right) = \tfrac{1}{3}M(b^2 + c^2),$$

$$I_{yO} = \int_0^a \int_0^b \int_0^c (z^2 + x^2)\rho\, dx\, dy\, dz = \rho\left(ab\frac{c^3}{3} + bc\frac{a^3}{3}\right) = \tfrac{1}{3}M(c^2 + a^2),$$

$$I_{zO} = \int_0^a \int_0^b \int_0^c (x^2 + y^2)\rho\, dx\, dy\, dz = \rho\left(bc\frac{a^3}{3} + ca\frac{b^3}{3}\right) = \tfrac{1}{3}M(a^2 + b^2).$$

5.5-3 Consider the rectangular box shown. Its mass density is ρ and its dimensions are a, b, and c. Develop expressions for the box's mass products about the point O.

Solution

The mass products of the box about the point O are calculated from the fifth column of Table 5.1-1, letting $dV = dx\, dy\, dz$:

$$I_{xyO} = -\int_0^a \int_0^b \int_0^c xy\rho\, dx\, dy\, dz = -\rho\left(\frac{a^2}{2}\frac{b^2}{2}c\right) = -\tfrac{1}{4}Mab,$$

$$I_{yzO} = -\int_0^a \int_0^b \int_0^c yz\rho\, dx\, dy\, dz = -\rho\left(a\frac{b^2}{2}\frac{c^2}{2}\right) = -\tfrac{1}{4}Mbc,$$

$$I_{zxO} = -\int_0^a \int_0^b \int_0^c zx\rho\, dx\, dy\, dz = -\rho\left(\frac{a^2}{2}b\frac{c^2}{2}\right) = -\tfrac{1}{4}Mca.$$

5.5-4 Consider the uniform cylinder shown. Its mass density is ρ and its dimensions are a and R. Develop expressions for the cylinder's mass and its mass center.

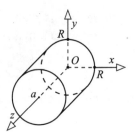

Examples 5.5-4 to 5.5-6

Solution

The mass products of the box about the point O are calculated from the fifth column of Table 5.1-1, letting $dV = r\,d\theta\,dr\,dz$:

$$M = \int_0^a \int_0^{2\pi} \int_0^R \rho r\,d\theta\,dr\,dz = \rho \int_0^a dz \int_0^{2\pi} d\theta \int_0^R r\,dr = \rho a 2\pi \frac{R^2}{2} = \rho a \pi R^2,$$

$$x_C = \frac{1}{M}\int_0^a \int_0^{2\pi} \int_0^R (r\cos\theta)\rho r\,d\theta\,dr\,dz = \frac{1}{M}\rho\int_0^a dz \int_0^{2\pi}\cos\theta\,d\theta \int_0^R r^2\,dr = 0,$$

$$y_C = \frac{1}{M}\int_0^a \int_0^{2\pi} \int_0^R (r\sin\theta)\rho r\,d\theta\,dr\,dz$$

$$= \frac{1}{M}\rho\int_0^a dz \int_0^{2\pi}\sin\theta\,d\theta \int_0^R r^2\,dr = 0,$$

$$z_C = \frac{1}{M}\int_0^a \int_0^{2\pi} \int_0^R z\rho r\,d\theta\,dr\,dz = \frac{1}{M}\rho\int_0^a z\,dz \int_0^{2\pi} d\theta \int_0^R r\,dr = \tfrac{1}{2}a.$$

5.5-5 Consider the uniform cylinder shown. Its mass density is ρ and its dimensions are a and R. Develop expressions for the cylinder's polar moments about the point O.

Solution

The polar moments of the cylinder about the point O are calculated from the fifth column of Table 5.1-1, letting $dV = r\,d\theta\,dr\,dz$:

$$I_{xO} = \int_0^a \int_0^{2\pi} \int_0^R (r^2\sin^2\theta + z^2)\rho r\,d\theta\,dr\,dz = M(\tfrac{1}{4}R^2 + \tfrac{1}{3}a^2),$$

$$I_{yO} = \int_0^a \int_0^{2\pi} \int_0^R (r^2\cos^2\theta + z^2)\rho r\,d\theta\,dr\,dz = M(\tfrac{1}{4}R^2 + \tfrac{1}{3}a^2),$$

$$I_{zO} = \int_0^a \int_0^{2\pi} \int_0^R r^2\rho r\,d\theta\,dr\,dz = \tfrac{1}{2}MR^2.$$

5.5-6 Consider the uniform cylinder shown. Its mass density is ρ and its dimensions are a and R. Develop expressions for the cylinder's mass products about the point O.

Solution

The mass products of the cylinder about the point O are calculated from the fifth column of Table 5.1-1, letting $dV = r\,d\theta\,dr\,dz$:

$$I_{xyO} = \int_0^a \int_0^{2\pi} \int_0^R (r\cos\theta)(r\sin\theta)\rho r\,d\theta\,dr\,dz = 0,$$

$$I_{yzO} = -\int_0^a \int_0^{2\pi} \int_0^R (r\sin\theta)(z)\rho r\,d\theta\,dr\,dz = 0,$$

$$I_{zxO} = -\int_0^a \int_0^{2\pi} \int_0^R (0)(r\cos\theta)\rho r\,d\theta\,dr\,dz = 0.$$

Problems

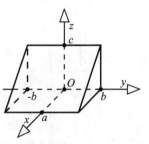

5.5-1 Consider the uniform wedge shown. Determine the wedge's mass and its mass center. ρ is given.

5.5-2 Consider the uniform wedge shown. Determine the wedge's polar moments about the point O. ρ is given.

Problems 5.5-1 to 5.5-3

5.5-3 Consider the uniform wedge shown. Determine the wedge's mass products about the point O. ρ is given.

5.5-4 Consider the uniform triangular wedge shown. Determine the wedge's mass and its mass center. ρ is given.

5.5-5 Consider the uniform triangular wedge shown. Determine the wedge's polar moments about the point O. ρ is given.

5.5-6 Consider the uniform triangular wedge shown. Determine the wedge's mass products about the point O. ρ is given.

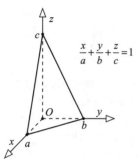

$$\frac{x}{a}+\frac{y}{b}+\frac{z}{c}=1$$

Problem 5.5-4 to 5.5-6

5.5-7 Consider the uniform cone shown. Determine expressions for the cone's mass and its mass center. ρ is given.

5.5-8 Consider the uniform cone shown. Determine expressions for the cone's polar moments about the point O. ρ is given.

5.5-9 Consider the uniform cone shown. Determine expressions for the cone's mass products. ρ is given.

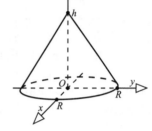

Problems 5.5-7 to 5.5-9

5.5-10 Consider the parabolic bowl shown. Determine expressions for the bowl's mass and its mass center. ρ is given.

5.5-11 Consider the parabolic bowl shown. Determine expressions for the bowl's polar moments about the point O. ρ is given.

5.5-12 Consider the parabolic bowl shown. Determine expressions for the bowl's mass products. ρ is given.

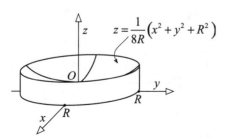

$$z=\frac{1}{8R}\left(x^{2}+y^{2}+R^{2}\right)$$

Problems 5.5-10 to 5.5-12

5.6 COMPOSITE BODIES

Theory

The previous section treated **simple** bodies. The mass integrals and the area integrals of many different types of simple bodies have been tabulated and are available in engineering reference books. (Some simple are tabulated at the end of this book). In practice, however, the body shapes of interest tend to be more complex than those treated in the reference books. More often, the body shape is a **composite** of simple shapes. The question arises then how to use the tables in those reference books when the body is a composite.

When dealing with composite bodies, one first divides the body into simple shapes. The integral in Eq. (5.1-1) is divided into parts associated with the simple shapes. For example, the shovel shown in Fig. 5.6-1 was divided into two

Figure 5.6-1

bodies: a handle and a blade. The handle can be taken to be a line body and the blade to be a surface body. Equation. (5.5-1) is rewritten as

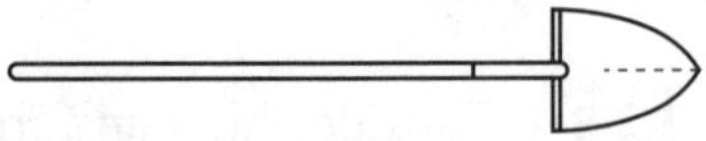

$$F = \sum_{i=1}^{n} \int_{D_i} f_i(x, y, z) \, dD_i, \qquad (5.6\text{-}1)$$

where n is the number of simple bodies that make up the composite body.

COMPOSITE MASS

In the case of mass, Eq. (5.6-1) reduces to a sum of the masses of the simple bodies; that is,

$$M = \sum_{i=1}^{n} \int_{D_i} \rho_i \, dV = \sum_{i=1}^{n} M_i \qquad (5.6\text{-}2)$$

where M_i denotes the mass of the ith simple body.

COMPOSITE MASS CENTER

From Eq. (5.6-1) and Table 5.1-1, the x component of the mass center of a composite body reduces to

$$x_C = \sum_{i=1}^{n} \int_{D_i} \left(\frac{x}{M}\right) \rho_i \, dV = \frac{1}{M} \sum_{i=1}^{n} M_i \int_{D_i} \left(\frac{x}{M_i}\right) \rho_i \, dV = \frac{1}{M} \sum_{i=1}^{n} M_i x_{Ci}, \quad (5.6\text{-}3)$$

where x_{Ci} denotes the x component of the mass center of the ith simple body. The formulas for the y and the z components of the mass center are the same as Eq. (5.6-3) if one replaces the x's with y's and z's, respectively. The mass center of a composite body is expressed in vector form as

$$\mathbf{r}_C = \frac{1}{M} \sum_{i=1}^{n} M_i \mathbf{r}_i, \qquad (5.6\text{-}4)$$

where $\mathbf{r}_i = x_i \mathbf{i} + y_i \mathbf{j} + z_i \mathbf{k}$ is the mass center of the ith simple body. Notice that the formula for the mass center of a system of particles is the same as the formula for the mass center of a composite body. Indeed, the masses of the simple bodies that compose a composite body can be lumped at their respective mass centers to form a system of particles that has the same mass center as the composite body.

COMPOSITE POLAR MOMENT

From Eq. (5.6-1) and Table 5.1-1,

$$I_O = \sum_{i=1}^{n} \int_{D_i} r^2 \rho_i \, dV = \sum_{i=1}^{n} I_{iO}, \qquad (5.6\text{-}5)$$

where I_{iO} denotes the polar moment of the ith simple body about an axis passing through the point O. Equation (5.6-5) is not very useful, because polar moments of simple bodies about every axis that lies outside the bodies are not available in reference books (it would be impossible to tabulate all the possibilities). On the other hand, polar moments of simple bodies about axes passing through their mass centers *are* available in reference books. The question is then how to determine the polar moments of simple bodies about arbitrary axes, from the polar moments of simple bodies about the axes through their mass centers. As it turns out, the two are related by the following rather simple formula (which will be derived shortly):

$$I_{iO} = I_{iC} + M_i d_i^2, \qquad (5.6\text{-}6)$$

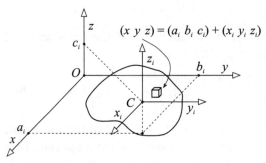

$(x \ y \ z) = (a_i \ b_i \ c_i) + (x_i \ y_i \ z_i)$

Figure 5.6-2

where I_{iO} is the polar moment of the ith simple body about an axis through the point O, I_{iC} is the polar moment of the ith simple body about the parallel axis through its mass center, and d_i is the perpendicular distance between the two axes. Equation (5.6-6) is derived as follows (see Fig. 5.6-2).

Without loss of generality, consider the polar moment about the z axis. Denoting the coordinate system of the composite body by (x, y, z) with origin at O, and denoting the coordinates of the ith simple body by (x_i, y_i, z_i) with origin at the mass center C of the ith simple body, the two coordinate systems are related by

$$x = x_i + a_i, \qquad y = y_i + b_i, \qquad z = z_i + c_i,$$

where a_i, b_i, and c_i, are the x-, y-, and z-component distances between the two origins. Thus

$$I_{iO} = \int_{V_i} r_i^2 \rho_i \, dV = \int_{V_i} (x^2 + y^2)\rho_i \, dV = \int_{V_i} [(x_i + a_i)^2 + (y_i + b_i)^2]\rho_i \, dV$$

$$= \int_{V_i} (x_i^2 + y_i^2)\rho_i \, dV + 2a_i M_i \left[\int_{V_i} \left(\frac{x_i}{M_i}\right)\rho_i \, dV \right]$$

$$+ 2b_i M_i \left[\int_{V_i} \left(\frac{y_i}{M_i}\right)\rho_i \, dV \right] + (a_i^2 + b_i^2) \int_{V_i} \rho_i \, dV,$$

where it is recognized that

$$I_{iC} = \int_{V_i} (x_i^2 + y_i^2)\rho_i \, dV, \qquad M = \int_{V_i} \rho_i \, dV, \qquad d_i^2 = a_i^2 + b_i^2,$$

and that the two terms in brackets [ ] are zero (these two terms are zero since they represent the position of the mass center measured in the ith simple body's coordinate system, which was taken to be its origin). Making these substitutions yields Eq. (5.6-6).

Equations (5.6-5) and (5.6-6) can now be combined to yield a useful formula for computing the polar moments of a composite body:

$$I_O = \sum_{i=1}^{n} (I_{iC} + M_i d_i^2).$$ (5.6-7)

In Eq. (5.6-7), $d_i^2 = x_i^2 + y_i^2$ when the polar moment is about the z axis, $d_i^2 = y_i^2 + z_i^2$ when the polar moment is about the x axis, and $d_i^2 = z_i^2 + x_i^2$ when the polar moment is about the y axis.

COMPOSITE MASS PRODUCT

From Eq. (5.6-1) and Table 5.1-1, I_{xy} can be written as

$$I_{xyO} = -\sum_{i=1}^{n} \int_{D_i} xy\rho_i \, dV = \sum_{i=1}^{n} I_{ixyO},$$ (5.6-8)

where I_{ixyO} denotes the mass product of the ith simple body about the point O. Eq. (5.6-8) is not very useful in itself, because mass products of simple bodies about general points that do not lie on the bodies are not available in reference books. On the other hand, mass products about their mass centers *are* available in reference books. The question is then how to determine the mass products of simple bodies about arbitrary points from the mass products of simple bodies about their mass centers. As it turns out, the two are related by the formula (which will be derived shortly)

$$I_{ixyO} = I_{ixyC} - M_i a_i b_i,$$ (5.6-9)

where I_{ixyO} is the mass product of the ith simple body about the point O, I_{ixyC} is the mass product of the ith simple body about its mass center, a_i is the x-component distance between the point O and the mass center C of the ith simple body, and b_i is the y-component distance between the point O and the mass center of the ith simple body. Equation (5.6-9) was determined as follows (again see Fig. 5.6-2):

$$
\begin{aligned}
I_{ixyO} &= -\int_{V_i} x_i y_i \rho_i \, dV = -\int_{V_i} (x_i + a_i)(y_i + b_i)\rho \, dV \\
&= -\int_{V_i} x_i y_i \rho_i \, dV - a_i M_i \left[\int_{V_i} \left(\frac{y_i}{M_i}\right) \rho_i \, dV \right] - b_i M_i \left[\int_{V_i} \left(\frac{x_i}{M_i}\right) \rho_i \, dV \right] \\
&\quad - a_i b_i \int_{V_i} \rho_i \, dV,
\end{aligned}
$$

where it is recognized that

$$I_{ixyC} = -\int_{V_i} x_i y_i \rho \, dV, \qquad M = \int_{V_i} \rho \, dV,$$

and that the two terms in brackets [ ] are zero (these two terms are zero since they represent the position of the mass center measured in the ith simple body's coordinate system, which was taken to be its origin). Making these substitutions yields Eq. (5.6-9).

Equations (5.6-8) and (5.6-9) can now be combined to yield a useful formula for computing the mass product of a composite body:

$$I_{xyO} = \sum_{i=1}^{n} (I_{ixyC} - M_i a_i b_i). \qquad (5.6\text{-}10)$$

The other mass products are determined in a similar way:

$$I_{yzO} = \sum_{i=1}^{n} (I_{iyzC} - M_i b_i c_i), \qquad (5.6\text{-}11)$$

$$I_{zxO} = \sum_{i=1}^{n} (I_{izxC} - M_i c_i a_i). \qquad (5.6\text{-}12)$$

Examples

5.6-1 The mailbox flag shown has a uniform mass density of 100 lb/ft³. Determine the flag's mass, and its mass center. Let $a = 2$ ft, $b = 0.1$ ft, and the thickness $t = 0.05$ ft.

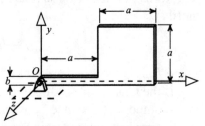

Examples 5.6-1 and 5.6-2

Solution
The body is divided into a line body and a surface body. The masses of each of these simple bodies are

$$M_1 = 100(2)(0.1)(0.05) = 1 \text{ slug},$$
$$M_2 = 100(2)(2)(0.05) = 20 \text{ slug}.$$

The mass centers of each of these bodies are located at

$$\mathbf{r}_1 = \mathbf{i}, \qquad \mathbf{r}_2 = 3\mathbf{i} + \mathbf{j}.$$

From Eqs. (5.6-2) and (5.6-4),

$M = 1 + 20 = 21$ slug,

$\mathbf{r}_1 = [(1)\mathbf{i} + (20)(3\mathbf{i} + \mathbf{j})]/21 = (61/21)\mathbf{i} + (20/21)\mathbf{j} = 2.90\mathbf{i} + 0.95\mathbf{j}$ ft.

5.6-2 The flag has a uniform mass density of 100 lb/ft³. Determine its polar moments and its mass products about the point O. Let $a = 2$ ft, $b = 0.1$ ft, and the thickness $t = 0.05$ ft.

Solution

The body is divided into a line body and a surface body. From Eqs. (5.6-7) and (5.6-10) through (5.6-12)

$$I_{xO} = 0 + \left[M_2 \frac{a^2}{12} + M_2 \left(\frac{a}{2} \right)^2 \right] = M_2 \frac{a^2}{3} = 26.67 \text{ slug·ft}^2,$$

$$I_{yO} = \left[M_1 \frac{a^2}{12} + M_1 \left(\frac{a}{2} \right)^2 \right] + \left[M_2 \frac{a^2}{12} + M_2 \left(\frac{3a}{2} \right)^2 \right] = M_1 \frac{a^2}{3} + M_2 \frac{7a^2}{3}$$

$$= 188 \text{ slug·ft}^2,$$

$$I_{zO} = \left[M_1 \frac{a^2}{12} + M_1 \left(\frac{a}{2} \right)^2 \right] + \left[M_2 \frac{a^2}{6} + M_2 \left[\left(\frac{3a}{2} \right)^2 + \left(\frac{a}{2} \right)^2 \right] \right]$$

$$= M_1 \frac{a^2}{3} + M_2 \frac{8a^2}{3} = 214.67 \text{ slug·ft}^2,$$

$$I_{xyO} = \left[0 - M_1 \left(\frac{a}{2} \right)(0) \right] + \left(0 - M_2 \frac{3a}{2} \frac{a}{2} \right) = -M_2 \frac{3a^2}{4} = -60 \text{ slug·ft}^2,$$

$$I_{yzO} = [0 - M_1(0)(0)] + \left[0 - M_2 \left(\frac{a}{2} \right)^2 (0) \right] = 0 \text{ slug·ft}^2,$$

$$I_{zxO} = \left[0 - M_1(0) \left(\frac{a}{2} \right) \right] + \left[0 - M_2(0) \left(\frac{3a}{2} \right) \right] = 0 \text{ slug·ft}^2.$$

5.6-3 The disk on the stand shown in the figure overleaf has a mass of M_1, the pole has a mass of M_2, and the block has a mass of M_3. Determine the mass center of the system. Let $L = 3$ m, $d = 4$ m, $R = 1$ m, $M_1 = 10$ kg, $M_2 = 1$ kg, and $M_3 = 25$ kg,

Solution

The body is divided into three simple bodies: the pole is a line body (body 2), and the block and disk are surface bodies (bodies 3 and 1, respectively). The mass centers of each of these bodies are located at

$$\mathbf{r}_1 = 4.0\mathbf{k}, \qquad \mathbf{r}_2 = 1.5\mathbf{k}, \qquad \mathbf{r}_3 = 0.$$

The mass center of the system is

$$r_C = \frac{(10)(4.0k) + (1)(1.5k) + (25)(0)}{10 + 1 + 25}$$

$$= \left(\frac{41.5}{36}\right)k = 1.15k \text{ m}.$$

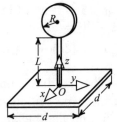

5.6-4 The disk on the block shown has a mass of M_1, the pole has a mass of M_2, and the block has a mass of M_3. Determine the polar moments and the mass products of the system about the point O. Let $L = 3$ m, $d = 4$ m, $R = 1$ m, $M_1 = 10$ kg, $M_2 = 1$ kg, and $M_3 = 25$ kg.

Examples 5.6-3 and 5.6-4

Solution

The body is divided into three simple bodies: the pole is a line body (body 2), and the block and disk are surface bodies (bodies 3 and 1, respectively). The mass centers of each of these bodies are located at

$$r_1 = 4.0k, \qquad r_2 = 1.5k, \qquad r_3 = 0.$$

Thus,

$$I_{xO} = \left[M_1 \frac{R^2}{2} + M_1 (L + R)^2 \right] + \left[M_2 \frac{L^2}{12} + M_2 \left(\frac{L}{2}\right)^2 \right] + \left(M_3 \frac{d^2}{12} \right)$$

$$= 201.33 \text{ kg·m}^2,$$

$$I_{yO} = \left[M_1 \frac{R^2}{4} + M_1 (L + R)^2 \right] + \left[M_2 \frac{L^2}{12} + M_2 \left(\frac{L}{2}\right)^2 \right] + \left(M_3 \frac{d^2}{12} \right)$$

$$= 198.83 \text{ kg·m}^2,$$

$$I_{zO} = \left[M_1 \frac{R^2}{4} + M_1 (0)^2 \right] + (0 + 0) + \left(M_3 \frac{d^2}{6} \right) = 69.17 \text{ kg·m}^2,$$

$$I_{xyO} = [0 - M_1(0)(0)] + [0 - M_2(0)(0)] + (0) = 0 \text{ kg·m}^2,$$

$$I_{yzO} = [0 - M_1(0)(L + R)] + \left[0 - M_2(0)\left(\frac{L}{2}\right) \right] + (0) = 0 \text{ kg·m}^2,$$

$$I_{zxO} = [0 - M_1(L + R)(0)] + \left[0 - M_2\left(\frac{L}{2}\right)(0) \right] + (0) = 0 \text{ kg·m}^2.$$

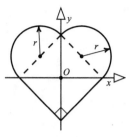

Problem 5.6-1

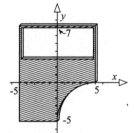

Problem 5.6-2

Problems

5.6-1 The thin heart-shaped chocolate box cover shown has a uniform mass distribution. Determine the position of its mass center, its polar moments, and its mass products about the point O. Let $M = 0.1$ kg and $r = 0.4$ m. You can divide the semicircular shapes into pie pieces tabulated in the back of the book.

5.6-2 The plastic part shown has a uniform mass distribution. The thickness of the part is 5 mm. The cross section is square in the part surrounding the rectangular hole. Determine the position of the mass center. Let $\rho = 10$ kg/m³.

5.6-3 The plastic part shown has a uniform mass distribution. The thickness of the part is 10 mm and the mass density is 10 kg/m³. Determine the position of the part's mass center, its polar moments, and its mass products about the point O.

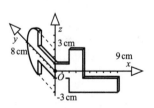

Problem 5.6-3

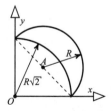

Problem 5.6-4

5.6-4 The uniform **lune** shown has a mass of M, an inside radius of $\sqrt{2}R$ and an outside radius of R. Determine the position of the lune's mass center, its polar moments, and its mass products about the point O. divide the semicircular shapes into pie pieces tabulated in the back of the book.

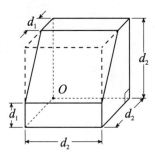

Problem 5.6-5

5.6-5 The part shown in the figure above has a uniform mass distribution. Determine the position of the part's mass center, its polar moments, and its mass moments about the point O. Let $\rho = 1000$ kg/m^3, $d_1 = 0.02$ m, and $d_2 = 0.1$ m.

CHAPTER 6

Newton's Laws of Motion

6.1 SOLVING DYNAMICS PROBLEMS

It is worth mentioning again that the process of solving a problem is really a process of **understanding**. The mathematical process consists of converting a problem into a set of mathematical equations, solving the resulting equations, and studying the solution.

As in statics, the process of understanding a dynamics problem is divided into five steps:

1. A coordinate system is selected for each object of concern.
2. The forces acting on each object are diagrammed.
3. Newton's Laws (or formulas derived from Newton's Laws) are applied to each object.
4. The equations are solved.
5. An understanding of the solution is sought.

The previous chapters dealt with statics problems in which the objects were in **static equilibrium**; that is, the objects were not moving. Chapters 6–9 deal with moving objects.

This chapter presents Newton's Laws. In Chapter 7, different types of coordinate systems will be studied. We'll examine rectangular coordinates, polar coordinates, cylindrical coordinates, tangential–normal coordinates, and spherical coordinates. In Chapter 8, equations that are derived from Newton's Laws will be introduced, along with the concepts of momentum and energy. The resulting momentum and energy equations can be used instead of Newton's Laws to solve some dynamics problems. The momentum and energy equations are particularly convenient to use when the system momentum and energy are conserved. Chapter 9 examines a special kind of system of particles, namely the rigid body. The rigid body, which can be thought of as a system composed of particles that translate and rotate together, has an associated angular velocity vector. That chapter expresses velocity, acceleration, angular momentum, and kinetic energy in terms of the rigid body's angular velocity, and, in doing so, develops equations that are needed for describing the motion of rigid bodies.

In terms of the five problem-solving steps listed earlier, Chapter 7 is used for Step 1. Chapter 6 provides you with the governing equations that are needed in Step 3. Chapter 8 provides you with alternate governing equations for Step 3, which can make the problem easier to solve. Chapter 9 provides you with governing equations for the situation in which the system is a rigid body. Notice that Steps 1 through 3 are associated with the **formulation of the problem**, that is, the conversion of a word statement into mathematical relationships. The equations are solved in Step 4 by inspecting them and determining their type. The method of solving the equations depends on their type, that is, whether they are linear algebraic, quadratic, linear differential, and so on. Regardless of the type of equations, the relationship between the number of equations and the number of independent unknowns is very important. In fact, as soon as Step 3 is completed, it is important to count the number of equations and independent unknowns to verify that they are equal to each other. There cannot be more equations than independent unknowns, and if there are fewer equations than independent unknowns then the mathematical formulation is incomplete. In other words, the physics of the problem has not been completely **captured** when the number of equations is fewer than the number of unknowns.

Understanding is achieved in Step 5. This understanding can be simple-minded or it can be comprehensive. The simplest type of understanding is a numerical answer. A comprehensive understanding is achieved by understanding the relationships among the parameters in the system. Engineering design often requires a comprehensive understanding of the parameters in the system.

One final point about problem solving is in order. When you perform the associated algebraic manipulations, it is good practice to manipulate the algebra and to solve the equations, **substituting numbers for the parameters at the last possible stage of the manipulation**. In fact, if possible, it is best to express the answers as general expressions, and then to substitute numbers for the parameters to obtain numerical answers. Doing this enables two important tests to be performed. First, by developing general expressions, the units in your answers can be verified to match as a way to test for careless errors. Secondly, the general expressions exhibit trends that can be inspected. The answers will be proportional to some of the parameters in the problem, and inversely proportional to other parameters in the problem. These trends are an important part of your comprehensive understanding of the system.

6.2 GOVERNING EQUATIONS

Theory

The field of **Newtonian mechanics** is governed by Newton's First, Second, and Third Laws and by Newton's Law of Gravitation.

Newton's First Law states:

A particle moves with constant speed and direction when the resultant force is zero.

Newton's Second Law states:

*A particle's acceleration is linearly proportional to the resultant force. The proportionality constant is called the particle's **mass**.*

Newton's Third Law states:

Two particles interact with each other through forces that are equal in magnitude, opposite in direction, and collinear.

Newton's Law of Gravitation states:

Two particles are attracted to each other through forces that are in linear proportion to their masses and in inverse proportion to the square of the distance between them.

Let's now examine a **system** of particles. The number of particles is n. Letting particles be distinguished from one another by subscripts, we pick out the ith particle and the jth particle and examine them (see Fig. 6.2-1). Newton's Second and Third Laws are expressed mathematically as

$$\mathbf{F}_i + (\mathbf{f}_{i1} + \mathbf{f}_{i2} + \ldots + \mathbf{f}_{in}) = m_i \mathbf{a}_i, \tag{6.2-1a}$$

$$\mathbf{f}_{ij} = -\mathbf{f}_{ji}, \tag{6.2-1b}$$

$$\mathbf{r}_i \times \mathbf{f}_{ij} = -\mathbf{r}_j \times \mathbf{f}_{ji} \tag{6.2-1c}$$

where $\mathbf{F}_i$ denotes the resultant **external** (to the system) force acting on the ith particle, $\mathbf{f}_{i1} + \mathbf{f}_{i2} + \ldots + \mathbf{f}_{in}$ is the resultant **internal** (to the system) force acting on the ith particle, $\mathbf{a}_i$ is the acceleration of the ith particle, m_i is the mass of the ith particle, $\mathbf{f}_{ij}$ is the force that the jth particle exerts on the ith particle, $\mathbf{f}_{ji}$ the force that the ith particle exerts on the jth particle, and $\mathbf{r}_i$ is the position of the ith particle. The left-hand side of Eq. (6.2-1a) is the resultant force (external and internal) acting on the ith particle. Thus, Eq. (6.2-1a) is the mathematical statement of Newton's Second Law (and the mathematical statement of Newton's First Law if $\mathbf{a}_i = \mathbf{0}$). Equation (6.2-1b) is the mathematical statement of Newton's Third Law, expressing that interacting forces are equal in magnitude and opposite in direction. Equation (6.2-1c) is the mathematical statement expressing that the associated interacting moments are

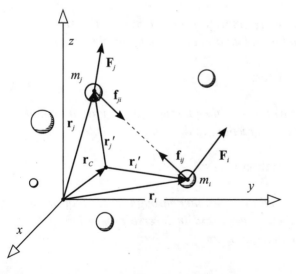

Figure 6.2-1

equal in magnitude and opposite in direction too. Although Newton's Third Law does not explicitly state that interacting moments are equal in magnitude and opposite in direction, this follows directly from this law, which states that the forces are equal in magnitude, opposite in direction, and colinear (see Example 3.1-6).

Let's now extend the equation of motion $\mathbf{F} = m\mathbf{a}$ for a single particle to $\mathbf{F} = m\mathbf{a}_C$ for a system of particles, in which $\mathbf{a}_C$ denotes the acceleration of the system's mass center. We start by defining the **position vector of the mass center** of a system of particles as the **weighted** average of the position vectors of the particles, written as

$$\mathbf{r}_C = \frac{1}{m}(m_1\mathbf{r}_1 + m_2\mathbf{r}_2 + \ldots m_n\mathbf{r}_n), \qquad m = m_1 + m_2 + \ldots + m_n,$$

where m denotes the **total mass** of the particles. Using an index notation, we can rewrite these two equations as

$$\mathbf{r}_C = \frac{1}{m}\sum_{i=1}^{n} m_i\mathbf{r}_i, \qquad m = \sum_{i=1}^{n} m_i. \tag{6.2-2}$$

Differentiating Eq. (6.2-2) twice with respect to time yields the velocity and acceleration vectors of the mass center, written as

$$\mathbf{v}_C = \frac{1}{m}\sum_{i=1}^{n} m_i\mathbf{v}_i, \qquad \mathbf{a}_C = \frac{1}{m}\sum_{i=1}^{n} m_i\mathbf{a}_i. \tag{6.2-3a,b}$$

Next, let's add together the equations governing the motion of each of the particles, Eqs. (6.2-1a). From the left-hand side of Eqs. (6.2-1a), we get

$$\sum_{i=1}^{n}\left(\mathbf{F}_i + \sum_{j=1}^{n}\mathbf{f}_{ij}\right) = \left(\sum_{i=1}^{n}\mathbf{F}_i\right) + \left(\sum_{i=1}^{n}\sum_{j=1}^{n}\mathbf{f}_{ij}\right) = \sum_{i=1}^{n}\mathbf{F}_i = \mathbf{F}, \tag{6.2-4}$$

where the double sum is zero since the internal forces are equal and opposite as stated in Newton's Third Law, Eq. (6.2-1b), and thus cancel out. $\mathbf{F}$ in Eq. (6.2-4) denotes the **resultant external force** acting on the system of particles. Substituting Eqs. (6.2-1a) and (6.2-4) into Eq. (6.2-3b), considering Eq. (6.2-4), and multiplying the result by m yields

$$\mathbf{F} = m\mathbf{a}_C. \tag{6.2-5}$$

Equation (6.2-5) states that the resultant external force acting on a system of particles is equal to the mass of the system multiplied by the acceleration of the system's mass center. The internal forces are not present in the equation. Thus, the behavior of the **macroscopic body** can be analyzed without regard to the internal forces acting inside the body. Since Eq. (6.2-5) looks so much like Newton's Second Law, it is called **Newton's Second Law for a System of Particles**.

Examples

6.2-1 A block is sliding to the right on a rough surface while being subjected to the force shown. Determine the acceleration of the block. Assume that the friction between the rough surface and the block is governed by a dry friction model that will be discussed shortly. The block has mass $m = 2$ slug, the kinetic friction coefficient is $\mu_k = 0.2$, $F = 15$ lb, and $\theta = 16°$.

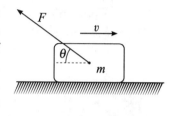

Example 6.2-1

Solution

The block can be regarded as a single particle. Let x be positive to the right and y be positive upward. Unless otherwise stated, the x axis is always taken positive to the right, and the y axis is always taken positive upward. Summing forces in the x and y directions yields

$$-F_f - F \cos \theta = ma_x,$$

$$N + F \sin \theta - mg = ma_y,$$

where F_f is the friction force. According to the **kinetic dry friction model**, **the friction force opposes the direction of the motion, and the magnitude of the force is proportional to the normal force N between the bodies (the block and the ground)**. The proportionality constant is called the **kinetic coefficient of friction** μ_k. The friction force is

$$F_f = \mu_k N. \qquad (6.2\text{-}6)$$

Also, the problem states that the block is sliding on the horizontal surface; there is no motion in the y direction, so the acceleration in the y direction is zero, that is, $a_y = 0$. Substituting $a_y = 0$ and Eq. (6.2-6) into the two governing equations for the block leaves the two unknowns a_x and N. Solving these two equations yields

$$a_x = -\frac{F}{m}[\cos\theta - \mu_k \sin\theta] - \mu_k g$$

$$= -\frac{15}{2}[\cos 16° - 0.2 \sin 16°] - 0.2(32.2) = -13.2 \text{ ft/s}^2,$$

$$N = m\left(g - \frac{F}{m}\sin\theta\right) = 2\left(32.2 - \frac{15}{2}\sin 16°\right) = 60.3 \text{ lb.}$$

Notice that the acceleration of the block a_x is composed of two contributions, one associated with the applied force and the other associated with the friction force. When the term in the brackets [ ] is positive, the applied force acts to slow down the block. The introduction of the applied force can also serve to lower the deceleration of the block when the term in brackets is negative. As the equations reveal, this can happen because the applied force causes the normal force to decrease, which in turn decreases the friction force, and hence the overall resistance to the motion. Also, notice that the assumption that the block **slides** on the surface is valid provided the normal force N given above is positive. For the block to slide to the right, we must have the condition $F < mg/\sin\theta$.

6.2-2 Assume that a block moves and accelerates down the incline shown. The incline angle θ is decreased until the block ceases to accelerate. Determine the critical angle at which the block ceases to accelerate. The kinetic coefficient of friction is $\mu_k = 0.32$.

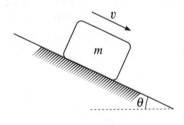

Solution
Let x denote the direction down the surface of the incline and let y denote the direction of the outward normal to the surface of the incline. Summing forces in the x and y directions yields

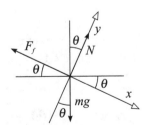

$$mg \sin\theta - F_f = ma_x,$$

$$N - mg \cos\theta = ma_y.$$

Example 6.2-2 and Problem 6.2-1

Since the block is sliding on the incline, $a_y = 0$, and, from the kinetic dry friction model, $F_f = \mu_k N$. Therefore, the two governing equations are expressed in terms of the two unknowns a_x and N. The solution is

$$a_x = g(\sin \theta - \mu_k \cos \theta),$$
$$N = mg \cos \theta.$$

The block ceases to accelerate when $a_x = 0$, which occurs when $\tan \theta = \mu_k$, so $\theta = \tan^{-1} 0.32 = 17.74°$. Notice that the critical angle is independent of the block's mass and the acceleration due to gravity, and it increases with the kinetic friction coefficient.

Example 6.2-3

6.2-3 Determine the position vector and the acceleration vector of the center of mass of the system shown, letting $m_1 = 4$ slug and $m_2 = 6$ slug. Assume that the mass of the connecting rod is negligible and that its length is 15 ft. Furthermore, assume that the system slides on a frictionless surface and that the applied forces are $F_1 = 120$ lb, $F_2 = 100$ lb, and $\theta = 30°$.

Solution
The position vector of the mass center is obtained by first considering a coordinate axis s having an origin located at m_2 directed toward m_1. From Eq. (6.2-2), the position of the mass center s_C is

$$s_C = \frac{1}{m}(m_1 s_1 + m_2 s_2) = \frac{1}{6+4}[4(15) + 6(0)] = 6 \text{ ft}.$$

Next, let's set up a coordinate system having an origin at m_2 with coordinates x directed to the right and y directed up the page. The position vector is

$$\mathbf{r}_C = -s_C \cos \theta \, \mathbf{i} + s_C \sin \theta \, \mathbf{j} = -6 \cos 30° \, \mathbf{i} + 6 \sin 30° \mathbf{j}$$
$$= -5.20\mathbf{i} + 3.00\mathbf{j} \text{ ft.}$$

The resultant force acting on the system is

$$\mathbf{F} = \mathbf{F}_1 + \mathbf{F}_2 = F_1(-\sin\theta\mathbf{i} - \cos\theta\,\mathbf{j}) + F_2\mathbf{i}$$
$$= 120(-\sin 30°\,\mathbf{i} - \cos 30°\,\mathbf{j}) + 100\mathbf{i} = 40\mathbf{i} + 103.9\mathbf{j}\,\text{lb}.$$

From Eq. (6.2-5),

$$a_C = \mathbf{F}/m = 3.33\mathbf{i} + 8.66\mathbf{j}\,\text{ft/s}^2.$$

Problems

6.2-1 Assume that the block in Example 6.2-2 is sliding **up** the incline. Determine the acceleration of the block. Let $m = 10\,\text{kg}$, $\mu_k = 0.20$, and $\theta = 30°$.

6.2-2 Determine the acceleration vector of the puck shown. The puck has mass $m = 0.1$ slug, and lies on a frictionless surface. The puck is subjected to forces $\mathbf{F}_1 = 17\mathbf{i} + 20\mathbf{j}\,\text{lb}$ and $\mathbf{F}_2 = -4\mathbf{i} + 25\mathbf{j}\,\text{lb}$.

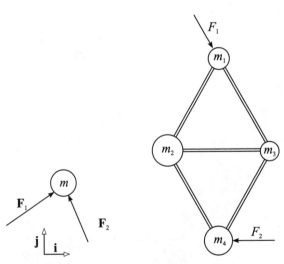

Problem 6.2-2 *Problem 6.2-3*

6.2-3 Determine the position vector and the acceleration vector of the mass center of the system shown. Let $m_1 = m_3 = 15\,\text{kg}$ and $m_2 = m_4 = 35\,\text{kg}$, let the lengths of the rods be 0.4 m each, and let $F_1 = 15\,\text{N}$ and $F_2 = 20\,\text{N}$.

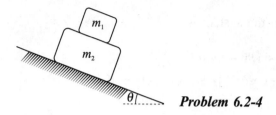

Problem 6.2-4

6.2-4 The blocks shown are released from rest. The static coefficient of friction between the surfaces is μ_s. The kinetic coefficient of friction between the surfaces is μ_k and the incline angle is θ. Determine conditions for which (a) the blocks remain at rest, (b) the two blocks slide together.

CHAPTER 7

Coordinate Systems

7.1 COORDINATE ROTATIONS

Theory

Frequently, more than one coordinate system is needed in order to describe the motion of a particle. Information may be given in one coordinate system, but **not** in the one that you wish to use. It may be preferable to use a **rotated** coordinate system because most of the forces in your free body diagram line up along its axes, or because the motion is directed along one of its axes. When this happens, it becomes advantageous to transform your quantities from the original coordinate system into the preferred coordinate system. When you have finished, it may be desirable to transform your quantities back to the original coordinate system.

Referring to Fig. 7.1-1, the position vector of a point is expressed in terms of the coordinates X and Y and the unit vectors $\mathbf{I}$ and $\mathbf{J}$ associated with one coordinate system. The same position vector is also expressed in terms of the quantities x, y, $\mathbf{i}$, and $\mathbf{j}$ associated with the rotated coordinate system. We write

$$\mathbf{r} = x\mathbf{i} + y\mathbf{j} \qquad (7.1\text{-}1a)$$

or

$$\mathbf{r} = X\mathbf{I} + Y\mathbf{J}. \qquad (7.1\text{-}1b)$$

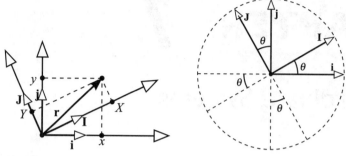

Figure 7.1-1   **Figure 7.1-2**

Our interest lies in expressing X and Y in terms of x and y, and vice versa. In order to do this, let's first express $\mathbf{I}$ and $\mathbf{J}$ in terms of $\mathbf{i}$ and $\mathbf{j}$, and vice versa. The relationships between the unit vectors are easy to see if they are collected and placed in a circle, as shown in Fig. 7.1-2. We get the unit vector transformations

$$\begin{array}{ll} \mathbf{i} = \cos\theta\,\mathbf{I} - \sin\theta\,\mathbf{J}, & \mathbf{I} = \cos\theta\,\mathbf{i} + \sin\theta\,\mathbf{j}, \\ \mathbf{j} = \sin\theta\,\mathbf{I} + \cos\theta\,\mathbf{J}, & \mathbf{J} = -\sin\theta\,\mathbf{i} + \cos\theta\,\mathbf{j}. \end{array} \qquad (7.1\text{-}2a,b)$$

Next, substituting Eqs. (7.1-2b) into Eq. (7.1-1b), yields

$$\begin{aligned} x\mathbf{i} + y\mathbf{j} &= X(\cos\theta\,\mathbf{i} + \sin\theta\,\mathbf{j}) + Y(-\sin\theta\,\mathbf{i} + \cos\theta\,\mathbf{j}) \\ &= (X\cos\theta - Y\sin\theta)\mathbf{i} + (X\sin\theta + Y\cos\theta)\mathbf{j}. \end{aligned}$$

It follows that

$$\begin{aligned} x &= \cos\theta\,X - \sin\theta\,Y, \\ y &= \sin\theta\,X + \cos\theta\,Y. \end{aligned} \qquad (7.1\text{-}3a)$$

Equations (7.1-3a) express x and y in terms of X and Y. Similarly, X and Y are expressed in terms of x and y by substituting Eqs. (7.1-2a) into Eq. (7.1-1a) to yield

$$\begin{aligned} X &= \cos\theta\,x + \sin\theta\,y, \\ Y &= -\sin\theta\,x + \cos\theta\,y. \end{aligned} \qquad (7.1\text{-}3b)$$

Notice that the transformation equations of the unit vectors and the transformation equations of the coordinates are of the same form. The coordinates x, y, X, and Y are interchangeable with the unit vectors $\mathbf{i}, \mathbf{j}, \mathbf{I}$, and $\mathbf{J}$, respectively. By interchanging these quantities, Eqs. (7.1-2a) become Eqs. (7.1-3a), and Eqs. (7.1-2b) become Eqs. (7.1-3b). Furthermore, notice that this entire derivation, rather than having started out with Eq. (7.1-1) expressing the position vector $\mathbf{r}$ in two ways, could have started out with $\mathbf{F} = F_x\mathbf{i} + F_y\mathbf{j}$ and $\mathbf{F} = F_X\mathbf{I} + F_Y\mathbf{J}$, expressing the force vector $\mathbf{F}$ in two ways, or could have started out with expressing any other vector in two ways. In other words, the transformation of coordinates given above is valid not only for the components of the position vector, but also for the components of **any** vector. For example, the transformations of the components of force are

$$
\begin{aligned}
F_x &= \cos\theta\, F_X - \sin\theta\, F_Y, \\
F_y &= \sin\theta\, F_X + \cos\theta F_Y,
\end{aligned}
\tag{7.1-4a}
$$

$$
\begin{aligned}
F_X &= \cos\theta\, F_x + \sin\theta\, F_y, \\
F_Y &= -\sin\theta\, F_x + \cos\theta\, F_y.
\end{aligned}
\tag{7.1-4b}
$$

Once you know how the unit vectors transform, you transform the components of any vector in the same way. In short, **the components of all of the vectors transform like the unit vectors**.

Examples

7.1-1 Determine the X and Y components of the position vector shown.

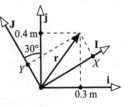

Example 7.1-1

Solution

Let's first determine the unit vector transformations. Referring to Fig. 7.1-2, we get

$$\mathbf{I} = \cos 30°\, \mathbf{i} + \sin 30°\, \mathbf{j}, \qquad \mathbf{J} = -\sin 30°\, \mathbf{i} + \cos 30°\, \mathbf{j}.$$

The coordinates transform like the unit vectors, so

$$
\begin{aligned}
X &= \cos 30°\, x + \sin 30°\, y = 0.866(0.3) + 0.5(0.4) = 0.46\,\text{m}, \\
Y &= -\sin 30°\, x + \cos 30°\, y = -0.5(0.3) + 0.866(0.4) = 0.20\,\text{m}.
\end{aligned}
$$

7.1-2 The resultant force acting on a three-stage Saturn I rocket is shown. Determine the x and y components of the force acting on the three-stage rocket. Also determine the X and Y components of the force. Let $F = 1\,620\,000$ lb. (The first stage is $1\,500\,000$ lb, the second stage is $80\,000$ lb, and the third stage is $40\,000$ lb.)

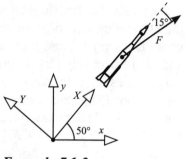

Example 7.1-2

Solution

We first determine the unit vector transformations, which are

$$\mathbf{i} = \cos 50° \, \mathbf{I} - \sin 50° \, \mathbf{J}, \qquad \mathbf{j} = \sin 50° \, \mathbf{I} + \cos 50° \, \mathbf{J}.$$

The X and Y components of the resultant force vector are given by $F_X = F \cos 15° = 1\,620\,000(0.966) = 1\,564\,920$ lb, and $F_Y = F \sin 15° = 1\,620\,000(0.259) = 419\,580$ lb. The forces transform like the unit vectors, so

$$F_x = \cos 50° F_X - \sin 50° \, F_Y = (0.643)1\,564\,920 - (0.766)419\,580$$
$$= 684\,845\,\text{lb},$$

$$F_y = \sin 50° \, F_X + \cos 50° \, F_Y = (0.766)1\,564\,920 + (0.643)419\,580$$
$$= 1\,468\,580 = 1\,468\,519\,\text{lb}.$$

Problems

7.1-1 Determine the X and Y components of the position vector $\mathbf{r}$ shown. Notice that the X–Y coordinates and the x–y coordinates have different origins; that is, the X–Y coordinates are rotated and **translated**.

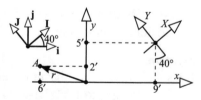

Problem 7.1-1

7.1-2 An airplane is subjected to a resultant force as shown. Determine the x and y components of the resultant force. Also determine the x and y components of the airplane velocity. Let $F = 2500$ N and $v = 25$ m/s.

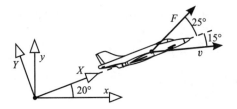

7.2 RECTANGULAR COORDINATES

Theory

Rectangular coordinates are the most basic type of coordinates. The position vector is written as

$$\mathbf{r}(t) = x(t)\mathbf{i} + y(t)\mathbf{j} + z(t)\mathbf{k}. \qquad (7.2\text{-}1)$$

Differentiating the position vector with respect to time yields the velocity vector, and differentiating the velocity vector with respect to time yields the acceleration vector, written

$$\mathbf{v}(t) = v_x(t)\mathbf{i} + v_y(t)\mathbf{j} + v_z(t)\mathbf{k} = \dot{x}(t)\mathbf{i} + \dot{y}(t)\mathbf{j} + \dot{z}(t)\mathbf{k}, \qquad (7.2\text{-}2a)$$

$$\mathbf{a}(t) = a_x(t)\mathbf{i} + a_y(t)\mathbf{j} + a_z(t)\mathbf{k} = \ddot{x}(t)\mathbf{i} + \ddot{y}(t)\mathbf{j} + \ddot{z}(t)\mathbf{k}, \qquad (7.2\text{-}2b)$$

where the over-dot represents time differentiation, that is $d([\ ])/dt = (\dot{\ })$. The over-dot is used for convenience. The rectangular components of velocity and acceleration are thus given by

$$
\begin{array}{lll}
v_x(t) = \dot{x}(t), & v_y(t) = \dot{y}(t), & v_z(t) = \dot{z}(t), \qquad (7.2\text{-}3) \\
a_x(t) = \ddot{x}(t), & a_y(t) = \ddot{y}(t), & a_z(t) = \ddot{z}(t). \qquad (7.2\text{-}4)
\end{array}
$$

The components of velocity are obtained from the components of position through time differentiation, and the components of acceleration are obtained from the components of velocity through time differentiation. Alternatively, the components of position are obtained from the components of velocity through time integration, and the components of velocity are obtained from the components of acceleration through time integration. Since Newton's Laws deal directly with acceleration components (since $\mathbf{a} = \mathbf{F}/m$), we frequently start with known acceleration components, integrate in time to obtain velocity components, and then integrate again to obtain position components. From Eqs. (7.2-3) and (7.2-4),

$$x(t) = x(t_1) + \int_{t_1}^{t} v_x(s) \, ds,$$

$$y(t) = y(t_1) + \int_{t_1}^{t} v_y(s) \, ds, \qquad (7.2\text{-}5)$$

$$z(t) = z(t_1) + \int_{t_1}^{t} v_z(s) \, ds,$$

and

$$v_x(t) = v_x(t_1) + \int_{t_1}^{t} a_x(s) \, ds,$$

$$v_y(t) = v_y(t_1) + \int_{t_1}^{t} a_y(s) \, ds, \qquad (7.2\text{-}6)$$

$$v_z(t) = v_z(t_1) + \int_{t_1}^{t} a_z(s) \, ds.$$

It is not always the case that the forces acting on an object (and hence the accelerations) are given as functions of time. For example, the force that a spring exerts on an object depends on the stretch of the spring. Rather than the force being an explicit function of time, it depends on time only implicitly, through the length of stretch in the spring. Thus, let's consider the situation in which acceleration is given as a function of position.

DETERMINATION OF $v_x(x)$ FROM $a_x(x)$

From the chain rule of differentiation

$$a_x(t) = \frac{dv_x}{dt} = \frac{dv_x}{dx}\frac{dx}{dt} = \frac{dv_x}{dx} v_x. \qquad (7.2\text{-}7)$$

Multiplying both sides by dx and integrating the result yields

$$\int_{x_1}^{x} a_x(x) \, dx = \int_{v_x(x_1)}^{v_x(x)} v_x \, dv_x = \tfrac{1}{2}v_x^2(x) - \tfrac{1}{2}v_x^2(x_1)$$

Rearranging terms yields

$$v_x^2(x) = v_x^2(x_1) + 2 \int_{x_1}^{x} a_x(x)\, dx \qquad (7.2\text{-}8)$$

Examples

7.2-1   Find the position x, velocity v, and acceleration a of the block shown at time $t = 30\,\text{s}$ given that the path is $x(t) = 2t^2 - 3t + 15$ m.

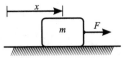

Examples 7.2-1 and 7.2-2

Solution
By time differentiation,

$$x(t) = 2t^2 - 3t + 15 = 2(30)^2 - 3(30) + 15 = 1725\,\text{m},$$
$$\dot{x}(t) = 4t - 3 = 4(30) - 3 = 117\,\text{m/s},$$
$$\ddot{x}(t) = 4\,\text{m/s}^2.$$

7.2-2   Find the position $x(t)$ and velocity $v(t)$ of the block shown given that the block is subjected to an applied force F. Initially the block's velocity is $v(0) = 3$ m/s and its position is $x(0) = 13$ m. Let the surface under the block be rough so that the block is subjected to a friction force. Assume that the friction force is governed by a dry friction model. The associated kinetic coefficient of friction is $\mu_k = 0.4$. Let $m = 20\,\text{kg}$ and $F = 100\,\text{N}$.

Solution
According to the dry friction model, the friction force between two surfaces in which there is relative motion is linearly proportional to the normal force between the surfaces. Denoting the normal force by N, the friction force between the two surfaces is $F_f = \mu_k N$. The direction of the friction force acting on the block is the opposite of the direction of the motion of the block; the friction force is directed to the left. Recalling the kinetic dry friction model and summing forces in the horizontal direction and the vertical direction yields

$$N - mg = ma_y,$$
$$F - F_f = ma_x,$$
$$F_f = \mu_k N$$

Since the block's motion is along the x direction, $a_y = 0$, and so, from the above,

$$a_x = \frac{F - \mu_k mg}{m} = \frac{100 - 0.4(20)9.81}{20} = 1.076 \, \text{m/s}^2.$$

Now, by integration

$$v_x = v_x(0) + \int_0^t 1.076 \, dt = 3 + 1.076t \, \text{m/s},$$

$$x(t) = x(0) + \int_0^t (3 + 1.076t) \, dt = 13 + 3t + 0.538t^2 \, \text{m}.$$

7.2-3   Find the relative velocity $v_{A/B}$ and the relative acceleration $a_{A/B}$ between the two blocks shown, given that block A follows the path $x_A(t) = 4t - 13 \, \text{m}$ and block B follows the path $x_B(t) = 12t^2 + 15t - 4 \, \text{m}$.

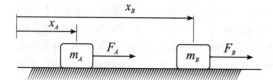

Examples 7.2-3 and 7.2-4

Solution
Relative velocity refers to the difference between two velocities; that is $v_{A/B} = v_A - v_B$. Verbally, one says **"the velocity of A relative to B."** Likewise, **relative acceleration** refers to $a_{A/B} = a_A - a_B$. Thus,

$$v_{A/B} = v_A - v_B = (4) - (24t + 15) = -24t - 11 \, \text{m/s},$$
$$a_{A/B} = a_A - a_B = (0) - (24) = -24 \, \text{m/s}^2.$$

7.2-4   The blocks A and B shown are subjected to the forces $F_A = 100e^{-t}$ lb and $F_B = -100e^{-t}$ lb.
Determine the relative position $x_{A/B}$, of the block if initially $x_A(0) = 3$ ft, $v_A(0) = 1$ ft/s, $x_B(0) = 100$ ft, and $v_B(0) = -1$ ft/s. Let $m_A = 1.5$ slug and $m_B = 1.2$ slug, assume a kinetic friction coefficient

between the blocks and the ground of $\mu_k = 0.15$. When do each of the blocks come to rest?

Solution
The free body diagrams of blocks A and B yield

$$N_A - m_A g = m_A a_{Ay},$$
$$F_A - \mu_k N_A = m_A a_{Ax},$$
$$N_B - m_B g = m_B a_{By},$$
$$F_B + \mu_k N_B = m_B a_{Bx}.$$

Notice that the friction force on block A is directed to the left and that the friction force on block B is directed to the right. We know the direction of the friction forces by the direction of the initial velocity. Also, since the blocks move in the x direction, $a_{Ay} = a_{By} = 0$. Solving these linear algebraic equations yields

$$a_{Ax} = \frac{F_A - \mu_k m_A g}{m_A} = \frac{100e^{-t} - 0.15(1.5)32.2}{1.5} = 66.7e^{-t} - 0.483,$$

$$a_{Bx} = \frac{F_B + \mu_k m_B g}{m_B} = \frac{-100e^{-t} + 0.15(1.2)32.2}{1.2} = -83.3e^{-t} + 0.483.$$

Next, by integration,

$$v_{Ax}(t) = v_{Ax}(0) + \int_0^t (66.7e^{-t} - 0.483) \, dt = 1 + 66.7(1 - e^{-t}) - 0.483t,$$

$$x_A(t) = x_A(0) + \int_0^t [1 + 66.7(1 - e^{-t}) - 0.483t] \, dt$$
$$= 3 + t + 66.7t - 66.7(1 - e^{-t}) - 0.2415t^2,$$

and

$$v_{Bx}(t) = v_{Bx}(0) + \int_0^t (-83.3e^{-t} + 0.483) \, dt = -1 - 83.3(1 - e^{-t}) + 0.483t,$$

$$x_B(t) = x_B(0) + \int_0^t [-1 - 83.3(1 - e^{-t}) + 0.483t] \, dt$$
$$= 100 - t - 83.3t + 83.3(1 - e^{-t}) + 0.21415t^2.$$

The times at which the blocks come to rest are found by letting $v_{Ax}(t_A) = 0$ and $v_B(t_B) = 0$ above. We get the two equations

$$0 = 1 + 66.7(1 - e^{-t_A}) - 0.483t_A,$$
$$0 = -1 - 83.3(1 - e^{-t_B}) + 0.483t_B.$$

These nonlinear algebraic equations cannot be solved in closed form, so we resort to an iterative approach. We guess answers, substitute them into the equations, and get closer and closer to the zeros on the left-hand side of the equations. We find

$$t_A = 140\,\text{s}, \qquad t_B = 174\,\text{s}.$$

Next, we determine $x_{A/B}$. We find

$$x_{A/B} = x_A - x_B = [3 + t + 66.7t - 66.7(1 - e^{-t}) - 0.2415t^2]$$
$$- [100 - t - 83.3t + 83.3(1 - e^{-t}) + 0.2415t^2]$$
$$= -97 + 152t + 150(1 - e^{-t}) - 0.483t^2\,\text{ft}.$$

These equations are valid throughout the time that block A moves to the right and block B moves to the left. If the block were to move to the left, the direction of the friction force would change, thereby changing the free body diagram and requiring a new set of equations.

 7.2-5   Find the magnitude of the velocity and the magnitude of the acceleration of a particle at time $t = 4\,\text{s}$ if its position is described by $\mathbf{r}(t) = \cos t\,\mathbf{i} + \sin(8t - 4)\,\mathbf{j}\,\text{m}$.

Solution
By time differentiation, we get

$$\mathbf{v}(t) = -\sin t\,\mathbf{i} + 8\cos(8t - 4)\,\mathbf{j}$$
$$\mathbf{a}(t) = -\cos t\,\mathbf{i} - 64\sin(8t - 4)\,\mathbf{j}$$

so, at $t = 4\,\text{s}$,

$$\mathbf{v}(t) = 0.757\mathbf{i} - 7.70\mathbf{j}$$
$$\mathbf{a}(t) = 0.654\mathbf{i} - 17.34\mathbf{j}$$

thus

$$v = 77.3\,\text{m/s}, \qquad a = 16.5\,\text{m/s}^2.$$

Note that your calculator should always be in radians unless the units specifically say degrees.

7.2-6 A football is thrown into the air with an initial speed of v_0 and with an initial angle of θ_0 (relative to the horizontal axis). Determine a formula for how far the football goes. Neglect air drag.

Solution
While in flight, the football is subject to the resultant force $F = mg$ directed downward. From Newton's Second Law, $a_x = 0$ and $a_y = -g$. The initial velocity vector is $\mathbf{v_0} = v_0(\cos\theta\,\mathbf{i} + \sin\theta\,\mathbf{j})$. Thus, by integration,

$$v_x(t) = v_x(0) + \int_0^t a_x\,dt = v_0\cos\theta_0 + 0 = v_0\cos\theta_0,$$

$$x(t) = x(0) + \int_0^t v_x(t)\,dt = 0 + \int_0^t v_0\cos\theta_0\,dt = v_0 t\cos\theta_0,$$

and

$$v_y(t) = v_y(0) + \int_0^t a_y\,dt = v_0\sin\theta_0 - gt,$$

$$y(t) = y(0) + \int_0^t v_y(t)\,dt = 0 + \int_0^t (v_0\sin\theta_0 - gt)\,dt = v_0 t\sin\theta_0 - \tfrac{1}{2}gt^2$$

$$= t(v_0\sin\theta_0 - \tfrac{1}{2}gt).$$

The football hits the ground at the final time t_f when $y(t_f) = 0 = v_0\sin\theta_0 - \tfrac{1}{2}gt_f$, so $t_f = (2v_0\sin\theta_0)/g$. Substituting this expression for t_f into $x(t)$ above yields the distance D that the football goes:

$$D = \frac{2v_0^2\sin\theta_0\cos\theta_0}{g} = \frac{v_0^2\sin 2\theta_0}{g}.$$

Notice that the football travels the farthest when $2\theta_0 = 90°$, that is when $\theta_0 = 45°$.

Problems

7.2-1 Find the velocity $v(t)$ and acceleration $a(t)$ of a particle that follows the path $x(t) = t^4 - 20t^2 + 60t - 8$ m.

7.2-2 A block is subjected to a force as shown in the figure for Examples 7.2-1 and 7.2-2. Determine the block's velocity $v(x)$ as a function of x. Let $m = 50\,\text{kg}$ and $\mu_k = 0.1$. Assume that the applied force is $F = 30(x^2 - 3)\,\text{N}$, and that the block starts out with $v(0) = 1\,\text{m/s}$ and $x(0) = 3\,\text{m}$.

7.2-3 Find the relative velocity $v_{A|B}$, and the relative acceleration $a_{A|B}$ of two particles A and B if particle A has velocity $v_A(t) = 17t - 2\,\text{m/s}$ and particle B has acceleration $a_B(t) = 3t\,\text{m/s}^2$, and let $x_A(0) = 1\,\text{m}$ and $v_B(0) = 6\,\text{m/s}$.

7.2-4 Find the relative position $x_{A|B}$, the relative velocity $v_{A|B}$, and the relative acceleration $a_{A|B}$ of the blocks shown in the figure for Examples 7.2-3 and 7.2-4. Let block A have mass $m_A = 1.5\,\text{slug}$ and be subjected to an applied force $F_A = 18\,\text{lb}$, and let block B have mass $m_B = 1.2$ slug and be subjected to a force $F_B = 16\,\text{lb}$. Let $\mu_k = 0.1$ between the surfaces. Initially, $x_A(0) = 3\,\text{ft}$, $v_A(0) = 1\,\text{ft/s}$, $x_B(0) = 10\,\text{ft}$, and $v_B(0) = -1\,\text{ft/s}$.

7.2-5 Find the height at which a projectile strikes a wall 30 ft away after being fired $45°$ from the horizontal with an initial velocity of $v(0) = 30\,\text{ft/s}$.

7.2-6 The particle shown slides on a planar surface while being subjected to a resultant applied force. Find the resultant applied force **F**. The particle has mass $m = 38\,\text{kg}$, and it follows the path $x(t) = 10\sin t\,\text{m}$, $y(t) = 2t^2 - 21t + 53\,\text{m}$.

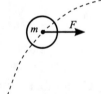

Problem 7.2-6

7.2-7 Determine the relative position vector $\mathbf{r}_{A|B}$, the relative velocity vector $\mathbf{v}_{A|B}$, and the relative acceleration vector $\mathbf{a}_{A|B}$ between objects A and B. Object A follows the path $\mathbf{r}_A = (3t^2 - 4t\cos t)\mathbf{i} + 5t\mathbf{j}$ and object B follows the path $\mathbf{r}_B = 16\mathbf{i} - 2t^3\mathbf{j}$.

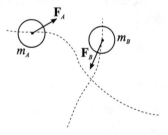

Problems 7.2-7 and 7.2-8

7.2-8 Two particles move in a horizontal plane as shown. Determine the velocity $\mathbf{v}_{A|B}$ of particle A relative to particle B. The resultant force acting on particle A is

$\mathbf{F}_A = (3t+2)\mathbf{i} + (4t+3)\mathbf{j}$, and particle B moves along the path $\mathbf{r}_B(t) = (12t-4)\mathbf{i} + (2t^2)\mathbf{j}$. Particle A is initially at rest, and $m_A = \frac{1}{2}$.

7.2-9 Find the path $\mathbf{r}(t)$ and velocity $\mathbf{v}(t)$ of a small particle flying in the air. The particle has mass $m = 0.125$ kg and is being influenced by the resultant force $\mathbf{F}(t) = (0.3t^2 - 0.4t + 0.8)\mathbf{i} + 0.1t\mathbf{j} + 0.5\cos\left(\frac{1}{30}\right)t\mathbf{k}$ N. The particle is initially located at the origin and at rest.

7.2-10 Find the relative position vector $\mathbf{r}_{R|Ed}(t)$ and the relative velocity vector $\mathbf{v}_{R|Ed}(t)$ of a model rocket as seen by Ed. Ed is traveling on a merry-go-round following the path $\mathbf{r}_{Ed}(t) = 50\sin t\,\mathbf{i} + 50\cos t\,\mathbf{j}$ ft and the rocket blasts off following the path $\mathbf{r}_R(t) = 200\mathbf{i} + 5t\mathbf{j} + [800 - 32(t-5)^2]\mathbf{k}$ ft.

7.2-11 Find the path of a small rocket $\mathbf{r}_{R|Ed}(t)$ as seen by Ed. Ed is now on a train traveling with a velocity of $\mathbf{v}_{Ed}(t) = 70\mathbf{i} + 40\mathbf{j}$ ft/s. The rocket, which was initially located at $\mathbf{r}_R(0) = 2000\mathbf{i}$ ft, is fired upward when Ed is at $\mathbf{r}_{Ed}(0) = -1000\mathbf{i}$ ft. For a short time, the rocket has mass $m_R = 300$ slug and a constant thrust of $\mathbf{F}_R(t) = 11\,000\mathbf{k}$ lb.

7.3 POLAR COORDINATES

Theory

We have seen that the **standard** way of expressing the position vector of a point is $\mathbf{r} = x\mathbf{i} + y\mathbf{j} + z\mathbf{k}$, in which $\mathbf{i} = (1 \quad 0 \quad 0)$, $\mathbf{j} = (0 \quad 1 \quad 0)$, and $\mathbf{k} = (0 \quad 0 \quad 1)$ denote **standard** unit vectors and where x, y, and z are called **rectangular coordinates**.

Another convenient way of expressing the position vector of a point is by writing it as a magnitude times a direction. This is called the **polar** form of the position vector. As shown in Fig. 7.3-1, the polar form of the position vector is

$$\mathbf{r} = r\mathbf{n}_r, \tag{7.3-1}$$

in which $r = |\mathbf{r}|$ denotes the magnitude of $\mathbf{r}$ and $\mathbf{n}_r$ denotes the unit vector in the direction of $\mathbf{r}$. Polar coordinates are used when the motion

of the point is restricted to a flat plane. They are
particularly convenient to use when the point
moves in a circular or circular-like path.

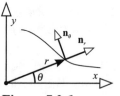

Figure 7.3-1

When polar coordinates are used, the vectors
are expressed in terms of the unit vector $\mathbf{n}_r$, called
the **radial unit vector**, and the unit vector $\mathbf{n}_\theta$,
called the **circumferential unit vector**. The
circumferential unit vector is perpendicular to $\mathbf{n}_r$.

Notice that the unit vectors $\mathbf{n}_r$ and $\mathbf{n}_\theta$ are merely **rotated** unit vectors,
like those we saw in Section 7.1. The $\mathbf{I}$ and $\mathbf{J}$ in that section are the same
as the $\mathbf{n}_r$ and $\mathbf{n}_\theta$ here. Therefore, $\mathbf{n}_r$ and $\mathbf{n}_\theta$ are related to $\mathbf{i}$ and $\mathbf{j}$ by

$$\mathbf{n}_r(t) = \cos\theta(t)\,\mathbf{i} + \sin\theta(t)\,\mathbf{j}, \qquad (7.3\text{-}2a)$$

$$\mathbf{n}_\theta(t) = -\sin\theta(t)\,\mathbf{i} + \cos\theta(t)\,\mathbf{j}. \qquad (7.3\text{-}2b)$$

Notice that the radial and circumferential unit vectors are functions of
time. Unlike the standard unit vectors, the radial and circumferential unit
vectors depend on the angle θ, which in turn depends on time t.
Differentiating Eq. (7.3-2a) with respect to time using the chain rule
yields

$$\dot{\mathbf{n}}_r = -\dot{\theta}\sin\theta\,\mathbf{i} + \dot{\theta}\cos\theta\,\mathbf{j} = \dot{\theta}(-\sin\theta\,\mathbf{i} + \cos\theta\,\mathbf{j}) = \dot{\theta}\mathbf{n}_\theta$$

Similarly, differentiating Eq. (7.3-2b) with respect to time yields

$$\dot{\mathbf{n}}_\theta(t) = -\dot{\theta}\mathbf{n}_r.$$

Thus, the derivatives of the unit vectors are given by

$$\boxed{\begin{aligned} \dot{\mathbf{n}}_r &= \dot{\theta}\mathbf{n}_\theta, \\ \dot{\mathbf{n}}_\theta &= -\dot{\theta}\mathbf{n}_r. \end{aligned}}$$

$$(7.3\text{-}3a)$$

$$(7.3\text{-}3b)$$

We are ready to differentiate $\mathbf{r}$ with respect to time to obtain the velocity
vector $\mathbf{v}$, and to differentiate $\mathbf{v}$ to obtain the acceleration vector $\mathbf{a}$. We

differentiate Eq. (7.3-1) with respect to time using the product rule for differentiation and Eq. (7.3-3a) to get

$$\mathbf{v} = v_r\mathbf{n}_r + v_\theta\mathbf{n}_\theta, \quad \text{where} \quad v_r = \dot{r}, \quad v_\theta = r\dot{\theta}. \tag{7.3-4}$$

Differentiating Eq. (7.3-4) with respect to time yields

$$\mathbf{a} = a_r\mathbf{n}_r + a_\theta\mathbf{n}_\theta, \quad \text{where} \quad a_r = \ddot{r} - r\dot{\theta}^2, \quad a_\theta = r\ddot{\theta} + 2\dot{r}\dot{\theta}. \tag{7.3-5}$$

Notice that the radial component of acceleration a_r is made up of two terms and that the circumferential component of acceleration is made up of two terms. The radial term $-r\dot{\theta}^2$ (which is always negative) is called the **centrifugal** term. The circumferential term $2\dot{r}\dot{\theta}$ is called the **Coriolis** term. A graphical representation of each of the acceleration terms is shown in Fig. 7.3-2(a–e). The figures show how each of the terms arises.

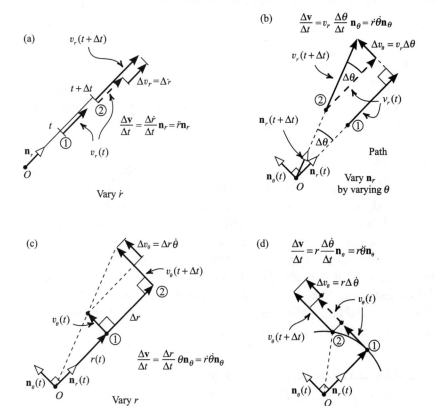

Figure 7.3-2(a–d)

(e) $\dfrac{\Delta \mathbf{v}}{\Delta t} = r\dot\theta \dfrac{\Delta\theta}{\Delta t}(-\mathbf{n}_r) = -r\dot\theta^2 \mathbf{n}_r$

Vary $\mathbf{n}_\theta$
by varying θ ***Figure 7.3-2(e)***

Notice that each figure shows an acceleration term arising by varying one of the variables and leaving the others fixed (which is what is done in the product rule for differentiation). In each figure, the particle moves from point 1 to point 2 over the infinitesimal time Δt.

Examples

 7.3-1 Let the position of a point be $\mathbf{r} = 10\cos t\,\mathbf{i} + 10\sin t\,\mathbf{j}$ m. Rewrite $\mathbf{r}$ using polar coordinates, and express $\mathbf{v}$ and $\mathbf{a}$ using polar coordinates, too. Describe the motion of the point.

Solution
By differentiation, the velocity vector and the acceleration vector expressed in rectangular coordinates are $\mathbf{v} = -10\sin t\,\mathbf{i} + 10\cos t\,\mathbf{j}$ m/s and $\mathbf{a} = -10\cos t\,\mathbf{i} - 10\sin t\,\mathbf{j}$ m/s^2. Let's now express them using polar coordinates. To do this, we first need to determine the radial and circumferential unit vectors. The radial unit vector is $\mathbf{n}_r = \mathbf{r}/r = [10\cos t\,\mathbf{i} + 10\sin t\,\mathbf{j}]/10 = \cos t\,\mathbf{i} + \sin t\,\mathbf{j}$. In order to determine $\mathbf{n}_\theta$, compare the $\mathbf{n}_r$ that we have just determined with the $\mathbf{n}_r$ given in Eqn. (7.3-2a), and notice that $\theta = t$. We get $\mathbf{n}_\theta = -\sin t\,\mathbf{i} + \cos t\,\mathbf{j}$.

With the radial and circumferential unit vectors in hand, the radial and circumferential components of $\mathbf{v}$ and $\mathbf{a}$ become

$$v_r = \mathbf{v}\cdot\mathbf{n}_r = (-10\sin t\,\mathbf{i} + 10\cos t\,\mathbf{j})\cdot(\cos t\,\mathbf{i} + \sin t\,\mathbf{j}) = 0,$$

$$v_\theta = \mathbf{v}\cdot\mathbf{n}_\theta = (-10\sin t\,\mathbf{i} + 10\cos t\,\mathbf{j})\cdot(-\sin t\,\mathbf{i} + \cos t\,\mathbf{j}) = 10,$$

$$a_r = \mathbf{a}\cdot\mathbf{n}_r = (-10\cos t\,\mathbf{i} - 10\sin t\,\mathbf{j})\cdot(\cos t\,\mathbf{i} + \sin t\,\mathbf{j}) = -10,$$

$$a_\theta = \mathbf{a}\cdot\mathbf{n}_\theta = (-10\cos t\,\mathbf{i} - 10\sin t\,\mathbf{j})\cdot(-\sin t\,\mathbf{i} + \cos t\,\mathbf{j}) = 0,$$

so $\mathbf{v} = 10\mathbf{n}_\theta$ m/s and $\mathbf{a} = -10\mathbf{n}_r$ m/s². Noticing that $r = 10$ m, we see that the point moves counterclockwise around a circle, starting out (at $t = 0$) 10 units to the right of the origin (at $x = 10$ m and $y = 0$ m).

7.3-2 Let the position of a point be $\mathbf{r} = e^{-t}\cos 5t\,\mathbf{i} + e^{-t}\sin 5t\,\mathbf{j}$ m. Rewrite r using polar coordinates, and express $\mathbf{v}$ and $\mathbf{a}$ using polar coordinates too. Describe the motion of the point.

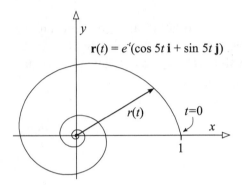

Example 7.3-2

Solution

By differentiation,

$$\mathbf{v} = e^{-t}(-\cos 5t - 5\sin 5t)\mathbf{i} + e^{-t}(-\sin 5t + 5\cos 5t)\mathbf{j}\text{ m,}$$

and

$$\mathbf{a} = e^{-t}(-24\cos 5t + 10\sin 5t)\,\mathbf{i} + e^{-t}(-24\sin 5t - 10\cos 5t)\mathbf{j}\text{ m/s}^2.$$

The magnitude of $\mathbf{r}$ is $r = [(e^{-t}\cos 5t)^2 + (e^{-t}\sin 5t)^2]^{1/2} = e^{-t}$ m and the direction is $\mathbf{n}_r = \mathbf{r}/r = \cos 5t\,\mathbf{i} + \sin 5t\,\mathbf{j}$. Comparing the $\mathbf{n}_r$ that we have just determined with the $\mathbf{n}_r$ given in Eq. (7.3-2a), we see that $\theta = 5t$, so that $\mathbf{n}_\theta = -\sin 5t\,\mathbf{i} + \cos 5t\,\mathbf{j}$.

Now, the radial and circumferential components of the velocity and acceleration are

$$v_r = \mathbf{v} \cdot \mathbf{n}_r = [e^{-t}(-\cos 5t - 5\sin 5t)\mathbf{i} + e^{-t}(-\sin 5t + 5\cos 5t)\mathbf{j}]$$
$$\cdot\,(\cos 5t\,\mathbf{i} + \sin 5t\,\mathbf{j}) = -e^{-t},$$

$$v_\theta = \mathbf{v} \cdot \mathbf{n}_\theta = [e^{-t}(-\cos 5t - 5\sin 5t)\mathbf{i} + e^{-t}(-\sin 5t + 5\cos 5t)\mathbf{j}]$$
$$\cdot\,(-\sin 5t\,\mathbf{i} + \cos 5t\,\mathbf{j}) = 5e^{-t},$$

$$a_r = \mathbf{a} \cdot \mathbf{n}_r = [e^{-t}(-24\cos 5t + 10\sin 5t)\mathbf{i} + e^{-t}(-24\sin 5t - 10\cos 5t)\mathbf{j}]$$
$$\cdot\,(\cos 5t\,\mathbf{i} + \sin 5t\,\mathbf{j}) = -24e^{-t},$$

$$a_\theta = \mathbf{a} \cdot \mathbf{n}_\theta = [e^{-t}(-24\cos 5t + 10\sin 5t)\mathbf{i} + e^{-t}(-24\sin 5t - 10\cos 5t)\mathbf{j}]$$
$$\cdot\,(-\sin 5t\,\mathbf{i} + \cos 5t\,\mathbf{j}) = -10e^{-t},$$

so $v = e^{-t}(-\mathbf{n}_r + 5\mathbf{n}_\theta)\,\text{m/s}$, and $\mathbf{a} = e^{-t}(-24\mathbf{n}_r - 10\mathbf{n}_\theta)\,\text{m/s}^2$. At $t = 0\,\text{s}$, the point is on the x axis at $x = 1\,\text{m}$ and the velocity is upward and slightly to the left. The angle between the position vector and the velocity vector is constant and slightly greater than $90°$. The point is spiraling in toward the origin, making one revolution every $\frac{2}{5}\pi\,\text{s}$. After each revolution, the radius of the spiral decreases by a factor of $e^{-2\pi/5} = 0.285$.

7.3-3 A block rests on a table that is rotating at a constant angular velocity $\dot\theta$ as shown. Determine the largest angular velocity that can be reached before the block begins to slide. The static coefficient of friction is μ_s and the distance from the origin is R.

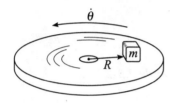

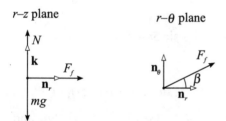

r–z plane r–θ plane

Example 7.3-3

Solution

In the vertical direction, the normal force balances with the gravitational force, so $N = mg$. In the plane of the surface, the direction of the friction force is unknown, as shown in the free body diagram. Summing forces in the radial and circumferential directions yields

$$\mu_s mg \cos\beta = ma_r,$$

$$\mu_s mg \sin\beta = ma_\theta,$$

in which

$$a_r = \ddot r - r\dot\theta^2 = -R\dot\theta^2,$$

$$a_\theta = r\ddot\theta + 2\dot r\dot\theta = 0$$

(since r and $\dot\theta$ are constant, $\dot r = 0$, $\ddot r = 0$, $\ddot\theta = 0$), and so

$$\mu_s mg \cos \beta = -mR\dot\theta^2,$$
$$\mu_s mg \sin \beta = 0.$$

Thus, $\beta = 180°$ and $\dot\theta = \sqrt{\mu_s g/R}$. Just before sliding, the friction force is directed toward the origin, and the block will slide outward.

| **7.3-4** | Find the orbital altitude h for which an object of mass $m_A = 5000$ kg remains in geosynchronous orbit. The Earth has mass $m_E = 5.97 \times 10^{24}$ kg and a mean radius of $R_E = 6370$ km. Also, recall that $G = 6.67 \times 10^{-11}$ m³/s²·kg. | 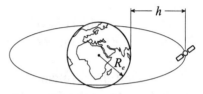 |

Example 7.3-4

Solution

A geosynchronous orbit is an orbit that is synchronized with the Earth's orbit, one revolution per day. Directly above the equator, an object in geosynchronous orbit appears to be fixed in the sky. From Newton's Law of Gravitation,

$$F = -G\frac{m_E m_A}{(R_E + h)^2} = m_A a_r = -m_A(R_E + h)\dot\theta^2.$$

Recognizing that the period of the Earth is $t = 86400$ seconds so the angular velocity of the Earth is $\dot\theta = 2\pi/T = 6.28/86\,400 = 7.27 \times 10^{-5}$ rad/s,

$$\frac{h}{R_E} = \left(\frac{Gm_E}{\dot\theta^2 R_E^3}\right)^{1/3} - 1 = 5.53.$$

so $h = 35860$ km (22280 miles).

Problems

| **7.3-1** | Find the unit vectors $\mathbf{n}_r$ and $\mathbf{n}_\theta$ for an object following the path $x(t) = 12 \cos t^2$ and $y(t) = t - 3$.

| **7.3-2** | Determine the resultant force exerted on a child who is spinning on a merry-go-round. Assume that the child is tightly gripping the

railings $r = 5$ ft from the center. The child weighs 60 lb and the merry-go-round is rotating at 0.25 rev/s.

7.3-3 Determine the difference in angles $\theta_{A|B}(t)$ and the time of impact t_{crash} of two cars traveling counterclockwise around a single-lane circular track of radius $R = 300$ ft. At time $t = 0$, car A is at $r_A = 300\mathbf{j}$ and car B is at $r_B = 300\mathbf{i}$. Car A travels at a constant $v_A = 45$ mph and car B travels at a constant $v_B = 37$ mph.

7.3-4 Let an object have velocity $\mathbf{v}(r) = \frac{1}{5}r\mathbf{n}_r + 6r^2\mathbf{n}_\theta$, $\mathbf{r}(0) = \mathbf{i} - 2\mathbf{j}$. Determine the position vector of the object at time $t = 5$ s.

7.3-5 Find the resultant force acting on a block of mass $m = 4$ slug that follows the path $r(t) = 2t^2 + 19$ ft and $\theta(t) = 4t$ rad.

7.3-6 Find the angular velocity of a freely swinging child's swing at $\theta = 45°$. The swing weighs 15 lb and has a chain length of $d = 8$ ft. At $\theta = 0$, the swing is at the bottom of its arc and has an angular velocity of 2 rad/s.

7.3-7 Determine the speed v of the gripper on the robotic arm shown. Segment 1 is rotating at $\dot{\theta}_1 = 1$ rad/s and segment 2 is rotating at $\dot{\theta}_2 = -2$ rad/s while $\theta_1 = 45°$ and $\theta_2 = 30°$. Let $d_1 = 2$ ft and $d_2 = 1.5$ ft.

7.3-8 A bead slides on a rod of length 2 ft, as shown. Determine the radial and circumferential components of the speed of the bead just before it reaches the tip of the rod. The bead is initially located at the center of the rod. The rod rotates at a constant angular velocity $\dot{\theta} = 5\pi$ rad/s (2.5 rev/s). Neglect the effect of gravity and neglect the friction between the bead and the rod.

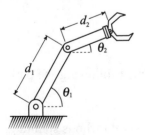

Problem 7.3-7

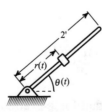

Problem 7.3-8

7.4 CYLINDRICAL COORDINATES

Theory

Cylindrical coordinates represent an extension of polar coordinates to three dimensions. The extension is directed perpendicular to the polar coordinate plane in the direction of the z axis. The common error that one makes when using cylindrical coordinates is to let r denote the distance between the origin and the tip of the position vector, as in polar coordinates. In cylindrical coordinates, however, r denotes the distance between the origin and the **projection** of the position vector onto the $r-\theta$ plane, as shown in Fig. 7.4-1. In cylindrical coordinates, $|\mathbf{r}| \neq r$. Denoting the projection of $\mathbf{r}$ onto the $r-\theta$ plane by $\mathbf{r}_p = r\mathbf{n}_r$, then $|\mathbf{r}_p| = r$. As shown, the position vector in cylindrical coordinates is

$$\mathbf{r} = r\mathbf{n}_r + z\mathbf{n}_z, \tag{7.4-1}$$

where $\mathbf{n}_z$ denotes the unit vector in the z direction. Differentiating Eq. (7.4-1) with respect to time yields the velocity vector, and differentiating the velocity vector with respect to time yields the acceleration vector, written as

$$\mathbf{v} = v_r\mathbf{n}_r + v_\theta\mathbf{n}_\theta + v_z\mathbf{n}_z, \quad \text{where} \quad v_r = \dot{r}, \quad v_\theta = r\dot{\theta}, \quad v_z = \dot{z}, \tag{7.4-2}$$

$$\mathbf{a} = a_r\mathbf{n}_r + a_\theta\mathbf{n}_\theta + a_z\mathbf{n}_z, \quad \text{where} \quad a_r = \ddot{r} - r\dot{\theta}^2, \quad a_\theta = r\ddot{\theta} + 2\dot{r}\dot{\theta}, \quad a_z = \ddot{z}. \tag{7.4-3}$$

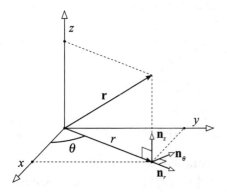

Figure 7.4-1

Notice that the radial and circumferential components of position, velocity, and acceleration in cylindrical coordinates and the corresponding components in polar coordinates are the same [see Eqs. (7.3-4) and (7.3-5)]. The differences lie in the z direction.

Examples

7.4-1 Determine the unit vectors $\mathbf{n}_r$, $\mathbf{n}_\theta$, and $\mathbf{n}_z$, and the cylindrical components of the position vector $\mathbf{r} = 4\mathbf{i} + 3\mathbf{j} + 12\mathbf{k}$ m.

Solution
The projection of the position vector onto the r–θ plane is $\mathbf{r}_p = 4\mathbf{i} + 3\mathbf{j}$. The magnitude of the projection of $\mathbf{r}$ is $r = \sqrt{4^2 + 3^2} = 5$. The radial unit vector is then $\mathbf{n}_r = \mathbf{r}_p/r = (4\mathbf{i} + 3\mathbf{j})/5 = 0.8\mathbf{i} + 0.6\mathbf{j}$. From the unit vector transformations, $\mathbf{n}_r = 0.8\mathbf{i} + 0.6\mathbf{j} = \cos\theta\,\mathbf{i} + \sin\theta\,\mathbf{j}$, so $\cos\theta = 0.8$ and $\sin\theta = 0.6$. Thus, $\mathbf{n}_\theta = -\sin\theta\,\mathbf{i} + \cos\theta\,\mathbf{j} = -0.6\mathbf{i} + 0.8\mathbf{j}$. The unit vector $\mathbf{n}_z$ is the same as the unit vector $\mathbf{k}$. We now verify the cylindrical components of the position vector. Taking dot products,

$$r = \mathbf{r} \cdot \mathbf{n}_r = (4\mathbf{i} + 3\mathbf{j} + 12\mathbf{k}) \cdot (0.8\mathbf{i} + 0.6\mathbf{j}) = 4(0.8) + 3(0.6) = 5\,\text{m},$$
$$z = \mathbf{r} \cdot \mathbf{n}_z = (4\mathbf{i} + 3\mathbf{j} + 12\mathbf{k}) \cdot \mathbf{k} = 12\,\text{m}.$$

so

$$\mathbf{r} = 5\mathbf{n}_r + 12\mathbf{n}_z\,\text{m}.$$

7.4-2 A bead is sliding down a cylindrical spiral having an incline angle of $\gamma = 10°$, as shown. Neglect the friction between the bead and the spiral, and assume that the bead starts from rest at the top of the spiral. Determine the position vector of the bead as a function of time.

Solution
As shown in the free body diagram, the bead is subjected to a gravitational force and a normal force, which has been broken down into two components that are perpendicular to the tangent of the spiral. Summing forces along the cylindrical directions, we get

$$N_r = ma_r = m(-R\dot{\theta}^2),$$
$$N_n \cos\gamma - mg = ma_z = m\ddot{z},$$
$$-N_n \sin\gamma = ma_\theta = mR\ddot{\theta},$$

where we notice that $\dot{r} = 0$ since r is constant. We also get the velocity vector in the form $\mathbf{v} = R\dot{\theta}\mathbf{n}_\theta + \dot{z}\mathbf{n}_z = +v\cos\gamma\,\mathbf{n}_\theta + v\sin\gamma\,\mathbf{n}_z$, taking v

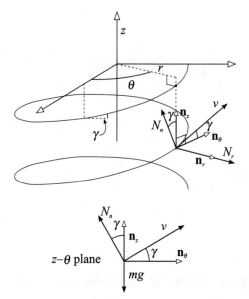

z–θ plane

Example 7.4-2

positive up the spiral. It follows that $R\dot\theta = v\cos\gamma$ and $\dot z = v\cos\gamma$.
Substituting $\ddot\theta = \dot v\cos\gamma/R$ and $\ddot z = \dot v\cos\gamma$ into the equations of motion
yields

$$N_r = -\frac{mv^2\cos^2\gamma}{R},$$

$$N_n\cos\gamma - mg = m\dot v\sin\gamma,$$

$$-N_n\sin\gamma = mR\ddot\theta = m\dot v\cos\gamma,$$

which represents three equations expressed in terms of the three
unknowns N_r, N_n, and $\dot v$. The solution is

$$N_r = -\frac{mv^2\cos^2\gamma}{R}, \qquad N_n = mg\cos\gamma \qquad \dot v = -g\sin\gamma.$$

By integration, $v = -gt\sin\gamma = R\dot\theta$, so

$$\theta(t) = \theta(0) - \int_0^t \frac{gt\sin\gamma}{R}\,dt = \theta(0) - \frac{g\sin\gamma}{2R}t^2,$$

$$z = z(0) + \int_0^t -gt\sin\gamma\cos\gamma\,dt = z(0) + \frac{-gt\sin\gamma\cos\gamma}{2}.$$

We get the position vector

$$r = R\cos\theta\,\mathbf{i} + R\sin\theta\,\mathbf{j} + \left[z(0) - \frac{gt^2\sin\gamma\cos\gamma}{2}\right]\mathbf{k},$$

in which

$$\theta(t) = \theta(0) - \frac{g\sin\gamma}{2R}t^2.$$

Problems

7.4-1 Determine the unit vectors $\mathbf{n}_r$, $\mathbf{n}_\theta$, and $\mathbf{n}_z$ for the position vector $r = 26\mathbf{i} + 17\mathbf{j} + 30\mathbf{k}$.

7.4-2 Determine the velocity v of a piece of chewing gum stuck on the bottom of a spinning chair seat as it rotates down along a threaded post. The gum is 6 inches from the center of the seat, which is rotating at 16 rev/s. The post has 7 threads per inch.

7.4-3 Determine the maximum velocity that a car can race down the spiral shown. The static coefficient of friction between the car and the incline is $\mu_s = 0.3$, $R = 80$ ft, and $\gamma = 20°$.

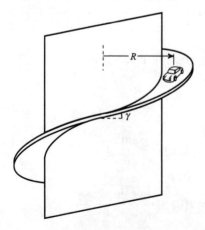

Problem 7.4-3

7.5 TANGENTIAL–NORMAL COORDINATES

Theory

When tangential–normal coordinates are used, the vector quantities are expressed in terms of components that act in the direction of the velocity vector and in the direction perpendicular, or **normal**, to the velocity vector. The direction of the velocity vector is also the tangent to the path that the point follows. This is why the coordinates used in this section are called tangential–normal coordinates. These coordinates are used when the motion is confined to a plane, and they are particularly convenient to use when the forces in the free body diagram act in the direction of the velocity vector and perpendicular to the velocity vector. This occurs in aerodynamic and hydrodynamic problems in which drag forces act along the velocity vector and lift forces act perpendicular to the velocity vector. It also occurs in certain mechanisms, such as in guided-rail problems, in which normal reactions act in the normal direction and friction forces act in the tangential direction. Notice, in contrast with polar coordinates, in which the vectors are based on position, that with tangential–normal coordinates the vectors are based on velocity.

Tangential–normal coordinates are set up by first writing the velocity vector as magnitude times direction. Referring to Fig. 7.5-1, we let

$$\mathbf{v} = v\mathbf{n}_t, \tag{7.5-1}$$

where v denotes the magnitude of the velocity vector and $\mathbf{n}_t$ denotes the **tangential unit vector**. The tangential unit vector is directed along the velocity vector. We also define the unit vector $\mathbf{n}_n$, called the **normal unit vector**. The normal unit vector is defined to be perpendicular to the tangential unit vector. Just as in the case of polar coordinates, the unit vectors $\mathbf{n}_t$ and $\mathbf{n}_n$ are **rotated** unit vectors, and thus are related to the standard unit vectors through transformation equations. From Eq. (7.3-3), replacing $\mathbf{n}_r$ and $\mathbf{n}_\theta$ with $\mathbf{n}_t$ and $\mathbf{n}_n$, we get

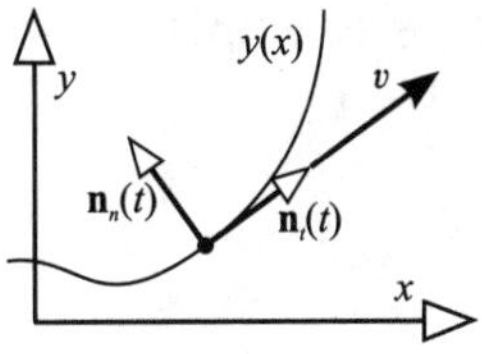

Figure 7.5-1

$$\dot{\mathbf{n}}_t = \dot{\theta}\mathbf{n}_n, \qquad \dot{\mathbf{n}}_n = -\dot{\theta}\mathbf{n}_t. \tag{7.5-2}$$

Next, referring to Fig. 7.5-2, we define the **radius of curvature** ρ of the path at time t, by intersecting the lines extending from $\mathbf{n}_t(t)$ and $\mathbf{n}_t(t + \Delta t)$ in which Δt is a small increment of time. Over the time increment Δt, the angle changes an incremental amount $\Delta\theta$ and the point moves an incremental amount Δs. Notice that $\rho\,\Delta\theta = \Delta s$, so dividing by Δt and letting the incremental quantities become infinitesimal yields

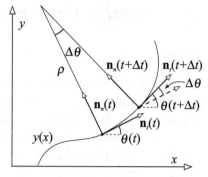

Figure 7.5-2

$$\rho\dot{\theta} = v, \qquad (7.5\text{-}3)$$

in which $d\theta/dt = \dot{\theta}$ and $ds/dt = v$. Now, differentiating $\mathbf{v}$ in Eq. (7.5-1) with respect to time yields

$$\mathbf{a} = a_t\mathbf{n}_t + a_n\mathbf{n}_n, \qquad \text{where} \qquad a_t = \dot{v}, \quad a_n = \frac{v^2}{\rho}. \qquad (7.5\text{-}4)$$

Notice in Eq. (7.5-4) that $|\mathbf{a}| \neq \dot{v}$.

With tangential–normal coordinates, it is useful to have a formula for the radius of curvature of a function $y(x)$. It can be shown from Fig. 7.5-3 (see Example 7.5-4) that

$$\rho = \frac{\left[1 + \left(\dfrac{dy}{dx}\right)^2\right]^{3/2}}{\dfrac{d^2y}{dx^2}}. \qquad (7.5\text{-}5)$$

It is interesting to notice that when the slope $dy/dx = \tan\theta$ of the function $y(x)$ is small (when $\tan\theta \ll 1$), the first derivative of the function is approximately $dy/dx = \theta$. From Eq. (7.5-5), when the slope of the function is small, the second derivative of the function is approximately $d^2y/dx^2 = 1/\rho$.

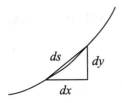

$$\frac{ds}{dx} = \sqrt{1 + \left(\frac{dy}{dx}\right)^2}$$

Figure 7.5-3

Examples

7.5-1 Find the radius of curvature of a particle that follows the path $y(x) = x^3 - 16x + 50$ m when $x = 4$.

Solution

Differentiating the function for the path yields

$$\frac{dy}{dx} = 3x^2 - 16 = 32, \qquad \frac{d^2y}{dx^2} = 6x = 24,$$

so, from Eq. (7.5-4),

$$\rho = \frac{(1 + 32^2)^{3/2}}{24} = 1367.3 \text{ m}.$$

7.5-2 Determine the polar form of the velocity vector $\mathbf{v} = 6\mathbf{n}_t$ m/s if the position vector is $\mathbf{r} = 5\mathbf{n}_r$ m, letting

$$\mathbf{n}_r = \frac{5}{13}\mathbf{i} - \frac{12}{13}\mathbf{j}, \qquad \mathbf{n}_t = \frac{7}{\sqrt{113}}\mathbf{i} + \frac{8}{\sqrt{113}}\mathbf{j}.$$

Solution

From the transformation equations for rotated unit vectors,

$$\mathbf{n}_\theta = \frac{12}{13}\mathbf{i} + \frac{5}{13}\mathbf{j}, \qquad \mathbf{n}_n = \frac{-8}{\sqrt{113}}\mathbf{i} + \frac{7}{\sqrt{113}}\mathbf{j}.$$

The polar components of the velocity vector are now easily determined by taking the dot products:

$$v_r = \mathbf{v} \cdot \mathbf{n}_r = 6\left(\frac{7}{\sqrt{113}}\mathbf{i} + \frac{8}{\sqrt{113}}\mathbf{j}\right) \cdot \left(\frac{5}{13}\mathbf{i} - \frac{12}{13}\mathbf{j}\right) = -2.65\,\text{m/s}.$$

$$v_\theta = \mathbf{v} \cdot \mathbf{n}_r = 6\left(\frac{7}{\sqrt{113}}\mathbf{i} + \frac{8}{\sqrt{113}}\mathbf{j}\right) \cdot \left(\frac{12}{13}\mathbf{i} + \frac{5}{13}\mathbf{j}\right) = 5.38\,\text{m/s}.$$

7.5-3 A car is traveling around a curved road with a radius of 50 ft. Determine the maximum speed the car can turn without slipping. Let $\mu_s = 0.3$.

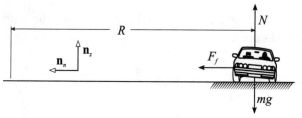

Example 7.5-3

Solution
Referring to the free body diagram, the normal force on the car balances with the gravity force on the car. Summing forces in the normal direction and applying Eq. (7.5-4) yields

$$F_f = \mu_s N = \mu_s mg = ma_n = m\frac{v^2}{R},$$

so

$$v = \sqrt{\mu_s g R} = 22\,\text{ft/s}$$

7.5-4 Derive Eq. (7.5-5).

Solution
Referring to Fig. 7.5-3, it follows from the Pythagorean Theorem that $ds = \sqrt{(dx^2 + (dy)^2}$, so

$$\frac{ds}{dx} = \sqrt{1 + \left(\frac{dy}{dx}\right)^2}.$$

From Eq. (7.5-3),

$$\rho = \frac{\dfrac{ds}{dt}}{\dfrac{d\theta}{dt}} = \frac{\dfrac{ds}{dx}}{\dfrac{d\theta}{dx}} = \frac{\sqrt{1 + \left(\dfrac{dy}{dx}\right)^2}}{\dfrac{d\theta}{dx}}.$$

We also know that

$$\frac{d^2y}{dx^2} = \frac{d}{dx}\left(\frac{dy}{dx}\right) = \frac{d\tan\theta}{dx} = \frac{d\tan\theta}{d\theta}\frac{d\theta}{dx} = (1 + \tan^2\theta)\frac{d\theta}{dx}$$

$$= \left[1 + \left(\frac{dy}{dx}\right)^2\right]\frac{d\theta}{dx},$$

so

$$\frac{d\theta}{dx} = \frac{\dfrac{d^2y}{dx^2}}{1 + \left(\dfrac{dy}{dx}\right)^2}.$$

Substituting this relationship into the above expression for ρ yields Eq. (7.5-5).

Problems

7.5-1 Determine the tangential and normal unit vectors if the projectile follows the path $y(x) = 160 - 0.1(x - 40)^2$ ft.

7.5-2 A child swings a bucket of water in a vertical circle. How fast does the child need to swing the bucket to prevent the water from spilling? Assume that the water is located 2 ft from the center of the circle.

7.5-3 Determine the tangential–normal form of the acceleration vector **a** of a particle that follows the path $r(t) = 2t$ m, $\theta(t) = 8t$ rad.

7.5-4 Express the relative velocity vector $\mathbf{v}_{A/B}(t)$ and the relative acceleration vector $\mathbf{a}_{A/B}(t)$ in terms of the tangential–normal coordinates of particle B given that $\mathbf{v}_A(t) = 2t\mathbf{i} - 4t^2\mathbf{j}$ and $\mathbf{v}_B(t) = -16\mathbf{i} + 4t\mathbf{j}$ at time $t = 10\,\text{s}$.

7.6 SPHERICAL COORDINATES

Theory

Spherical coordinates are used when objects move along spherical or spherical-like paths. As in polar coordinates, in spherical coordinates, the position vector is expressed as a magnitude times a direction. The unit vector in the direction of the position vector is called the **radial unit vector**. The angle ϕ that the position vector makes relative to the z axis is called the **azimuthal angle**. The unit vector in the direction that the position vector moves as the azimuthal angle is increased is called the **azimuthal unit vector**. The angle θ between the x axis and the projection of the position vector onto the x–y plane is the called the **polar angle**. The unit vector in the direction that the projection of the position vector moves as the polar angle is increased is called the **polar unit vector**. Vector quantities expressed in terms of spherical coordinates are broken down into components along the radial, azimuthal, and polar directions.

Using spherical coordinates, the position vector is expressed as

$$\mathbf{r} = r\mathbf{n}_r, \qquad (7.6\text{-}1)$$

where r is the magnitude of $\mathbf{r}$, and where $\mathbf{n}_r$ is the radial unit vector. From Fig. 7.6-1, the radial, azimuthal, and polar unit vectors are related to the standard unit vectors by

$$\mathbf{n}_r = \sin\phi\cos\theta\,\mathbf{i} + \sin\phi\sin\theta\,\mathbf{j} + \cos\phi\mathbf{k}, \qquad (7.6\text{-}2\text{a})$$

$$\mathbf{n}_\phi = \cos\phi\cos\theta\,\mathbf{i} + \cos\phi\sin\theta\,\mathbf{j} - \sin\phi\,\mathbf{k}, \qquad (7.6\text{-}2\text{b})$$

$$\mathbf{n}_\theta = -\sin\theta\,\mathbf{i} + \cos\theta\,\mathbf{j}. \qquad (7.6\text{-}2\text{c})$$

The radial, azimuthal, and polar unit vectors are perpendicular to each other, which can be deduced either from Fig. 7.6-1 or by checking

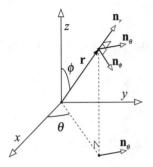

Figure 7.6-1

that $\mathbf{n}_r \cdot \mathbf{n}_\phi = \mathbf{n}_\phi \cdot \mathbf{n}_\theta = \mathbf{n}_\theta \cdot \mathbf{n}_r = 0$. As with $\mathbf{i}, \mathbf{j}$, and $\mathbf{k}$, the spherical unit vector cross products are $\mathbf{n}_r \times \mathbf{n}_\phi = \mathbf{n}_\theta$, $\mathbf{n}_\phi \times \mathbf{n}_\theta = \mathbf{n}_r$, and $\mathbf{n}_\theta \times \mathbf{n}_r = \mathbf{n}_\phi$.

Next, the spherical unit vectors given in Eqs. (7.6-2) are differentiated with respect to time and expressed in terms of the spherical unit vectors, to obtain

$$\dot{\mathbf{n}}_r = \dot{\phi}\,\mathbf{n}_\phi + \dot{\theta}\sin\phi\,\mathbf{n}_\theta, \tag{7.6-3a}$$

$$\dot{\mathbf{n}}_\phi = -\dot{\phi}\,\mathbf{n}_r + \dot{\theta}\cos\phi\,\mathbf{n}_\theta, \tag{7.6-3b}$$

$$\dot{\mathbf{n}}_\theta = -\dot{\theta}\sin\phi\,\mathbf{n}_r - \dot{\theta}\cos\phi\,\mathbf{n}_\phi. \tag{7.6-3c}$$

The position vector given in Eq. (7.6-1) can now be differentiated twice with respect to time to yield the velocity and acceleration vectors:

$$\mathbf{v} = v_r\mathbf{n}_r + v_\phi\mathbf{n}_\phi + v_\theta\mathbf{n}_\theta,$$

$$v_r = \dot{r}, \qquad v_\phi = r\dot{\phi}, \qquad v_\theta = r\dot{\theta}\sin\phi, \tag{7.6-4}$$

$$\mathbf{a} = a_r\mathbf{n}_r + a_\phi\mathbf{n}_\phi + a_\theta\mathbf{n}_\theta,$$

$$a_r = \ddot{r} - r\dot{\phi}^2 - r\dot{\theta}^2\sin^2\phi,$$

$$a_\phi = r\ddot{\phi} + 2\dot{r}\dot{\phi} - r\dot{\theta}^2\sin\phi\cos\phi, \tag{7.6-5}$$

$$a_\theta = r\ddot{\theta}\sin\phi + 2\dot{r}\dot{\theta}\sin\phi + 2r\dot{\theta}\dot{\phi}\cos\phi.$$

Examples

7.6-1 Determine the spherical unit vectors $\mathbf{n}_r$, $\mathbf{n}_\phi$, and $\mathbf{n}_\theta$ for the position vector $\mathbf{r}(t) = 3\mathbf{i} + 4\mathbf{j} + 12\mathbf{k}$.

Solution

Referring to Fig. 7.6-1, we take dot products to get

$$\cos\phi = (\mathbf{r}/|\mathbf{r}|) \cdot \mathbf{k} = 12/\sqrt{169} = 12/13 = 0.923,$$
$$\cos\theta = (\mathbf{r}_p/|\mathbf{r}_p|) \cdot \mathbf{i} = 3/5 = 0.6.$$

We could also take cross products to get

$$|\sin\phi| = |(\mathbf{r}/|\mathbf{r}|) \times \mathbf{k}| = |-3\mathbf{j} + 4\mathbf{i}|/\sqrt{169} = 5/13 = 0.385,$$
$$|\sin\theta| = |(\mathbf{r}_p/|\mathbf{r}_p|) \times \mathbf{i}| = 4/5 = 0.8,$$

so $\theta = 53.13°$ and $\phi = 22.63°$. Now, from Eqs. (7.6-2),

$$\mathbf{n}_r = (0.385)0.6\mathbf{i} + (0.385)0.8\mathbf{j} + 0.923\mathbf{k} = 0.231\mathbf{i} + 0.308\mathbf{j} + 0.923\mathbf{k},$$
$$\mathbf{n}_\phi = (0.923)0.6\mathbf{i} + (0.923)0.8\mathbf{j} - 0.385\mathbf{k} = 0.554\mathbf{i} + 0.738\mathbf{j} - 0.385\mathbf{k}$$
$$\mathbf{n}_\theta = -0.8\mathbf{i} + 0.6\mathbf{j}.$$

7.6-2 A bead slides down a smooth rotating ring starting from rest just beside the point A, as shown. The ring rotates about the z axis at a constant angular velocity $\dot{\theta}$. Determine an expression for the velocity of the bead when it reaches the point B on the ring.

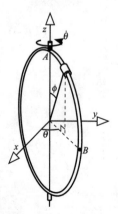

Example 7.6-2

Solution

The free body diagram of the bead consists of a gravitational force and a normal force perpendicular to the tangent of the spiral. In vector form, the sum of the forces is

$$-mg\mathbf{k} + N_r\mathbf{n}_r + N_\theta\mathbf{n}_\theta = m\mathbf{a}.$$

Substituting $\mathbf{k} = \cos\phi\,\mathbf{n}_r - \sin\phi\,\mathbf{n}_\phi$ and Eq. (7.6-5) into the above equation yields the equations that govern the motion of the bead:

$$-mg\cos\phi + N_r = m(-R\dot{\phi}^2 - R\dot{\theta}^2\sin^2\phi),$$
$$N_\theta = m(R\ddot{\theta}\sin\phi + 2R\dot{\theta}\dot{\phi}\cos\phi),$$
$$mg\sin\phi = m(R\ddot{\phi} - R\dot{\theta}^2\sin\phi\cos\phi).$$

These three equations are expressed in terms of the three independent unknowns N_r, N_θ, and ϕ. The third of these equations must be solved to obtain $\dot{\phi}$ as a function of ϕ, and the result substituted into the first two equations to obtain N_r and N_ϕ. Our interest lies in determining the velocity at $\phi = \frac{1}{2}\pi$, so we'll solve the third equation and leave the determination of the normal forces as an exercise for the reader. The third equation is rewritten as

$$\ddot{\phi} = \frac{g}{R}\sin\phi + \dot{\theta}^2\sin\phi\cos\phi.$$

From the chain rule, we know that

$$\ddot{\phi} = \frac{d\dot{\phi}}{dt} = \frac{d\dot{\phi}}{d\phi}\frac{d\phi}{dt} = \frac{d\dot{\phi}}{d\phi}\dot{\phi},$$

so

$$\dot{\phi}\,d\dot{\phi} = \left(\frac{g}{R}\sin\phi + \dot{\theta}^2\sin\phi\cos\phi\right)d\phi.$$

Integrating both sides of the equation from the point A to the point B yields

$$\int_0^{\dot\phi} \dot\phi \, d\dot\phi = \frac{\dot\phi^2}{2}, \qquad \int_0^{\pi/2} \left(\frac{g}{R}\sin\phi + \dot\theta^2 \sin\phi \cos\phi\right) d\phi = \frac{g}{R} + \frac{\dot\theta^2}{2},$$

so, at the point B,

$$\dot\phi^2 = \frac{2g}{R} + \dot\theta^2.$$

From Eq. (7.6-4), the velocity of the bead at the point B is

$$\mathbf{v} = R\dot\phi\mathbf{n}_\phi + R\dot\theta\mathbf{n}_\theta = \sqrt{2gR + (R\dot\theta)^2}\,\mathbf{n}_\phi + R\dot\theta\mathbf{n}_\theta.$$

The magnitude of the velocity vector is

$$v = \sqrt{2gR + 2(R\dot\theta)^2}.$$

Problems

7.6-1   Determine the azimuthal angle and the polar angle for the position vector $\mathbf{r} = -17\mathbf{i} + 22\mathbf{j} - 43\mathbf{k}$.

7.6-2   The base of the spherically mounted gripper shown rotates at a constant rate of $2\,\text{rad/s}$ as the rod rotates downward at a constant angular rate of $3\,\text{rad/s}$, and telescopes outward at a constant rate of

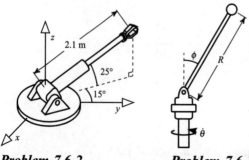

Problem 7.6-2   **Problem 7.6-3**

5 m/s. Determine the position vector, the velocity vector, and the acceleration vector of the gripper in the position shown in rectangular coordinates.

7.6-3 The spinning inverted pendulum shown is released from rest at $\phi(t) = 0$. The polar angle of the hinge is prescribed to be $\theta(t) = (\sqrt{g/R})t$ and the azimuthal angle rotates freely. Determine the velocity of the tip of the pendulum when it reaches $\phi = \frac{1}{2}\pi$.

CHAPTER 8

Energy and Momentum

8.1 WORK AND KINETIC ENERGY

Theory

As discussed in Chapter 6, dynamics problems are solved in five steps. The third step consists of applying Newton's Laws, or formulas derived from these laws, to each object. In Chapter 7, Newton's Laws were applied to each object directly. This chapter derives formulas based on Newton's Laws that can be applied to each object. The development given in this section deals with the **spatial** integration of Newton's Laws, which leads us to the concepts of work and kinetic energy.

We begin by defining the **work** done to move particle from point 1 to point 2 by the resultant force **F** acting on the particle as

$$U_{1-2} = \int_1^2 \mathbf{F} \cdot d\mathbf{r}.$$

(8.1-1)

Substituting $\mathbf{F} = m\mathbf{a}$ into Eq. (8.1-1) yields

$$U_{1-2} = \int_1^2 m\mathbf{a} \cdot d\mathbf{r} = \int_{t_1}^{t_2} m\mathbf{a} \cdot \mathbf{v}\, dt = \int_{t_1}^{t_2} \frac{d}{dt}\left(\tfrac{1}{2} m\mathbf{v} \cdot \mathbf{v}\right) dt,$$

since $d\mathbf{r} = \mathbf{v}\,dt$ and

$$\frac{d}{dt}\left(\tfrac{1}{2}m\mathbf{v}\cdot\mathbf{v}\right) = \tfrac{1}{2}m\mathbf{a}\cdot\mathbf{v} + \tfrac{1}{2}m\mathbf{v}\cdot\mathbf{a} = m\mathbf{a}\cdot\mathbf{v}.$$

It follows that

$$U_{1-2} = \int dT = T_2 - T_1,$$

where T is defined as **kinetic energy**; that is,

$$\boxed{U_{1-2} = T_2 - T_1, \qquad T = \tfrac{1}{2}mv^2,} \qquad (8.1\text{-}2)$$

where $v^2 = \mathbf{v}\cdot\mathbf{v}$, with v denoting the magnitude of the velocity vector. Equation (8.1-2) states that **the work done on a particle by the resultant force as it moves from point 1 to point 2 is equal to the change in the particle's kinetic energy.**

The expression for work given in Eq. (8.1-2) can be modified as follows:

$$U_{1-2} = \int_{t_1}^{t_2} \mathbf{F}\cdot\mathbf{v}\,dt = \int_{t_1}^{t_2} P\,dt, \qquad \text{where} \qquad P = \mathbf{F}\cdot\mathbf{v}, \qquad (8.1\text{-}3a,b)$$

in which we recognize that

$$d\mathbf{r} = \frac{d\mathbf{r}}{dt}\,dt = \mathbf{v}\,dt$$

and where P is called the **power**. Equation (8.1-3a) states that the work done on a particle by a force as it moves from point 1 to point 2 is equal to the power integrated over the time interval of the action.

In the case of a **system** of particles, we can let Eq. (8.1-2) represent the work–energy equation for the ith particle. Introducing the subscript i in Eq. (8.1-2) in the expressions for work and kinetic energy, and summing these equations, yields Eq. (8.1-2) again, in which

$$U_{1-2} = \sum_{i=1}^{n} U_{1-2i}, \qquad (8.1\text{-}4a)$$

$$T = \sum_{i=1}^{n} T_i, \qquad (8.1\text{-}4b)$$

where U_{1-2} now represents the work done on the system of particles, and where T now represents the kinetic energy of the system of particles.

Indeed, Eq. (8.1-2) applies either to a single particle or to a system of particles. Thus, **the work done on a system of particles by the resultant forces acting on each of the particles as the particles move from state 1 to state 2 is equal to the change in the kinetic energy of the system of particles.**

Examples

8.1-1 Determine the work done by the resultant force acting on the block shown as the block slides down the incline from the point A to the point B. Let $mg = 100$ N, $h = 2$ m, $\theta = 45°$, and $\mu_k = 0.2$.

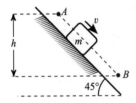

Examples 8.1-1 and 8.1-2

Solution
The resultant force acting on the block is

$$\mathbf{F} = N\mathbf{J} - F_f\mathbf{I} - mg\mathbf{j} = (N - mg \cos \theta)\mathbf{J} + (-F_f + mg \sin \theta)\mathbf{I}.$$

Since the block travels in the X direction from $X = 0$ (at the point A) to $X = h/\sin \theta$ (at the point B), it follows that $d\mathbf{r} = dX\,\mathbf{I}$ and

$$U_{A-B} = \int_A^B \mathbf{F} \cdot d\mathbf{r} = \int_0^{h/\sin \theta} [(N - mg \cos \theta)\mathbf{J} + (-F_f + mg \sin \theta\mathbf{I})] \cdot dX\,\mathbf{I}$$

$$= \int_0^{h/\sin \theta} (-F_f + mg \sin \theta)\,dX = \frac{(-F_f + mg \sin \theta)h}{\sin \theta}$$

$$= -\frac{F_f h}{\sin \theta} + mgh.$$

From Newton's Second Law, $N - mg\cos\theta = ma_Y = 0$, so that $N = mg\cos\theta$. Furthermore, according to the dry friction model, $F_f = \mu_k N = \mu_k\,mg\cos\theta$, so the work done by the resultant force acting on the block is

$$U_{A-B} = mgh(1 - \mu_k\cot\theta) = 160 \text{ N·m}.$$

8.1-2 Assuming the block in Example 8.1-1 starts from rest at the point A, determine the velocity of the block at the point B.

Solution
From Example 8.1-1 and Eq. (8.1-2),

$$U_{A-B} = mgh(1 - \mu_k\cot\theta) = \tfrac{1}{2}mv_B^2 - \tfrac{1}{2}mv_A^2 = \tfrac{1}{2}mv_B^2,$$

since $v_A = 0$. Thus,

$$v_B = \sqrt{2gh(1 - \mu_k\cot\theta)} = 2.80 \text{ m/s}.$$

8.1-3 The pendulum shown consists of a tip mass connected to a bar of negligible mass. If the pendulum is released from rest at $\theta = 0°$, determine the speed of the tip mass when $\theta = 90°$. Let $R = 3$ m and $m = 0.2$ kg.

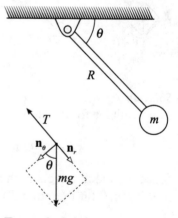

Solution
Referring to the free body diagram of the pendulum, the tension force T does no work, since it is perpendicular to the motion. This leaves the gravitational force. The differential position, which is in the circumferential direction, is

Example 8.1-3

$d\mathbf{r} = R\,d\theta\,\mathbf{n}_\theta$. The circumferential component of the gravitational force is $mg\cos\theta$, so

$$U_{1-2} = \int_0^{\pi/2} mg\cos\theta\,R\,d\theta = mgR.$$

From Eq. (8.1-2),

$$U_{1-2} = mgR = \tfrac{1}{2}mv_2^2 - \tfrac{1}{2}mv_1^2 = \tfrac{1}{2}mv_2^2,$$

so

$$v_2 = \sqrt{2gR} = 7.67 \text{ m/s}.$$

Problems

8.1-1 A block that slides on a rough horizontal surface is subjected to an applied force F as shown. Determine the velocity of the block after the block has traveled 10 m if the block starts from rest. Let $F = 50$ N, $m = 20$ kg, $\mu_k = 0.2$, and $\theta = 30°$.

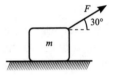

Problem 8.1-1

8.1-2 Find the initial velocity necessary to move a shuffleboard puck precisely 14.3 ft before it comes to rest. The puck weighs 0.8 lb and has a friction coefficient of $\mu_k = 0.07$.

8.1-3 Determine the maximum velocity that a 1600 lb car with a 49 horsepower engine can maintain as it travels up a 15° mountain incline. Neglect internal energy losses.

8.1-4 Express the power and energy consumed by each of the motors shown in terms of the mass m, the distance traveled D, and the time of travel T. Neglect the weight of the pulley. How do the differences between the configurations affect the requirements of the motor?

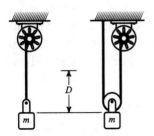

Problem 8.1-4

8.2 CONSERVATIVE AND NON-CONSERVATIVE FORCES

Theory

Forces can be recognized as being one of two types: **conservative** or **non-conservative**. Conservative forces $\mathbf{F}_C$ are forces that are defined in terms of a potential energy function V as $\mathbf{F}_C = -\nabla V$ (the function ∇V will be defined shortly). If such a function V is found or exists, then the force is conservative. Otherwise, the force is regarded as non-conservative.

By distinguishing between conservative forces and non-conservative forces, certain advantages arise. The work expression developed in the previous section can be simplified. The work done by the forces is divided into the work done by the conservative forces and the work done by the non-conservative forces. The work done by the conservative forces does not need to be obtained by integration. Instead, we only need to take the difference between the values of the potential energy at the limits of integration to determine the work done by the conservative forces. The second advantage that arises by distinguishing between conservative forces and non-conservative forces is that the associated potential energy function can be added to the kinetic energy function to obtain what is called the total energy. In the absence of non-conservative forces, it will be shown that the total energy is constant. This is called **conservation of energy**. Conservative forces conserve the total energy of the system, which is why they're called conservative forces.

Let's begin by breaking up the resultant force into the conservative part and the non-conservative part as

$$\mathbf{F} = \mathbf{F}_C + \mathbf{F}_{NC}, \tag{8.2-1a}$$

$$\mathbf{F}_C = -\nabla V, \tag{8.2-1b}$$

where $\mathbf{F}_C$ denotes the conservative part of the force and $\mathbf{F}_{NC}$ denotes the non-conservative part of the force. The ∇ symbol represents the **gradient** of a scalar function (in our case, the function V). In planar rectangular coordinates, in which $\mathbf{r} = x\mathbf{i} + y\mathbf{j}$, the gradient of $V(x, y)$ is defined as

$$\nabla V = \frac{\partial V}{\partial x}\mathbf{i} + \frac{\partial V}{\partial y}\mathbf{j}.$$

Notice that

$$\nabla V \cdot d\mathbf{r} = \left(\frac{\partial V}{\partial x}\mathbf{i} + \frac{\partial V}{\partial y}\mathbf{j}\right) \cdot (dx\,\mathbf{i} + dy\,\mathbf{j}) = \frac{\partial V}{\partial x}\,dx + \frac{\partial V}{\partial y}\,dy = dV, \quad (8.2\text{-}2)$$

in which dV is the differential change in $V(x, y)$ as a result of the differential changes dx and dy. (See Fig. 8.2-1.) In fact, for any type of coordinate system, the definitions for ∇V are determined so that $\nabla V \cdot d\mathbf{r} = dV$. The definitions of ∇V for rectangular, cylindrical, and spherical coordinate systems are given in Table 8.2-1.

Let's now return to the problem at hand, and learn how conservative forces help us. For a single particle, substituting Eqs. (8.2-1) and (8.2-2) into Eq. (8.1-2) yields

$$U_{1-2} = \int_1^2 (\mathbf{F}_C + \mathbf{F}_{NC}) \cdot d\mathbf{r} = \int_1^2 \mathbf{F}_C \cdot d\mathbf{r} + \int_1^2 \mathbf{F}_{NC} \cdot d\mathbf{r} = U_{C1-2} + U_{NC1-2},$$

$$(8.2\text{-}3)$$

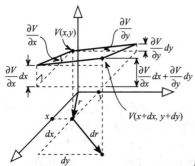

Figure 8.2-1

Table 8.2-1   The Gradient Function

V	∇V	$d\mathbf{r}$
Rectangular		
$V(x, y, z)$	$\dfrac{\partial V}{\partial x}\mathbf{i} + \dfrac{\partial V}{\partial y}\mathbf{j} + \dfrac{\partial V}{\partial z}\mathbf{k}$	$dx\,\mathbf{i} + dy\,\mathbf{j} + dz\,\mathbf{k}$
Cylindrical		
$V(r, \theta, z)$	$\dfrac{\partial V}{\partial r}\mathbf{n}_r + \dfrac{1}{r}\dfrac{\partial V}{\partial \theta}\mathbf{n}_\theta + \dfrac{\partial V}{\partial z}\mathbf{n}_z$	$dr\,\mathbf{n}_r + r\,d\theta\,\mathbf{n}_\theta + dz\,\mathbf{n}_z$
Spherical		
$V(r, \phi, \theta)$	$\dfrac{\partial V}{\partial r}\mathbf{n}_r + \dfrac{1}{r}\dfrac{\partial V}{\partial \phi}\mathbf{n}_\phi + \dfrac{1}{r\sin\phi}\dfrac{\partial V}{\partial \theta}\mathbf{n}_\theta$	$dr\,\mathbf{n}_r + r\,d\phi\,\mathbf{n}_\phi + r\sin\phi\,d\theta\,\mathbf{n}_z$

where U_{C1-2} denotes the work done by the conservative forces and U_{NC1-2} denotes the work done by the non-conservative forces. From Eq. (8.2-2),

$$U_{C1-2} = \int_1^2 \mathbf{F}_C \cdot d\mathbf{r} = -\int_1^2 \nabla V \cdot dr = -\int_1^2 dV = V_1 - V_2, \quad (8.2\text{-}4a)$$

$$U_{NC1-2} = \int_1^2 \mathbf{F}_{NC} \cdot d\mathbf{r}. \quad (8.2\text{-}4b)$$

Equation (8.2-4a) states that **the work done by the conservative force is the difference between the potential energies at points 1 and 2.** From Eq. (8.1-2), we get

$$U_{1-2} = U_{NC1-2} + V_1 - V_2 = T_2 - T_1,$$

from which

$$\boxed{U_{NC1-2} = E_2 - E_1, \quad E = T + V,} \quad (8.2\text{-}5a,b)$$

where E is called the **total energy** (kinetic and potential). Equation (8.2-5) states that **the work done by the non-conservative forces is equal to the change in the total energy of the particle.** If there are no non-conservative forces acting on the particle then the total energy is conserved.

Now, consider a **system** of particles. The work done by the resultant forces (external and internal) acting on the ith particle is

$$U_{1-2i} = \int \left(\mathbf{F}_i + \sum_{j=1}^n \mathbf{f}_{ij} \right) \cdot d\mathbf{r}_i = \int \mathbf{F}_i \cdot d\mathbf{r}_i + \sum_{j=1}^n \int \mathbf{f}_{ij} \cdot d\mathbf{r}_i.$$

The first term on the right-hand side of this equation is the work done by the resultant external force, and the second term is the work done by the resultant internal force. For the ith particle, we know that

$$U_{1-2i} = T_{2i} - T_{1i} \quad (8.2\text{-}6)$$

where T_{2i} is the kinetic energy of the ith particle at the final time and T_{1i} is the kinetic energy of the ith particle at the initial time. Summing over the particles yields

$$U_{1-2} = \sum_{i=1}^{n} U_{1-2i} = \sum_{i=1}^{n} \int (\mathbf{F}_{Ci} + \mathbf{F}_{NCi}) \cdot d\mathbf{r}_i + \sum_{i=1}^{n} \sum_{j=1}^{n} \int \mathbf{f}_{ij} \cdot d\mathbf{r}_i$$

$$= \sum_{i=1}^{n} \int \mathbf{F}_{Ci} \cdot d\mathbf{r}_i + \sum_{i=1}^{n} \int \mathbf{F}_{NCi} \cdot d\mathbf{r}_i + \sum_{i=1}^{n} \sum_{j=1}^{n} \int \mathbf{f}_{ij} \cdot d\mathbf{r}_i$$

$$= U_{C1-2} + U_{NC1-2} + U_{I1-2} = \sum_{i=1}^{n} (T_{2i} - T_{1i}); \qquad (8.2\text{-}7)$$

that is,

$$U_{C1-2} + U_{NC1-2} + U_{I1-2} = T_2 - T_1, \qquad (8.2\text{-}8)$$

in which

$$U_{C1-2} = \sum_{i=1}^{n} \int \mathbf{F}_{Ci} \cdot d\mathbf{r}_i, \qquad U_{NC1-2} = \sum_{i=1}^{n} \int \mathbf{F}_{NCi} \cdot d\mathbf{r}_i, \quad (8.2\text{-}9a, b)$$

$$U_{I1-2} = \sum_{i=1}^{n} \sum_{j=1}^{n} \int \mathbf{f}_{ij} \cdot d\mathbf{r}_i, \qquad (8.2\text{-}9c)$$

$$T = \sum_{i=1}^{n} T_i. \qquad (8.2\text{-}9d)$$

From Eq. (8.2-4a), $U_{C1-2} = V_1 - V_2$, so

$$\boxed{U_{NC1-2} + U_{I1-2} = E_2 - E_1, \qquad E = T + V.} \qquad (8.2\text{-}10)$$

The sum of the work done by the non-conservative external forces and the work done by the internal forces acting on a system of particles is equal to the change in the total (kinetic and potential) energy of the system of particles over the time interval of the action.

In the case of a rigid body, the work done by the internal forces is identically zero (see Problem 8.2-5). However, no body is perfectly rigid, even if it is idealized as such. The work done by the internal forces causes internal **vibration**. This vibrational energy is a form of heat.

If only a few forces in nature had the property that they were conservative, then the value of Eq. (8.2-10) would be marginal. However, both the mechanical forces (gravitational) and the electrical forces (electrostatic) in nature are conservative. In fact, it is the conservative nature of these forces that explains why planets can orbit around the Sun and why electrons can spin around the nucleus.

Examples

8.2-1 Determine the potential energy function for the gravitational force $\mathbf{F} = -mg\mathbf{j}$.

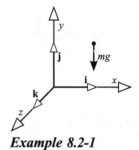

Example 8.2-1

Solution
From Eq. (8.2-1b) and Table 8.2-1, the gravitational force is

$$\mathbf{F} = -mg\mathbf{j} = -\frac{\partial V}{\partial x}\mathbf{i} - \frac{\partial V}{\partial y}\mathbf{j} - \frac{\partial V}{\partial z}\mathbf{k},$$

so

$$\frac{\partial V}{\partial x} = 0, \quad \frac{\partial V}{\partial z} = 0, \quad \frac{\partial V}{\partial y} = mg.$$

It follows that V is not a function of x, nor is it a function of z. By integration, we get

$$V = mgy + C,$$

where C is a constant relative to x, y, and z. Changes in potential energy are calculated in the work–energy equations (8.2-5) and (8.2-10). Thus, the constant C, whatever its value, cancels out. For no other reason than for simplicity, we let $C = 0$. The potential energy function associated with the gravitational force is

$$\boxed{V = mgy.} \qquad (8.2\text{-}11)$$

8.2-2 Determine the potential energy function for the spring force $\mathbf{F} = -ks\mathbf{n}$, where s is the stretch in the spring, k is the spring constant or stiffness, and $\mathbf{n}$ is the unit vector in the direction of the stretch, as shown.

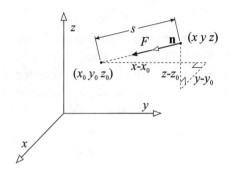

Example 8.2-2

Solution

From Eq. (8.2-1b) and Table 8.2-1, the spring force is given by

$$F = -ks\left(\frac{x - x_0}{s}i + \frac{y - y_0}{s}j + \frac{z - z_0}{s}k\right) = -\frac{\partial V}{\partial x}i - \frac{\partial V}{\partial y}j - \frac{\partial V}{\partial z}k,$$

so

$$\frac{\partial V}{\partial x} = k(x - x_0), \qquad \frac{\partial V}{\partial y} = k(y - y_0), \qquad \frac{\partial V}{\partial z} = k(z - z_0).$$

By integration, it follows that

$$V = \tfrac{1}{2}k(x - x_0)^2 + f(y, z) = \tfrac{1}{2}k(y - y_0)^2 + g(x, z) = \tfrac{1}{2}k(z - z_0)^2 + h(x, y),$$

for some functions $f(x, y)$, $g(x, z)$, and $h(y, z)$. The only possible form of V is

$$V = \tfrac{1}{2}k(x - x_0)^2 + \tfrac{1}{2}k(y - y_0)^2 + \tfrac{1}{2}k(z - z_0)^2 + C,$$

where C is a constant. From the figure and the Pythagorean Theorem, $s^2 = k[(x - x_0)^2 + (y - y_0)^2 + (z - z_0)^2]$, and, letting $C = 0$, we get

$$\boxed{V = \tfrac{1}{2}ks^2.} \tag{8.2-12}$$

8.2-3   A block of mass $m = 24$ slug slides down a smooth incline starting from rest $h = 10$ ft from the ground. Determine the speed when it reaches the ground. Let $\theta = 30°$.

Solution

Referring to the figure for Example 8.1-1, the forces acting on the block consist of a normal force and a gravity force. There is no friction force in this example. The normal force is perpendicular to the motion, and hence does no work. The gravitational force is also conservative; its potential energy function is given by Eq. (8.2-11), so Eq. (8.2-10) yields

$$U_{NC1-2} = E_2 - E_1,$$
$$U_{NC1-2} = 0,$$
$$E_2 = T_2 + V_2 = \tfrac{1}{2}mv_2^2 + mgy_2 = \tfrac{1}{2}mv_2^2,$$
$$E_1 = T_1 + V_1 = \tfrac{1}{2}mv_1^2 + mgy_1 = mgh,$$

so

$$v_2 = \sqrt{2gh} = 25.38 \text{ ft/s}.$$

8.2-4 A pin slides down the smooth track shown. At the point A, the velocity of the pin is $v_A = 2 \text{ m/s}$ to the right. Determine the velocity of the pin at the point B. Let $k = 4 \text{ N/m}$, $m = 9 \text{ kg}$, and the unstretched length of the spring be $x_0 = 1.5 \text{ m}$.

Solution

The pin is subjected to a normal force, which does no work, along with a spring force and a gravitational force; the gravitational force and the

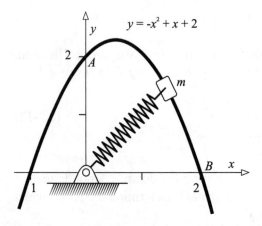

$$y = -x^2 + x + 2$$

Example 8.2-4

spring force are conservative. Using Eq. (8.2-12), the terms in Eq. (8.2-10) are,

$$U_{NCA-B} = 0,$$

$$T_A = \tfrac{1}{2}mv_A^2, \qquad V_A = mgy_A + \tfrac{1}{2}ks_A^2,$$

$$T_B = \tfrac{1}{2}mv_B^2, \qquad V_B = mgy_B + \tfrac{1}{2}ks_B^2,$$

so

$$0 = [\tfrac{1}{2}9v_B^2 + \tfrac{1}{2}4(0.5)^2] - [\tfrac{1}{2}9(2)^2 + 9(9.81)2 + \tfrac{1}{2}4(0.5)^2].$$

We finally solve for v_B to get

$$v_B = 6.58 \, \text{m/s}.$$

Problems

8.2-1 A Teflon (polytetrafluoroethylene)-coated collar travels over a rod at $v_A = 15$ ft/s when $\theta = 50°$. Determine the angle θ_B at which the collar momentarily comes to rest before rebounding. The unstretched length of the spring is equal to the separation distance $d = 2.5$ ft. The mass of the collar is 15 slug and the stiffness of the spring is $k = 1.5$ lb/ft.

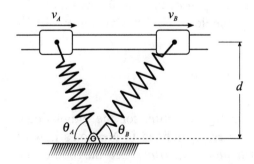

Problem 8.2-1

8.2-2 The block shown in the figure overleaf, of weight 220 lb, is dropped with an initial slackness in the rope of $h = 4$ ft. The spring has a spring constant of $k = 400$ lb/ft. Determine the maximum stretch in the spring. Neglect internal energy losses.

Problem 8.2-2

8.2-3 Find the least compressed setting on the toy racetrack shown so that a car of weight $W = 0.2$ lb makes it around the loop of diameter $D = 12$ inches. The launching mechanism is in the uncompressed position at

Problem 8.2-3

setting 0. The spring constant is $k = 3$ lb/inch and the spring latches in $\frac{1}{2}$ inch increments. Neglect friction.

8.2-4 An amusement park ride consists of a carpeted barrel that can spin in either the horizontal or vertical directions as shown. The passengers are held in place by friction. Find the minimum angular velocity that keeps the passengers held against the barrel. Let $\mu_s = 0.2$ and $d = 20$ ft.

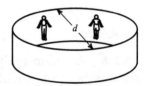

Problem 8.2-4

8.2-5 Show in the case of a rigid body that the work done by the internal forces is zero. *Hint:* Examine a pair of internal forces $\mathbf{f}_{ij}$, and $\mathbf{f}_{ji}$, and notice that for a rigid body the length of the vector $\mathbf{r}_i - \mathbf{r}_j$ is constant, so

$$\frac{d}{dt}|\mathbf{r}_i - \mathbf{r}_j| = 0.$$

8.2-6 Show that the potential energy function for the gravitational force of attraction between two bodies, as defined in Newton's Law of Gravitation, is given by $V = Gm_A m_B / r_{AB}$. *Hint:* Let the bodies be located in a rectangular coordinate system, with one body at the origin of the coordinate system and the second body at (x, y, z). The gravitational force acting on the second body is $\mathbf{F} = F\mathbf{n}$, in which the magnitude of the force is $F = Gm_A m_B / r_{AB}^2$ and the unit vector is $\mathbf{n} = -(x/r_{AB})\mathbf{i} - (y/r_{AB})\mathbf{j} - (z/r_{AB})\mathbf{k}$, where $r_{AB}^2 = x^2 + y^2 + z^2$.

8.2-7 At what angle θ does the boy shown start his swing in order to land on a pile of leaves 12 ft from the swing? As shown, assume that the boy jumps from the swing at 30°. Also assume that the boy's mass center is located at the swing seat.

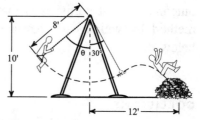

Problem 8.2-7

8.3 LINEAR IMPULSE AND LINEAR MOMENTUM

Theory

Let's now examine the integral with respect to time of $\mathbf{F} = m\mathbf{a}_C$, Eq. (6.2-5). Toward this end, the **linear momentum** of the ith particle is defined as $\mathbf{L}_i = m_i\mathbf{v}_i$, where $\mathbf{v}_i$ is the ith particle's velocity. The linear momentum of the system of particles is the sum of the linear momenta of the particles, that is, $\mathbf{L} = \sum_{i=1}^{n} \mathbf{L}_i$. Integrating Eq. (6.2-5) with respect to time yields

$$\mathbf{G}_{1-2} = \int_{t_1}^{t_2} \mathbf{F}\, dt = \int_{t_1}^{t_2} m\mathbf{a}_C\, dt = m\mathbf{v}_C(t_2) - m\mathbf{v}_C(t_1) = \mathbf{L}_2 - \mathbf{L}_1,$$

where $\mathbf{L}$ denotes the linear momentum of the system and $\mathbf{G}_{1-2} = \int_{t_1}^{t_2} \mathbf{F}\, dt$ denotes the **linear impulse** acting on the system imparted over the time interval in question. Thus,

$$\boxed{\mathbf{G}_{1-2} = \mathbf{L}_2 - \mathbf{L}_1,} \qquad (8.3\text{-}1a)$$

in which

$$\mathbf{L} = \sum_{i=1}^{n} m_i\mathbf{v}_i, \qquad \mathbf{G}_{1-2} = \int_{t_1}^{t_2} \mathbf{F}\, dt. \qquad (8.3\text{-}1b,c)$$

Equation (8.3-1) states that **the linear impulse acting on the system is equal to the change in the system's linear momentum over the time interval in question**. Equation (8.3-1) is the time integral of Eq. (6.2-5).

Collision problems are an important class of problems that employ linear impulse principles. When two approaching particles collide, the outcome of that collision is that each particle's velocity changes. The method by which each particle's velocity is determined is described below.

COLLISIONS

Figure 8.3-1 shows two particles. The collision forces between the particles are equal and opposite and are assumed to act in the x direction. Thus, the x direction is taken to be the **line of force**. The plane that is perpendicular to the line of force is called the **plane of impact**. The collision forces between the particles are assumed to be **impulsive**, by which it is meant that the impulse $\int F\, dt$ that they create is finite. In the impulse integral, a very large force acts over a very small collision time, thereby producing a finite impulse. Other forces might occur during the collision that are not considered to be impulsive. These are smaller finite forces that occur over the infinitesimal collision time, resulting in infinitesimal impulses that can be neglected. Gravitational forces and spring forces are examples of non-impulsive forces.

As shown, particle A is situated to the left of particle B and the x axis is positive to the right. In this situation, a collision is imminent provided $v_{Ax} > v_{Bx}$, where v_{Ax} and v_{Bx} denote the components of the velocities of the particles in the x direction **just before** the collision. The components of the velocities of the particles in the x direction **just after** the collision are denoted by v'_{Ax} and v'_{Bx}. A prime indicates a quantity just after the collision; quantities without primes are those just before the collision. Just after the collision, $v'_{Bx} \geq v'_{Ax}$; otherwise, the particles would travel through each other.

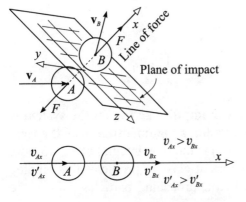

Figure 8.3-1

Let's first examine the effects of the collision in the plane of impact (the y and z directions). In the y direction, no impulsive forces act on either particle over the time of the collision, so

$$0 = G_{Ay} = \int -F_y \, dt = L'_{Ay} - L_{Ay},$$

$$0 = G_{By} = \int F_y \, dt = L'_{By} - L_{By}.$$

It follows that the y components of the velocities of the particles do not change over the time of the collision. By the same reasoning, the z components of the velocities of the particles also do not change, so

$$\boxed{v_{Ay} = v'_{Ay}, \quad v_{Az} = v'_{Az}, \quad v_{By} = v'_{By}, \quad v_{Bz} = v'_{Bz}.} \tag{8.3-2}$$

In the x direction, however, the particle velocities **do** change. Recognizing that the resultant impulse exerted on the **system** of particles over the time of the collision is zero (since the collision forces are internal), the linear momentum of the system is conserved over the time of the collision. In the x direction, we get

$$\boxed{m_A v_{Ax} + m_B v_{Bx} = m_A v'_{Ax} + m_B v'_{Bx}.} \tag{8.3-3}$$

Equation (8.3-3) is one equation expressed in terms of two unknown velocities just after the collision. An additional equation is needed in order to determine the two unknown velocities just after the collision. This additional equation is associated with the energy loss during the collision, which, in turn, depends on the elasticity and the plasticity of the colliding objects. Although it is beyond the scope of this book to analyze these elasticity–plasticity effects, the results from this analysis are quite interesting. It turns out that, for a given pair of objects, there exists a characteristic parameter, called the **coefficient of restitution**, which is a constant of the collision. The coefficient of restitution is given by

$$\boxed{e = \frac{v'_{Bx} - v'_{Ax}}{v_{Ax} - v_{Bx}},} \tag{8.3-4}$$

which is the relative velocity of the particles after the collision divided by the relative velocity of the particles before the collision. Equations (8.3-3) and (8.3-4) are two equations from which the two unknown velocities, v'_{Ax} and v'_{Bx}, are determined. It can be shown that e is always between 0 and 1. When $e = 1$, energy is conserved during the collision, and the collision is called **purely elastic**. When $e = 0$, it follows from Eq. (8.3-4) that the velocities of the two particles just after the collision are identical, in which case the particles stay together. That case is called a **purely plastic** collision. In general, e is between 0 and 1, and the collisions are referred to as **elastic–plastic**.

Examples

8.3-1 Collisions are common in sports. One class of sports collisions is set up as follows. Object A has mass m_A and is moving with speed v_A just before it collides with object B. Object B has mass m_B, and is at rest before the collision. Furthermore, assume that this class of collisions is purely elastic; that is, let $e = 1$. Determine the speeds v'_A and v'_B of the objects just after the collision. Consider three cases. In case 1, let m_A/m_B be very small, as when object A is a racketball and object B is a wall. In case 2, let $m_A/m_B = 1$, as when objects A and B are billiard balls or croquet balls. In case 3, let m_A/m_B be very large, as when object A is a bat and object B is a baseball on a tee.

Before $\overset{v_A}{\longrightarrow}$ $v_B = 0$

$----\left(m_A\right)--------\left(m_B\right)--\overset{x}{\longrightarrow}$

After v'_A v'_B Line of force

Example 8.3-1

Solution
From Eqs. (8.3-3) and (8.3-4),

$$m_A v_A = m_A v'_A + m_B v'_B,$$
$$v_A = v'_B - v'_A,$$

which are two equations expressed in terms of the unknowns v'_A and v'_B. The solution is

$$v'_A = \frac{(m_A/m_B) - 1}{(m_A/m_B) + 1} v_A, \qquad v'_B = \frac{2(m_A/m_B)}{(m_A/m_B) + 1} v_A.$$

In case 1, let $m_A/m_B \ll 1 \approx 0$, in which case $v'_A = -v_A$ and $v'_B = 0$. As a result of the collision, the ball changes direction and the wall does not move.

In case 2, let $m_A/m_B = 1$, in which case $v'_A = 0$ and $v'_B = v_A$. As a result of the collision, the moving billiard ball stops and the other billiard ball moves with the speed of the first ball.

In case 3, let $m_A/m_B \gg 1 \approx \infty$, in which case $v'_A = v_A$ and $v'_B = 2v_A$. As a result of the collision, the speed of the base ball does not change, and the ball takes off at twice the speed of the bat.

8.3-2 Two identical pucks A and B approach each other on a smooth surface, as shown. Determine their velocities just after the collision, letting $e = 0.4$.

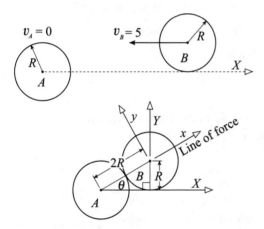

Example 8.3-2

Solution

This is an oblique impact problem involving motion along the line of force as well as motion in the plane of impact. Let's first determine the line of force. From the figure, the line of force x is the line that is $30°$ counterclockwise to the X direction. Indeed, the triangle reveals that the angle between the X axis and the x axis is $\sin^{-1}(R/2R) = \sin^{-1}(1/2) = 30°$. Let's solve this problem using x–y coordinates, and when we're done transform our quantities back to X–Y coordinates. Just before the collision,

$$v_{Ax} = 0, \quad v_{Ay} = 0, \quad v_{Bx} = -5\cos 30° = -\frac{5\sqrt{3}}{2}, \quad v_{By} = 5\sin 30° = \frac{5}{2}.$$

From Eqs. (8.3-2),

$$v'_{Ay} = 0, \quad v'_{By} = \frac{5}{2},$$

From Eqs. (8.3-3) and (8.3-4),

$$mv_{Bx} = mv'_{Ax} + mv'_{Bx}, \qquad e(0 - v_{Bx}) = v'_{Bx} - v'_{Ax},$$

so

$$v'_{Ax} = \frac{1+e}{2} v_{Bx} = -3.03, \qquad v'_{Bx} = \frac{1-e}{2} v_{Bx} = -1.30.$$

Now, transforming the velocity components back to the $X-Y$ coordinate system yields

$$v'_{AX} = \cos\theta\, v'_{Ax} - \sin\theta\, v'_{Ay} = 0.866(-3.03) - 0.5(0) = -2.62\,\text{m/s},$$
$$v'_{AY} = \sin\theta\, v'_{Ax} + \cos\theta\, v'_{Ay} = 0.5(-3.03) + 0.866(0) = -1.52\,\text{m/s},$$
$$v'_{BX} = \cos\theta\, v'_{Bx} - \sin\theta\, v'_{By} = 0.866(-1.30) - 0.5(2.5) = -2.38\,\text{m/s},$$
$$v'_{BY} = \sin\theta\, v'_{Bx} + \cos\theta\, v'_{By} = 0.5(-1.30) + 0.866(2.5) = 1.52\,\text{m/s},$$

Notice that the Y components of the velocities of the two pucks are equal and opposite just after the collision.

Problems

8.3-1 Determine the velocity $v(t_1)$ of the block shown at time $t_1 = 18$ s. The block of mass $m = 3$ kg begins at rest and is pushed by a force $F = 16e^{-t/20}$ N over a smooth surface.

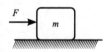

Problem 8.3-1

8.3-2 Find the time t_1 it takes for the block shown to reach a speed of 16 mph. The block of mass $m = 0.2$ slug has a speed of 1 mph uphill at $t = 0$. A horizontal force of $F = 10$ lbs pushes the block up the 25° incline with coefficient of friction of $\mu_k = 0.11$.

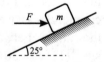

Problem 8.3-2

8.3-3 A block of wood of mass $m_1 = 2$ kg is suspended by strings of length 3.5 m as shown. The stationary block is struck by a bullet of mass $m_2 = 0.01$ kg traveling at $v_2 = 300$ m/s. The bullet is quickly lodged in the wood. Find the height h that the block of wood travels upward.

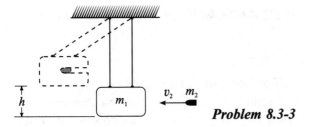

v_2 m_2

m_1

h

Problem 8.3-3

8.3-4 Two balls, one of mass $m_1 = 5m$ and the other of mass $m_2 = m$, are dropped from a height h_1, as shown. Determine the height h_2 that the second ball reaches after bouncing off the first ball. Let $e = 1$. Hint: There are two collisions

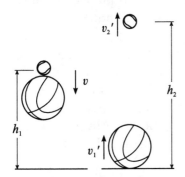

v_2'

v

h_1

h_2

v_1'

Problem 8.3-4

8.3-5 A car of mass $m_1 = 1000\,\text{kg}$ is traveling gracefully sideways across an iced-over bridge at $v_1 = 25\,\text{m/s}$. It strikes a second car of mass $m_2 = 1200\,\text{kg}$ that was standing still, and the two cars lock fenders. Determine the speeds of the cars after the collision.

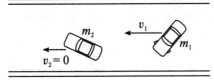

m_2

v_1

m_1

$v_2 = 0$

Problem 8.3-5

8.3-6 Show that energy is conserved during a purely elastic collision ($e = 1$). *Hint:* There is no change in potential energy during the collision, because particles cannot be displaced a finite amount in an infinitesimal amount of time. It is then sufficient to show that kinetic energy is conserved during the collision. Also, note that you'll need to use the fact that the linear momentum of the system is conserved during the collision. (The only impulsive forces acting on the system are internal.)

8.4 ANGULAR IMPULSE AND ANGULAR MOMENTUM

Theory

The previous sections developed equations that could be used instead of Newton's Second Law. In Sections 8.1 and 8.2, Newton's Second Law was integrated over the path (spatial integration) to produce a work–energy equation and associated work–energy principles. In Section 8.3, Newton's Second Law was integrated with respect to time to produce linear-impulse–linear-momentum equations and associated linear-impulse–linear-momentum principles. In this section, Newton's Second Law is pre-cross-multiplied by the position vector to produce angular-impulse–angular-momentum equations and associated principles.

Pre-cross-multiplying a vector by the position vector is frequently referred to as taking the **moment of the vector**. Pre-cross-multiplying the force **F** by the position vector yields $\mathbf{M}_O = \mathbf{r} \times \mathbf{F}$, which is called the **moment of the force about the point O**, or simply the **moment about the point O**, where the point O is the origin of the position vector. The time integral of the moment of force is called the **angular impulse about the point O**. The moment of the linear momentum, $\mathbf{H}_O = \mathbf{r} \times \mathbf{L}$, is called the **angular momentum about the point O**.

The point O about which we take moments can be a fixed point, the mass center, or any other moving point. However, when we select the point O to be a fixed point or the mass center, the relationship between the resultant moment of force and the angular momentum simplifies. In fact, in the absence of a resultant external moment acting on the system of particles about either a fixed point or the mass center, we'll see shortly that the angular momentum of the system about that point is constant; that is, the angular momentum is conserved.

We begin by defining the **angular momentum of a system of n particles about a point A** as

$$\mathbf{H}_A = \sum_{i=1}^{n} \mathbf{H}_{Ai} = \sum_{i=1}^{n} \mathbf{r}_{i/A} \times \mathbf{L}_i = \sum_{i=1}^{n} \mathbf{r}_{i/A} \times m_i \mathbf{v}_i. \tag{8.4-1}$$

For now, the point A is any point. It can be a fixed point, the mass center, or any other point. The resultant external moment about the point A is

$$\mathbf{M}_A = \sum_{i=1}^{n} \mathbf{M}_{Ai} = \sum_{i=1}^{n} \mathbf{r}_{i/A} \times \mathbf{F}_i. \tag{8.4-2}$$

Differentiating $\mathbf{H}_A$ with respect to time yields

$$\dot{\mathbf{H}}_A = \sum_{i=1}^{n} (\mathbf{v}_{i/A} \times m_i \mathbf{v}_i + \mathbf{r}_{i/A} \times m_i \mathbf{a}_i) = \sum_{i=1}^{n} [(\mathbf{v}_i - \mathbf{v}_A) \times m_i \mathbf{v}_i + \mathbf{r}_{i/A} \times m_i \mathbf{a}_i]$$

$$= -\mathbf{v}_A \times \sum_{i=1}^{n} m_i \mathbf{v}_i + \sum_{i=1}^{n} \mathbf{r}_{i/A} \times \left(\mathbf{F}_i + \sum_{j=1}^{n} \mathbf{f}_{ij} \right)$$

$$= -\mathbf{v}_A \times \mathbf{L} + \sum_{i=1}^{n} \mathbf{r}_{i/A} \times \mathbf{F}_i + \sum_{i=1}^{n} \sum_{j=1}^{n} \mathbf{r}_{i/A} \times \mathbf{f}_{ij}$$

$$= -\mathbf{v}_A \times \mathbf{L} + \sum_{i=1}^{n} \mathbf{r}_{i/A} \times \mathbf{F}_i.$$

Thus,

$$\boxed{\dot{\mathbf{H}}_A = -\mathbf{v}_A \times \mathbf{L} + \mathbf{M}_A,} \qquad (8.4\text{-}3)$$

in which we have recognized that $\mathbf{v}_i \times m_i \mathbf{v}_i = \mathbf{0}$, since the cross product of parallel vectors is zero, and that $\sum_{i=1}^{n} \sum_{j=1}^{n} \mathbf{r}_{i/A} \times \mathbf{f}_{ij} = \mathbf{0}$, since the internal moments are equal and opposite and hence their sum cancels out.

If the point A is fixed then $\mathbf{v}_A = \mathbf{0}$ in Eq. (8.4-3), and if the point A is the mass center C then $\mathbf{v}_A \times \mathbf{L} = \mathbf{v}_C \times m\mathbf{v}_C = \mathbf{0}$. In both cases, whether the point A represents a fixed point O or the mass center C, Eq. (8.4-3) reduces to

$$\boxed{\dot{\mathbf{H}}_A = \mathbf{M}_A,} \qquad (8.4\text{-}4)$$

which states that **the time derivative of the angular momentum about a fixed point O (or the mass center C) is equal to the resultant external moment about the fixed point O (or the mass center C).**

Integrating Eq. (8.4-4) with respect to time yields

$$\mathbf{N}_{A1-2} = \int_{t_1}^{t_2} \mathbf{M}_A \, dt = \int_{t_1}^{t_2} \dot{\mathbf{H}}_A \, dt = \mathbf{H}_A(t_2) - \mathbf{H}_A(t_1) = \mathbf{H}_{A2} - \mathbf{H}_{A1}, \quad (8.4\text{-}5)$$

in which $N_{A1-2} = \int_{t_1}^{t_2} M_A \, dt$ denotes the **angular impulse** imparted over the time interval in question. Thus,

$$N_{A1-2} = H_{A2} - H_{A1}.$$ (8.4-6)

Equation (8.4-6) states that **the angular impulse acting on the system about the fixed point O (or the mass center C) is equal to the change in the system's angular momentum about the point O (or the mass center C) over the time interval in question.** Equation (8.4-6) is the time integral of Eq. (8.4-4).

Examples

8.4-1 Determine the angular momentum of the system of particles shown about the point A, and then about the mass center C. Let $m_1 = 3\,\text{kg}$, $m_2 = 1\,\text{kg}$, $m_3 = 5\,\text{kg}$, $v_1 = -10j$, $v_2 = 10k$, and $v_3 = -4j + 12k$.

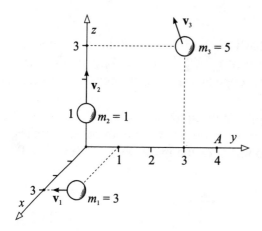

Example 8.4-1

Solution
From the figure,

$$r_1 = 3i + 2j, \qquad r_2 = k, \qquad r_3 = 3j + 3k, \qquad r_A = 4j,$$

so

$$r_{1/A} = 3i - 2j, \qquad r_{2/A} = k - 4j, \qquad r_{3/A} = -j + 3k.$$

The angular momentum of the system about the point A is

$$
\begin{aligned}
H_A &= \mathbf{r}_{1/A} \times m_1 \mathbf{v}_1 + \mathbf{r}_{2/A} \times m_2 \mathbf{v}_2 + \mathbf{r}_{3/A} \times m_3 \mathbf{v}_3 \\
&= (3\mathbf{i} - 2\mathbf{j}) \times 3(-10\mathbf{j}) + (\mathbf{k} - 4\mathbf{j}) \times 1(10\mathbf{k}) \\
&\quad + (-\mathbf{j} + 3\mathbf{k}) \times 5(-4\mathbf{j} + 12\mathbf{k}) \\
&= -90\mathbf{k} - 40\mathbf{i} + 0 = -40\mathbf{i} - 90\mathbf{k} \ \text{kg·m}^2/\text{s}.
\end{aligned}
$$

The mass center is located at

$$
\mathbf{r}_C = \frac{1}{3 + 1 + 5}[3(3\mathbf{i} + 2\mathbf{j}) + 1(\mathbf{k}) + 5(3\mathbf{j} + 3\mathbf{k})] = \mathbf{i} + \tfrac{7}{3}\mathbf{j} + \tfrac{16}{9}\mathbf{k},
$$

so

$$
\mathbf{r}_{1/C} = 2\mathbf{i} - \tfrac{1}{3}\mathbf{j} - \tfrac{16}{9}\mathbf{k}, \qquad \mathbf{r}_{2/C} = -\mathbf{i} - \tfrac{7}{3}\mathbf{j} - \tfrac{7}{9}\mathbf{k}, \qquad \mathbf{r}_{3/C} = -\mathbf{i} + \tfrac{2}{3}\mathbf{j} + \tfrac{11}{9}\mathbf{k}.
$$

The angular momentum of the system about the mass center C is

$$
\begin{aligned}
H_C &= \mathbf{r}_{1/C} \times m_1 \mathbf{v}_1 + \mathbf{r}_{2/C} \times m_2 \mathbf{v}_2 + \mathbf{r}_{3/C} \times m_3 \mathbf{v}_3 \\
&= (2\mathbf{i} - \tfrac{1}{3}\mathbf{j} - \tfrac{16}{9}\mathbf{k}) \times 3(-10\mathbf{j}) + (-\mathbf{i} - \tfrac{7}{3}\mathbf{j} - \tfrac{7}{9}\mathbf{k}) \times 1(10\mathbf{k}) \\
&\quad + (-\mathbf{i} + \tfrac{2}{3}\mathbf{j} + \tfrac{11}{9}\mathbf{k}) \times 5(-4\mathbf{j} + 12\mathbf{k}) \\
&= (-\tfrac{160}{3}\mathbf{i} - 60\mathbf{k}) + (-\tfrac{70}{3}\mathbf{i} + 10\mathbf{j}) + \tfrac{382}{9}\mathbf{i} + 60\mathbf{j} + 20\mathbf{k}) \\
&= -34.22\mathbf{i} + 70\mathbf{j} - 40\mathbf{k} \ \text{kg·m}^2/\text{s}.
\end{aligned}
$$

| **8.4-2** | A block slides on a smooth table while being pulled inward by a rope, as shown in the figure overleaf. At point 1, the block's circumferential component of velocity is $v_{\theta 1}$. At point 2, the rope has been pulled in halfway, and the block's circumferential component of velocity is $v_{\theta 2}$. Determine $v_{\theta 2}/v_{\theta 1}$.

Solution
In the plane of the motion, the block is subjected to only one force, which is the rope force. The rope force is always directed toward the point O, thereby not producing a moment about O. Referring to Eq. (8.4-4) or (8.4-6), in the absence of a moment, the angular momentum about the point O is conserved. Thus, $H_{A1} = H_{A2}$, that is, $r_1 m v_{\theta 1} = r_2 m v_{\theta 2}$, so $v_{\theta 2}/v_{\theta 1} = r_1/r_2 = 2$.

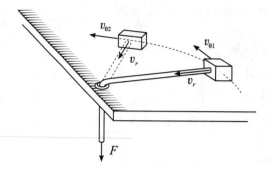

Example 8.4-2

8.4-3 A block slides on a smooth conical surface, as shown. Determine the block's speed v and angle β at the point B. Let $v_A = 0.6 \, \text{m/s}$, $r_A = 0.5 \, \text{m}$, and $r_B = 0.2 \, \text{m}$.

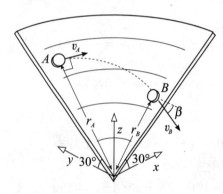

Example 8.4-3

Solution

The forces acting on the block consist of a gravitational force and a normal force. The gravitational force is a conservative force. The normal force does no work, nor does it produce a moment about the z axis (the line of the normal force intersects the z axis). Therefore, as the block moves from the point A to the point B, both energy and the z component of angular momentum about the point O are conserved. These quantities are given by

$$E_A = \tfrac{1}{2}mv_A^2 + mgh_A, \qquad E_B = \tfrac{1}{2}mv_B^2 + mgh_B,$$
$$H_{OA} = mr_A v_A, \qquad H_{OB} = mr_B v_B \cos \beta,$$

so, from $E_A = E_B$, we get

$$v_B = \sqrt{v_A^2 + 2g(h_A - h_B)} = 1.82 \, \text{m/s}.$$

From $H_{OA} = H_{OB}$, we get

$$\beta = \cos^{-1}\left(\frac{r_A v_A}{r_B v_B}\right) = 34.5°.$$

8.4-4 Children, from about the age of two, learn how to swing on a playground swing set. They learn how to shift their body mass center to elevate their swing. Develop a simple formula for the increase in swing angle after each half-cycle of swing. Denote the radius of the swing by R, the mass center shift by ΔR, and the increase in the swing angle by $\Delta \theta$. (See the diagram.)

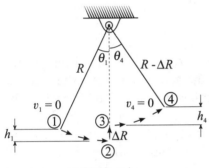

Half-cycle   *Example 8.4-4*

Solution
The path of the child's mass center over the half-cycle starts at point 1, and continues to points 2, 3, and 4, as shown. Point 1 is the top of the swing, where $v_1 = 0$. Point 2 is the bottom of the swing. The optimal swing is one in which the mass center is shifted at the bottom of the swing. Thus, point 3 represents the mass center of the child just after the mass center was shifted upward. Point 4 is the top of the swing at the end of the half-cycle, where again $v_4 = 0$. In order to solve this problem, the motion is divided into three phases. The first phase is when the child is between points 1 and 2, the second phase is when the child is between points 2 and 3, and the third phase is when the child is between points 3 and 4. During the first and third phases, energy is conserved. During the second phase, angular momentum about the point O is conserved. Thus,

$$mgh_1 = \tfrac{1}{2}mv_2^2,$$

$$Rmv_2 = (R - \Delta R)mv_3,$$

$$\tfrac{1}{2}mv_3^2 = mgh_4,$$

in which

$$h_1 = R(1 - \cos\theta_1),$$
$$h_4 = (R - \Delta R)(1 - \cos\theta_4).$$

The unknowns in these five equations are $h_1, h_2, v_2, v_3,$ and θ_4. Substituting the last two equations into the first and third, and then substituting the first and third equations into the second, yields

$$\cos\theta_4 = 1 - \left(\frac{R}{R - \Delta R}\right)^3 (1 - \cos\theta_1).$$

The change in the swing angle is $\Delta\theta = \theta_4 - \theta_1$. So, the change in the swing angle can be determined from the above equation. Although the problem is solved, the formula is rather complicated. The formula can be simplified and an appreciation for the relationships among the parameters can be found by making two small-number approximations. The numbers $\Delta\theta$ and $\Delta R/R$ are recognized as being small compared with 1 (they are usually less than 0.2). Therefore, as a simplification, we can perform the following first-order (retaining only the first two terms) Taylor series approximations:

$$\cos\theta_4 = \cos(\theta_1 + \Delta\theta_1) = \cos\theta_1 - \sin\theta_1 \, \Delta\theta,$$
$$\left(\frac{R}{R - \Delta R}\right)^3 = \left(1 - \frac{\Delta R}{R}\right)^{-3} = 1 + \frac{3\Delta R}{R}.$$

Substituting these approximations into the expression for $\cos\theta_4$ yields

$$\cos\theta_1 - \sin\theta_1 \, \Delta\theta = 1 - \left(1 + \frac{3\Delta R}{R}\right)(1 - \cos\theta_1),$$

so

$$\Delta\theta = 3\frac{\Delta R}{R}\frac{(1 - \cos\theta_1)}{\sin\theta_1} = 3\frac{\Delta R}{R}\frac{2\sin^2\frac{1}{2}\theta_1}{2\sin\frac{1}{2}\theta_1 \cos\frac{1}{2}\theta_1}$$

and so

$$\Delta\theta = 3\frac{\Delta R}{R}\tan(\tfrac{1}{2}\theta_1).$$

From this relatively simple formula, we see that the change in the swing angle is proportional both to the change in the mass center and to the angle of the swing. When the swing angle is large, it is easier to elevate than when the swing angle is small.

Problems

8.4-1 Determine the angular momentum $\mathbf{H}_A$ of the system of particles shown about the point A. Let $\mathbf{v}_1 = 10\mathbf{i} + 10\mathbf{j}\,\text{m/s}$, $\mathbf{v}_2 = -10\mathbf{i} - 10\mathbf{j}\,\text{m/s}$, $\mathbf{v}_3 = -5\mathbf{i} + 20\mathbf{j}\,\text{m/s}$, $m_1 = 3\,\text{kg}$, $m_2 = 6\,\text{kg}$, and $m_3 = 9\,\text{kg}$.

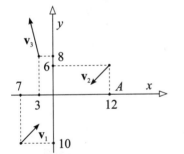

Problem 8.4-1

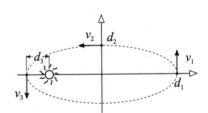

Problem 8.4-2

8.4-2 A comet orbits the Sun in an ellipse as shown. The semi-major axis is $d_1 = 3.24$ gigamiles, the semi-minor axis is $d_2 = 2.00$ gigamiles, and the perihelion is $d_3 = 0.1255$ gigamiles. The velocity of the comet at point 2 is 2.3 gigamiles per year. Determine the velocity of the comet when it is closest and farthest from the Sun.

8.4-3 An ice cube of mass $m = 30$ g initially lands on a funnel with a velocity of $v = 0.3$ m/s at an angle of $30°$ measured relative to the horizontal, as shown in the figure overleaf. Find the vertically downward distance h that the cube has traveled after reaching a speed of $v = 3$ m/s.

8.4-4 Find the maximum constant rotational speed of the motor $\dot{\theta}$ (in revolutions per minute, rpm) imposed by the governor shown in the

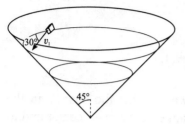

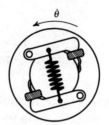

Problem 8.4-3   **Problem 8.4-4**

figure above. Also find the angular momentum H of the arms. The motor governor limits the speed of the motor by swinging out the arms at a specific rpm, thus breaking the circuit that drives the motor. The arms have a mass of $m = 0.5$ grams and are held by a spring exerting a force of 0.1 N. The center of mass of the arms is $d = 0.01$ m from the center of the motor.

CHAPTER 9

Rigid Body Dynamics

9.1 BODY-FIXED COORDINATES

Theory

The **angular velocity vector** is used to describe three-dimensional motion of rigid bodies. The direction of the angular velocity vector is taken to be the axis of rotation, and the magnitude of the angular velocity vector is taken to be the angular rate of the motion.

Let $\mathbf{i}_1$, $\mathbf{i}_2$, and $\mathbf{i}_3$ denote the unit vectors of a rotating coordinate system. The angular velocity vector of the coordinate system is denoted by $\boldsymbol{\omega}$ (up until now we called it $\dot{\theta}$). If we assume that a rigid body is fixed to this coordinate system, then the coordinate system is referred to as a **body-fixed coordinate system**, the associated unit vectors, $\mathbf{i}_1$, $\mathbf{i}_2$, and $\mathbf{i}_3$, are called **body-fixed unit vectors**, and the angular velocity vector is then associated with the rigid body. (See Fig. 9.1-1.)

Without loss of generality, we first look at the rotational motion of the unit vector $\mathbf{i}_1$. As shown in Fig. 9.1-2, $\mathbf{i}_1$ rotates around the axis of the angular velocity vector. The time derivative of $\mathbf{i}_1$ is expressed as magnitude times direction as

$$\frac{d\mathbf{i}_1}{dt} = \left|\frac{d\mathbf{i}_1}{dt}\right|\mathbf{n}, \tag{9.1-1}$$

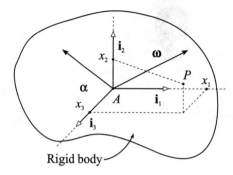

Figure 9.1-1

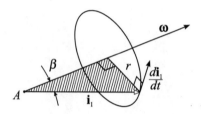

Figure 9.1-2

where the unit vector **n** is directed perpendicular to the shaded plane in the figure. Since **n** is a unit vector that is perpendicular to both $\boldsymbol{\omega}$ and $\mathbf{i}_1$, it follows that

$$\mathbf{n} = \frac{\boldsymbol{\omega} \times \mathbf{i}_1}{|\boldsymbol{\omega} \times \mathbf{i}_1|} = \frac{\boldsymbol{\omega} \times \mathbf{i}_1}{|\boldsymbol{\omega}| \sin \beta}. \tag{9.1-2}$$

Also, recognize that $|d\mathbf{i}_1/dt|$ is a circumferential component of velocity; that is,

$$\left|\frac{d\mathbf{i}_1}{dt}\right| = |\boldsymbol{\omega}|r = |\boldsymbol{\omega}| \sin \beta. \tag{9.1-3}$$

Substituting Eqs. (9.1-2) and (9.1-3) into Eq. (9.1-1) yields

$$\frac{d\mathbf{i}_1}{dt} = \boldsymbol{\omega} \times \mathbf{i}_1. \tag{9.1-4}$$

These steps can be repeated to obtain similar expressions for $\mathbf{i}_2$ and $\mathbf{i}_3$. Thus, the time derivatives of the body-fixed unit vectors and the angular velocity vector are related by

$$\boxed{\frac{d\mathbf{i}_1}{dt} = \boldsymbol{\omega} \times \mathbf{i}_1, \qquad \frac{d\mathbf{i}_2}{dt} = \boldsymbol{\omega} \times \mathbf{i}_2, \qquad \frac{d\mathbf{i}_1}{dt} = \boldsymbol{\omega} \times \mathbf{i}_1.} \tag{9.1-5}$$

Now that we can differentiate in time the body-fixed unit vectors, let's look at differentiating in time any vector that is expressed in terms of body-fixed unit vectors. For example, the rotating coordinate system could be the Earth's rotating axes, and we could be observing an airplane. The airplane's motion relative to our Earth-fixed coordinate system is different than the motion that would be observed relative to another coordinate system. The rotating coordinate system is also called the **body-fixed frame**.

First define a vector in the body-fixed frame as

$$\mathbf{A} = A_1\mathbf{i}_1 + A_2\mathbf{i}_2 + A_3\mathbf{i}_3. \tag{9.1-6}$$

The unit vectors $\mathbf{i}_1$, $\mathbf{i}_2$, and $\mathbf{i}_3$ are directed along the coordinate axes of the body-fixed frame. The vector $\mathbf{A}$ could represent position, velocity, force, or any other vector. The vector $\mathbf{A}$ is differentiated with respect to time using the product rule and Eq. (9.1-5), to get

$$
\begin{aligned}
\dot{\mathbf{A}} &= \left(\dot{A}_1\mathbf{i}_1 + A_1\frac{d\mathbf{i}_1}{dt}\right) + \left(\dot{A}_2\mathbf{i}_2 + A_2\frac{d\mathbf{i}_2}{dt}\right) + \left(\dot{A}_3\mathbf{i}_3 + A_3\frac{d\mathbf{i}_3}{dt}\right) \\
&= (\dot{A}_1\mathbf{i}_1 + \dot{A}_2\mathbf{i}_2 + \dot{A}_3\mathbf{i}_3) + \left(A_1\frac{d\mathbf{i}_1}{dt} + A_2\frac{d\mathbf{i}_2}{dt} + A_3\frac{d\mathbf{i}_3}{dt}\right) \\
&= (\dot{A}_1\mathbf{i}_1 + \dot{A}_2\mathbf{i}_2 + \dot{A}_3\mathbf{i}_3) + (A_1\boldsymbol{\omega} \times \mathbf{i}_1 + A_2\boldsymbol{\omega} \times \mathbf{i}_2 + A_3\boldsymbol{\omega} \times \mathbf{i}_3) \\
&= (\dot{A}_1\mathbf{i} + \dot{A}_2\mathbf{i}_2 + \dot{A}_3\mathbf{i}_3) + \boldsymbol{\omega} \times (A_1\mathbf{i}_1 + A_2\mathbf{i}_2 + A_3\mathbf{i}_3); \tag{9.1-7}
\end{aligned}
$$

that is,

$$\boxed{\dot{\mathbf{A}} = \dot{\mathbf{A}}_{bf} + \boldsymbol{\omega} \times \mathbf{A},} \tag{9.1-8}$$

where $\dot{\mathbf{A}}_{bf} = \dot{A}_1\mathbf{i}_1 + \dot{A}_2\mathbf{i}_2 + \dot{A}_3\mathbf{i}_3$ is called the time derivative **in** the body-fixed frame of $\mathbf{A}$. Notice in Eq. (9.1-8) that $\dot{\mathbf{A}} = \dot{\mathbf{A}}_{bf}$ when the body is not rotating ($\boldsymbol{\omega} = 0$), and that $\dot{\mathbf{A}} = \boldsymbol{\omega} \times \mathbf{A}$ when $\mathbf{A}$ is fixed **in** the body-fixed frame ($\dot{\mathbf{A}}_{bf} = 0$).

Example

 9.1-1   A boy walks on top of a see-saw, as shown in the figure overleaf. When the boy is 2 ft to the right of the center of the see-saw, he is walking 1 ft/s to the right (in the see-saw's reference frame) while the see-saw is rotating clockwise 3 rad/s. Assuming that the see-saw is horizontal, determine the velocity of the boy at this instant.

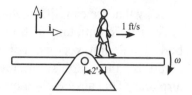

Solution

Let the see-saw represent the rotating body. The angular velocity vector of the see-saw is $\boldsymbol{\omega} = -3\mathbf{k}$. Also, let $\mathbf{A}$ in Eq. (9.1-8) represent the position vector of the boy measured from the center of the see-saw; that is, $\mathbf{r} = \mathbf{A} = 2\mathbf{i}$ m. Thus, in Eq. (9.1-8), $\mathbf{v} = \dot{\mathbf{A}}$ and $\mathbf{v}_{bf} = \dot{\mathbf{A}}_{bf}$, so, from Eq. (9.1-8), $\mathbf{v} = \mathbf{v}_{bf} + \boldsymbol{\omega} \times \mathbf{r}$, in which $\mathbf{v}_{bf} = \mathbf{i}$. We get

Example 9.1-1

$$\mathbf{v} = \mathbf{i} + -3\mathbf{k} \times 2\mathbf{i} = \mathbf{i} - 6\mathbf{j} \text{ ft/s.}$$

Notice in the answer that the first term is the velocity of the boy **in** the frame of the see-saw, and that the second term is the velocity of the boy had he been standing still in the rotating frame.

Problems

9.1-1 Find the derivative of the vector $\mathbf{A} = 7t\mathbf{i}_1 - 4\mathbf{i}_2 + 12(t-2)\mathbf{i}_3$ at $t = 10$, given that the indicated unit vectors rotate at an angular velocity $\boldsymbol{\omega} = 3\mathbf{i}_1 + 11\mathbf{i}_2 - 10\mathbf{i}_3$.

9.1-2 Find the derivative of the position vector $\mathbf{r} = 21t^2\mathbf{i}_1 + 16\mathbf{i}_3$ at $t = 5$, given that the indicated unit vectors are spinning with an angular velocity $\boldsymbol{\omega} = 12\mathbf{i}_1 - 14\mathbf{i}_3$.

9.2 COMPONENTS OF VELOCITY AND ACCELERATION IN A BODY-FIXED FRAME

Theory

The previous section provided us with the tools needed to differentiate in time the position vector of an object located in a body-fixed frame. Upon differentiation, we get the velocity vector, and upon differentiation once more, we get the acceleration vector. The acceleration vector will have four components: the relative acceleration effect, the Coriolis effect, the angular acceleration effect, and the centrifugal effect.

First, let $\mathbf{A} = \mathbf{r}$ in Eq. (9.1-8) to get

$$\mathbf{v} = \mathbf{v}_{bf} + \boldsymbol{\omega} \times \mathbf{r}. \qquad (9.2\text{-}1)$$

Equation (9.2-1) expresses the velocity of a point as the sum of two parts, the first being the velocity observed in the body-fixed frame and the second being the velocity of the point had it been fixed in the body-fixed frame.

Next, let's differentiate Eq. (9.2-1) with respect to time to get the acceleration vector. As we perform the differentiation, we'll again use Eq. (9.1-8), letting $\mathbf{A} = \mathbf{v}_{bf}$, and we'll use the product rule. Letting $\dot{\mathbf{r}}_{bf} = \mathbf{v}_{bf}$ and $\ddot{\mathbf{r}}_{bf} = \mathbf{a}_{bf}$, we get

$$\mathbf{a} = (\mathbf{a}_{bf} + \boldsymbol{\omega} \times \mathbf{v}_{bf}) + (\dot{\boldsymbol{\omega}} \times \mathbf{r} + \boldsymbol{\omega} \times \mathbf{v}), \qquad (9.2\text{-}2)$$

from which (letting $\boldsymbol{\alpha} = \dot{\boldsymbol{\omega}}$)

$$\mathbf{a} = \mathbf{a}_{bf} + 2\boldsymbol{\omega} \times \mathbf{v}_{bf} + \boldsymbol{\alpha} \times \mathbf{r} + \boldsymbol{\omega} \times (\boldsymbol{\omega} \times \mathbf{r}). \qquad (9.2\text{-}3)$$

The first term is the acceleration observed **in** the body-fixed coordinate system. The second term is called the **Coriolis** term. The third term is the result of the body's angular acceleration. The fourth term is called the **centrifugal** term.

Examples

| 9.2-1 | The brace shown rotates about the axis AB at a rate of 13 rad/s such that the z component of the velocity of the point C is positive. Determine the velocity vector of the point C.

Example 9.2-1

Solution
The brace ABC is a rotating body. The point A is a fixed point and the point C is a point that is not fixed in inertial space, although it is fixed **in** the rotating body ABC. From Eq. (9.2-1), the velocity vector of the point C is

$$\mathbf{v}_C = \mathbf{v}_{bf} + \boldsymbol{\omega} \times \mathbf{r}_{C/A},$$

in which

$$\mathbf{v}_{bf} = \mathbf{0}\,\text{m/s},$$

$$\boldsymbol{\omega} = \pm 13\frac{\mathbf{r}_{A/B}}{|\mathbf{r}_{A/B}|} = \pm 13\frac{-10\mathbf{i} - 12\mathbf{j} - 2\mathbf{k}}{\sqrt{10^2 + 12^2 + 2^2}}$$

$$= \mp 0.8255(10\mathbf{i} + 12\mathbf{j} + 2\mathbf{k})\,\text{rad/s},$$

$$\mathbf{r}_{C/A} = 12\mathbf{j}\,\text{m}.$$

Notice that the sign in front of the unit vector associated with the angular velocity vector is not yet known (this is why the $\pm$ symbol was used). Substituting these quantities into the equation for $\mathbf{v}_C$ yields

$$\mathbf{v}_C = \mathbf{0} \mp 0.8255(10\mathbf{i} + 12\mathbf{j} + 2\mathbf{k}) \times 12\mathbf{j} = \mp 0.8255(-24\mathbf{i} + 120\mathbf{k})$$

$$= \mp(-19.81\mathbf{i} + 99.06\mathbf{k})\,\text{m/s}.$$

The $\mathbf{k}$ component of the velocity vector is positive provided the plus part of the minus/plus symbol is selected. Thus, the angular velocity of ABC and the velocity of C are given by

$$\boldsymbol{\omega} = 0.8255(10\mathbf{i} + 12\mathbf{j} + 2\mathbf{k})\,\text{rad/s}, \qquad \mathbf{v}_C = -19.81\mathbf{i} + 99.06\mathbf{k}\,\text{m/s}.$$

Another way to calculate $\mathbf{v}_C$ is to notice that the point B is a fixed point too, in which case we could have performed the calculation the following way:

$$\mathbf{v}_C = \mathbf{v}_{bf} + \boldsymbol{\omega} \times \mathbf{r}_{C/B},$$

in which $\mathbf{r}_{C/B} = -10\mathbf{i} - 2\mathbf{k}\,\text{m}$. Then

$$\mathbf{v}_C = \mathbf{0} \mp 0.8255(10\mathbf{i} + 12\mathbf{j} + 2\mathbf{k}) \times (-10\mathbf{i} - 2\mathbf{k})$$

$$= \mp 0.8255(-24\mathbf{i} + 120\mathbf{k})$$

$$= \mp(-19.81\mathbf{i} + 99.06\mathbf{k})\,\text{m/s}.$$

Again, we see that the $\mathbf{k}$ component of the velocity is positive provided the plus part of the minus/plus symbol is selected, as expected. As before, the angular velocity of ABC and the velocity of C are given by

$$\boldsymbol{\omega} = 0.8255(10\mathbf{i} + 12\mathbf{j} + 2\mathbf{k})\,\text{rad/s}, \qquad \mathbf{v}_C = -19.81\mathbf{i} + 99.06\mathbf{k}\,\text{m/s}.$$

9.2-2 A block slides in a channel at a constant 4 m/s (in the body frame of the disk), as shown. Determine the velocity and acceleration of the block as it passes the point C. Let $r = 0$, $\omega = 3$ rad/s and $\alpha = -5$ rad/s^2.

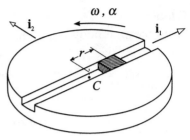

Example 9.2-2

Solution
Consider Eqs. (9.2-1) and (9.2-3), in which $r = 0$ m, $v_{bf} = 4i_1$ m/s, $\omega = 3i_3$ rad/s, and $\alpha = -5i_3$ rad/s^2. Thus,

$$v = 4i_1 + 3i_3 \times 0 = 4i_1 \text{ m/s},$$
$$a = 0 + 2(3i_3) \times 4i_1 + -5i_3 \times 0 + 3i_3 \times (3i_3 \times 0) = 24i_2 \text{ m/s}^2.$$

9.2-3 A bead is flicked from a pen, as shown. The bead was initially at rest 3 inches from the pivot point O. The flick was produced by flicking (rotating) the pen 90° in $\frac{1}{20}$ s at a constant rate of $[90°(\pi/180 \text{ rad/deg})]/(\frac{1}{20} \text{ s}) = 31.4$ rad/s. Determine the velocity of the bead when it flies off the end of the pen. Neglect friction between the bead and the pen and the effect of gravity.

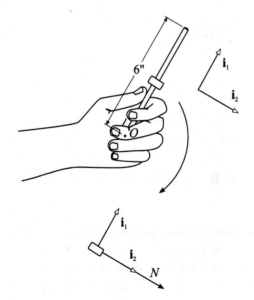

Example 9.2-3

Solution

From the free body diagram,

$$N\mathbf{i}_2 = m\mathbf{a} = m[\mathbf{a}_{bf} + 2\boldsymbol{\omega} \times \mathbf{v}_{bf} + \boldsymbol{\omega} \times (\boldsymbol{\omega} \times \mathbf{r}) + \boldsymbol{\alpha} \times \mathbf{r}]$$
$$= m[\ddot{r}\mathbf{i}_1 + 2\omega\mathbf{i}_3 \times \dot{r}\mathbf{i}_1 + \omega\mathbf{i}_3 \times (\omega\mathbf{i}_3 \times r\mathbf{i}_1) + \alpha\mathbf{i}_3 \times r\mathbf{i}_1]$$
$$= m[(\ddot{r} - \omega^2 r)\mathbf{i}_1 + (r\alpha + 2\omega\dot{r})\mathbf{i}_2],$$

so

$$0 = \ddot{r} - \omega^2 r,$$
$$N = r\alpha + 2\omega\dot{r}.$$

The first of these equations expresses $\ddot{r}$ as a function of r. The second equation is used to determine N, which we don't need. From the chain rule, $\ddot{r} = \dot{r}\, d\dot{r}/dr$, so the first equation becomes

$$\int_3^6 r\omega^2 \, dr = \int_0^{\dot{r}} \dot{r} \, d\dot{r};$$

so when the bead flies off the end of the pencil, the outward (radial) speed of the bead is

$$\dot{r} = 3\sqrt{3}\omega.$$

The velocity vector is

$$\mathbf{v} = \mathbf{v}_{bf} + \omega\mathbf{i}_3 \times r\mathbf{i}_1 = 3\sqrt{3}\omega\mathbf{i}_1 + 6\omega\mathbf{i}_2$$
$$= 163.16\mathbf{i}_1 + 188.4\mathbf{i}_2 \text{ inches/s}.$$

Notice that the bead flies outward off the end of the pencil at 86.6% of the speed of the tip of the pencil.

9.2-4 Show that the four-term acceleration expression given in Eq. (9.2-3) can be regarded as the extension of polar coordinates to three-dimensional problems. To show this, collapse the four-term expression for acceleration into the polar coordinate expression for acceleration.

Solution

Let's replace the body-fixed unit vectors with the polar coordinate unit vectors, and express the angular velocity vector and the angular acceleration vector in polar forms, as well. We substitute

$$\boldsymbol{\omega} = \dot{\theta}\mathbf{k}, \quad \boldsymbol{\alpha} = \ddot{\theta}\mathbf{k}, \quad \mathbf{r} = r\mathbf{n}_r, \quad \mathbf{v}_{bf} = \dot{r}\mathbf{n}_r, \quad \mathbf{a}_{bf} = \ddot{r}\mathbf{n}_r$$

into the four terms in Eq. (9.2-3) to yield

$$2\boldsymbol{\omega} \times \mathbf{v}_{bf} = 2\dot{\theta}\dot{r}\mathbf{k}, \quad \boldsymbol{\alpha} \times \mathbf{r} = r\ddot{\theta}\mathbf{k}, \quad \boldsymbol{\omega} \times (\boldsymbol{\omega} \times \mathbf{r}) = -r\dot{\theta}^2\mathbf{k}.$$

When these are substituted into Eq. (9.2-3), the following polar coordinate expression for the acceleration is obtained:

$$\mathbf{a} = (\ddot{r} - r\dot{\theta}^2)\mathbf{n}_r + (r\ddot{\theta} + 2\dot{r}\dot{\theta})\mathbf{n}_\theta.$$

Notice that in planar problems, both the acceleration in the body frame and the centrifugal component act in the radial direction, and that both the angular acceleration component and the Coriolis component act in the circumferential direction. On the other hand, in general three-dimensional problems, the four components of acceleration can act in four different directions.

9.2-5 The link AB rotates at an angular velocity $\omega = 5$ rad/s and an angular acceleration $\alpha = 2$ rad/s^2, as shown. Determine the velocity and acceleration of the block in the channel at this instant. The length of the link AB is 0.5 m, and the length of the link BC is 0.4 m.

Example 9.2-5

Solution

The position vector of the point C (measured from the point A) is

$$\mathbf{r}_C = \mathbf{r}_B + \mathbf{r}_{C/B}.$$

By differentiating with respect to time, we get the expressions

$$\mathbf{v}_C = \mathbf{v}_B + \mathbf{v}_{C/B}, \quad \mathbf{a}_C = \mathbf{a}_B + \mathbf{a}_{C/B}.$$

The points A and B are fixed to the body AB, and the points B and C are fixed to the body BC, so, from Eqs. (9.2-1) and (9.2-3),

$$
\begin{aligned}
\mathbf{v}_B &= \boldsymbol{\omega} \times \mathbf{r}_B = 5\mathbf{k} \times 0.5(\cos 40° \, \mathbf{i} + \sin 40° \, \mathbf{j}) = -1.607\mathbf{i} + 1.915\mathbf{j} \, \text{m/s}, \\
\mathbf{v}_{C/B} &= \boldsymbol{\omega}_{BC} \times \mathbf{r}_{C/B} = \omega_{BC}\mathbf{k} \times 0.4(\cos 20° \, \mathbf{i} + \sin 20° \, \mathbf{j}) \\
&= \omega_{BC}(-0.137\mathbf{i} + 0.376\mathbf{j}) \, \text{m/s}, \\
\mathbf{a}_B &= \boldsymbol{\alpha} \times \mathbf{r}_B + \boldsymbol{\omega} \times (\boldsymbol{\omega} \times \mathbf{r}_B) = 2\mathbf{k} \times 0.5(\cos 40° \, \mathbf{i} + \sin 40° \, \mathbf{j}) \\
&\quad + 5\mathbf{k} \times [5\mathbf{k} \times 0.5(\cos 40° \, \mathbf{i} + \sin 40° \, \mathbf{j})] = -10.22\mathbf{i} + 8.80\mathbf{j} \, \text{m/s}^2, \\
\mathbf{a}_{C/B} &= \boldsymbol{\alpha}_{BC} \times \mathbf{r}_{C/B} + \boldsymbol{\omega}_{BC} \times (\boldsymbol{\omega}_{BC} \times \mathbf{r}_{C/B}) \\
&= \alpha_{BC}\mathbf{k} \times 0.4(\cos 20° \, \mathbf{i} + \sin 20° \, \mathbf{j}) \\
&\quad + 5\mathbf{k} \times [5\mathbf{k} \times 0.4(\cos 20° \, \mathbf{i} + \sin 20° \, \mathbf{j})] \\
&= (-0.137\alpha_{BC} - 0.376\omega_{BC}^2)\mathbf{i} + (0.376\alpha_{BC} - 0.137\omega_{BC}^2)\mathbf{j} \, \text{m/s}^2
\end{aligned}
$$

Substituting for $\mathbf{v}_B$ and $\mathbf{v}_{C/B}$ in the expression for $\mathbf{v}_C$ yields

$$
\mathbf{v}_C = v_C\mathbf{j} = -1.607\mathbf{i} + 1.915\mathbf{j} + \omega_{BC}(-0.137\mathbf{i} + 0.376\mathbf{j}),
$$

which represents the two equations

$$
\begin{aligned}
0 &= 1.915 + 0.376\omega_{BC}, \\
v_C &= 1.607 + 0.137\omega_{BC}.
\end{aligned}
$$

The solution is $\omega_{BC} = -5.09$ rad/s and $v_C = 0.909$ m/s. Now, turning to the expression for the acceleration of the point C, we get

$$
\begin{aligned}
\mathbf{a}_C = a_C\mathbf{j} &= -10.22\mathbf{i} + 8.80\mathbf{j} + (-0.137\alpha_{BC} - 0.376\omega_{BC}^2)\mathbf{i} \\
&\quad + (0.376\alpha_{BC} - 0.137\omega_{BC}^2)\mathbf{j},
\end{aligned}
$$

which represents the two equations

$$
\begin{aligned}
0 &= -10.22 - 0.137\alpha_{BC} - 0.376(-5.09)^2, \\
a_C &= 8.80 + 0.376\alpha_{BC} - 0.137(-5.09)^2.
\end{aligned}
$$

The solution is $\alpha_{BC} = -145.7$ rad/s^2 and $a_C = -49.53$ m/s^2.

Problems

9.2-1 Determine the velocity and acceleration vectors of the point A on the triangular plate shown, given that the plate rotates at an angular velocity of 12 rad/s and an angular acceleration of -6 rad/s² at the instant shown.

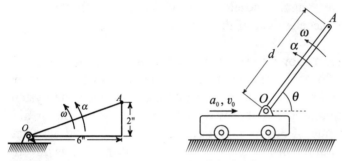

Problem 9.2-1   **Problem 9.2-2**

9.2-2 At a given instant the cart shown has a velocity of 4 m/s and an acceleration of 5 m/s². The bar has an angular velocity of -4 rad/s and an angular acceleration of -7.5 rad/s. Determine the velocity and acceleration vectors of the point A. Let $d = 0.8$ m and $\theta = 50°$.

9.2-3 The block A shown rotates counterclockwise at a constant rate of 10 m/s. The rod that is attached to the block rotates at an angular rate of 10 rad/s and at an angular acceleration of 5 rad/s². Determine the velocity and acceleration vectors of the point B.

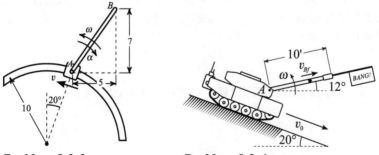

Problem 9.2-3   **Problem 9.2-4**

9.2-4 A projectile is launched from the barrel of a super-tank, as shown. The projectile leaves the barrel at a speed of 100 m/s (relative to the barrel) while the barrel is rotating upward at a rate of 1 rad/s while the tank is moving forward at a rate of 10 m/s. Determine the velocity of the projectile as it leaves the barrel.

9.2-5 Determine the acceleration vector of the point C in Example 9.2-1. Let the axis AB rotate with an angular velocity of 13 m/s such that the z component of the velocity of the point C is positive, and let the axis AB rotate with an angular acceleration of 10 m/s^2 such that the z component of the acceleration of the point C is positive.

9.2-6 The system shown can represent either a compressor, which is driven by a motor located at the point O, or an engine, which is driven by a piston at the point A. Consider the first case, in which the system represents a compressor. Let the point B be driven by the motor at a constant angular rate of 30 rad/s. Determine the velocity and acceleration of the point A at the instant shown. Notice that the point B is directly above the point O, and let $R = 0.4$ m and $L = 1.2$ m.

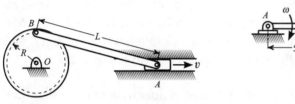

Problem 9.2-6   **Problem 9.2-7**

9.2-7 The bar AB has a constant angular velocity of 5 rad/s in the direction shown. Determine the angular velocity and angular acceleration of the bar BC. Also determine the velocity and acceleration of the collar B **in** the body frame of the bar BC.

9.3 ANGULAR IMPULSE AND ANGULAR MOMENTUM

Theory

In general, the motion of a single particle is governed by three independent equations corresponding to three independent degrees of freedom. Newton's Second Law $\mathbf{F} = m\mathbf{a}$ gives us the three needed equations. Instead of using these three equations, some or all of them can be replaced by a work–energy equation, some linear-impulse–linear-momentum equations, and some angular-impulse–angular-momentum equations. Regardless of which equations are used, the number of equations is equal to the number of independent degrees of freedom.

In the case of a system of n particles, the most general motion consists of $3n$ equations and $3n$ independent degrees of freedom: three for each particle. When the system of particles is a rigid body, the relative motion of the particles is constrained, leaving at most six independent degrees of freedom. Three of the independent degrees of freedom correspond to **translations**, and the other three correspond to **rotations**. The three equations corresponding to the translations are Eqs. (6.2-5), $\mathbf{F} = m\mathbf{a}_C$, and the three equations corresponding to the rotations are Eqs. (8.4-4), $\mathbf{M}_A = \dot{\mathbf{H}}_A$. Some or all of Eqs. (6.2-5) and (8.4-4) can be replaced by a work–energy equation, some linear-impulse–linear-momentum equations, and some angular-impulse–angular-momentum equations, that is, Eqs. (8.2-10), (8.3-1), and (8.4-6).

In the case of a rigid body, the body-fixed components of angular momentum have a special form wherein they can be expressed as a linear combination of the body-fixed components of the angular velocity. For convenience the coefficients in the linear combination are collected into a 3×3 matrix called the **inertia matrix**. The diagonal terms of the inertia matrix are called **moments of inertia** and its off-diagonal terms are called **products of inertia**.

For simplicity, we'll first develop the relationship between angular momentum and angular velocity for planar motion, and then treat the general case of three-dimensional motion.

PLANAR MOTION

We first consider the planar motion of a rigid body that rotates about a fixed point, after which we'll consider the general planar motion of a rigid body. So, let the rigid body rotate about a point O about the z axis, thereby confining the motion to the x–y plane. From Eqs. (8.4-1) and (9.2-1), the angular momentum of the rigid body is

$$\mathbf{H}_O = \sum \mathbf{r}_i \times m_i \mathbf{v}_i = \int \mathbf{r} \times \mathbf{v} \, dm = \int \mathbf{r} \times (\boldsymbol{\omega} \times \mathbf{r}) \, dm. \qquad (9.3\text{-}1)$$

Letting $\mathbf{r} = x\mathbf{i} + y\mathbf{j} + z\mathbf{k}$ and $\boldsymbol{\omega} = \omega\mathbf{k}$, we get $\mathbf{H}_O = H_{Ox}\mathbf{i} + H_{Oy}\mathbf{j} + H_{Oz}\mathbf{k}$, in which

$$H_{Ox} = \left(-\int xz \, dm\right)\omega, \quad H_{Oy} = \left(-\int yz \, dm\right)\omega, \quad H_{Oz} = \left[\int\int (x^2 + y^2) \, dm\right]\omega.$$

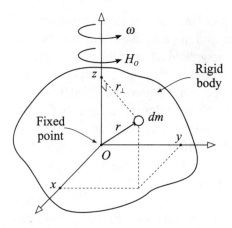

Figure 9.3-1

Let's focus on the third equation, and rewrite it as

$$H_O = I_O\omega, \quad \text{in which} \quad I_O = \int r_\perp^2 \, dm \qquad (9.3\text{-}2a,b)$$

denotes the **rigid body's moment of inertia about the z axis**, where $r_\perp = \sqrt{x^2 + y^2}$ denotes the distance between the differential mass dm and the z axis (see Fig. 9.3-1). From Eq. (8.4-4), the time derivative of the angular momentum is given by

$$M_O = I_O\alpha. \qquad (9.3\text{-}3)$$

Equation (9.3-3) governs the rotational motion of a rigid body just as $\mathbf{F} = m\mathbf{a}_C$ governs the translational motion of a rigid body. Indeed, just as mass resists translational motion ($\mathbf{a}_C = \mathbf{F}/m$), the moment of inertia resists rotational motion ($\alpha = M_O/I_O$).

Next, let's assume that the rigid body does **not** rotate about a fixed point, although it continues to rotate in the x–y plane (see Fig. 9.3-2). From Eqs. (8.4-1) and (9.2-1),

$$\mathbf{H}_C = \int \mathbf{r}' \times \mathbf{v} \, dm = \int \mathbf{r}' \times (\mathbf{v}_C + \boldsymbol{\omega} \times \mathbf{r}') \, dm,$$

in which $\mathbf{r}' = \mathbf{r} - \mathbf{r}_C$ is the position of dm relative to the position of the mass center C. Then

$$\mathbf{H}_C = \int \mathbf{r}' \, dm \times \mathbf{v}_C + \int \mathbf{r}' \times (\boldsymbol{\omega} \times \mathbf{r}') \, dm.$$

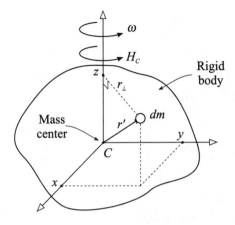

Figure 9.3-2

Notice that the first term is zero; that is, $\int \mathbf{r}' \, dm = \int (\mathbf{r} - \mathbf{r}_C) \, dm = \int \mathbf{r} \, dm - \mathbf{r}_C \int dm = m\mathbf{r}_C - \mathbf{r}_C m = \mathbf{0}$. Thus,

$$\mathbf{H}_C = \int \mathbf{r}' \times (\boldsymbol{\omega} \times \mathbf{r}') \, dm, \tag{9.3-4}$$

which is the same as the expression for $\mathbf{H}_O$ in Eq. (9.3-1) except that $\mathbf{r}$ has been replaced by $\mathbf{r}'$. Letting $\mathbf{r}' = x\mathbf{i} + y\mathbf{j} + z\mathbf{k}$ and $\boldsymbol{\omega} = \omega\mathbf{k}$, just as Eq. (9.3-1) led to Eqs. (9.3-2) and (9.3-3), Eq. (9.3-4) leads to the two equations

$$H_C = I_C \omega, \tag{9.3-5}$$

$$M_C = I_C \alpha. \tag{9.3-6}$$

Equations (9.3-2), (9.3-3), (9.3-5), and (9.3-6) can be written compactly as

$$\boxed{H_A = I_A \omega, \qquad M_A = I_A \alpha.} \tag{9.3-7a,b}$$

The point A denotes either a fixed point or the mass center, and

$$\boxed{I_A = \int r_\perp^2 \, dm} \tag{9.3-8}$$

is the moment of inertia of the rigid body about the point A, in which $r_\perp$ denotes the distance between the differential mass dm and the z axis of rotation (which passes through the point A).

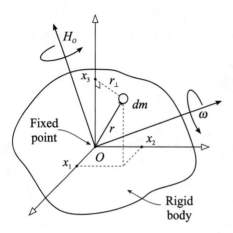

Fixed point

Rigid body

Figure 9.3-3

THREE-DIMENSIONAL MOTION

First consider the angular momentum of a rigid body about a fixed point O (see Fig. 9.3-3). The angular momentum vector is given by Eq. (9.3-1), in which

$$\mathbf{H}_O = H_{O1}\mathbf{i}_1 + H_{O2}\mathbf{i}_2 + H_{O3}\mathbf{i}_3, \qquad \mathbf{r} = x_1\mathbf{i}_1 + x_2\mathbf{i}_2 + x_3\mathbf{i}_3,$$
$$\boldsymbol{\omega} = \omega_1\mathbf{i}_1 + \omega_2\mathbf{i}_2 + \omega_3\mathbf{i}_3.$$

Thus,

$$
\begin{aligned}
H_{O1}&\mathbf{i}_1 + H_{O2}\mathbf{i}_1 + H_{O3}\mathbf{i}_3 \\
&= \int (x_1\mathbf{i}_1 + x_2\mathbf{i}_2 + x_3\mathbf{i}_3) \\
&\quad \times [(\omega_1\mathbf{i}_1 + \omega_2\mathbf{i}_2 + \omega_3\mathbf{i}_3) \times (x_1\mathbf{i}_1 + x_2\mathbf{i}_2 + x_3\mathbf{i}_3)]\, dm \\
&= \int (x_1\mathbf{i}_1 + x_2\mathbf{i}_2 + x_3\mathbf{i}_3) \\
&\quad \times [(\omega_2 x_3 - \omega_3 x_2)\mathbf{i}_1 + (\omega_3 x_1 - \omega_1 x_3)\mathbf{i}_2 + (\omega_1 x_2 - \omega_2 x_1)\mathbf{i}_3]\, dm \\
&= \left[\int (x_2^2 + x_3^2)\, dm\, \omega_1 - \int x_1 x_2\, dm\, \omega_2 - \int x_1 x_3\, dm\, \omega_3 \right]\mathbf{i}_1 \\
&\quad + \left[-\int x_1 x_2\, dm\, \omega_1 + \int (x_3^2 + x_1^2)\, dm\, \omega_2 - \int x_2 x_3\, dm\, \omega_3 \right]\mathbf{i}_2 \\
&\quad + \left[-\int x_3 x_1\, dm\, \omega_1 - \int x_3 x_2\, dm\, \omega_2 + \int (x_1^2 + x_2^2)\, dm\, \omega_3 \right]\mathbf{i}_3.
\end{aligned}
$$

Each of the components of the angular momentum vector can now be expressed as a linear combination of the angular velocity components. In matrix–vector form, we get

$$
\begin{bmatrix} H_{O1} \\ H_{O2} \\ H_{O3} \end{bmatrix} = \begin{bmatrix} I_{11} & I_{12} & I_{13} \\ I_{12} & I_{22} & I_{23} \\ I_{13} & I_{23} & I_{33} \end{bmatrix} \begin{bmatrix} \omega_1 \\ \omega_2 \\ \omega_3 \end{bmatrix}, \tag{9.3-10}
$$

in which

$$
I_{11} = \int (x_2^2 + x_3^2)\, dm, \qquad I_{22} = \int (x_3^2 + x_1^2)\, dm, \qquad I_{33} = \int (x_1^2 + x_2^2)\, dm,
$$

$$
I_{12} = -\int x_1 x_2\, dm, \qquad I_{13} = -\int x_1 x_3\, dm, \qquad I_{23} = -\int x_2 x_3\, dm.
$$

$$
\tag{9.3-11}
$$

The quantities I_{11}, I_{22}, and I_{33} are the **moments of inertia** of the rigid body about the $\mathbf{i}_1$, $\mathbf{i}_2$, and $\mathbf{i}_3$ axes, respectively, and I_{12}, I_{13}, and I_{23} are the **products of inertia** of the rigid body. Notice that the integrand of I_{11} is equal to the squared distance between the differential mass dm and the $\mathbf{i}_1$ axis. Similarly, the integrand of $\mathbf{I}_{22}$ is equal to the squared distance between the differential mass dm and the $\mathbf{i}_2$ axis, and the integrand of I_{33} is equal to the squared distance between the differential mass dm and the $\mathbf{i}_3$ axis.

Next consider the angular momentum of a rigid body about the mass center (see Fig. 9.3-4). Begin with the definition of the angular momentum of a rigid body about the mass center, which is given in Eq. (9.3-3). The steps followed to obtain Eqs. (9.3-10) and (9.3-11) from Eq. (9.3-1) for the angular momentum about a fixed point are the same as the steps followed from Eq. (9.3-4) to obtain

$$
\begin{bmatrix} H_{C1} \\ H_{C2} \\ H_{C3} \end{bmatrix} = \begin{bmatrix} I_{11} & I_{12} & I_{13} \\ I_{12} & I_{22} & I_{23} \\ I_{13} & I_{23} & I_{33} \end{bmatrix} \begin{bmatrix} \omega_1 \\ \omega_2 \\ \omega_3 \end{bmatrix}. \tag{9.3-12}
$$

Comparing Eqs. (9.3-10) and (9.3-12), we see that the angular momentum about a fixed point and the angular momentum about the mass

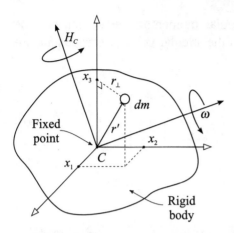

Figure 9.3-4

center are related to the angular velocity by the same equations. Thus, in matrix–vector form, Eqs. (9.3-10) and (9.3-12) are written as

$$[\mathbf{H}_A] = [\mathbf{I}_A][\boldsymbol{\omega}] \qquad (9.3\text{-}13)$$

in which **the point A is either a fixed point or the mass center**, and $[\mathbf{I}_A]$ is called the **inertia matrix**. Notice that $[\mathbf{H}_A]$ and $[\boldsymbol{\omega}]$ are column vectors, whereas

$$\mathbf{H}_A = H_{A1}\mathbf{i}_1 + H_{A2}\mathbf{i}_2 + H_{A3}\mathbf{i}_3, \qquad \boldsymbol{\omega} = \omega_1\mathbf{i}_1 + \omega_2\mathbf{i}_2 + \omega_3\mathbf{i}_3.$$

Since the derivative of the angular momentum about the point A with respect to time is equal to the external moment about A, we get from Eqs. (8.4-4) and (9.1-8)

$$\mathbf{M}_A = \dot{\mathbf{H}}_A = \dot{\mathbf{H}}_{Abf} + \boldsymbol{\omega} \times \mathbf{H}_A. \qquad (9.3\text{-}14)$$

This equation governs the rotational motion of a rigid body. Equation (9.3-14) can also be written in matrix–vector form, as

$$[\mathbf{I}_A][\boldsymbol{\alpha}] + [\boldsymbol{\omega}] \times ([\mathbf{I}_A][\boldsymbol{\omega}]) = [\mathbf{M}_A]. \qquad (9.3\text{-}15)$$

The moments of inertia and the products of inertia are quantities associated with a given rigid body. They characterize its rotational behavior. Unlike mass, the moments of inertia depend on the axis of rotation. Indeed, the moments of inertia and the products of inertia depend on the location of the origin of the coordinate system and its orientation. In fact, for any rigid body, there exists a coordinate system for which the rigid body's products of inertia are all zero. This type of body-fixed coordinate system is called a **principal coordinate system**. If principal coordinates are used, Eq. (9.3-13) reduces to

$$H_{A1} = I_{11}\omega_1, \quad H_{A2} = I_{22}\omega_2, \quad H_{A3} = I_{33}\omega_3, \qquad (9.3\text{-}16)$$

and Eq. (9.3-15) reduces to

$$
\begin{aligned}
M_{A1} &= I_{11}\dot{\omega}_1 + (I_{33} - I_{22})\omega_2\omega_3, \\
M_{A2} &= I_{22}\dot{\omega}_2 + (I_{11} - I_{33})\omega_3\omega_1, \\
M_{A3} &= I_{33}\dot{\omega}_3 + (I_{22} - I_{11})\omega_2\omega_3.
\end{aligned}
\qquad (9.3\text{-}17)
$$

Equations (9.3-17) are a set of three coupled nonlinear differential equations describing the rotational behavior of the rigid body. Notice that the nonlinear terms $(I_{33} - I_{22})\omega_2\omega_3$, $(I_{11} - I_{33})\omega_1\omega_3$, and $(I_{22} - I_{11})\omega_2\omega_1$ are responsible for the **coupling** in the system. Otherwise, the equations would be independent of each other. These nonlinear terms are responsible for some interesting **wobbling** effects observed in bodies undergoing three-dimensional motion.

Whereas the solution of these equations generally requires numerical integration, which is not treated in this book, two types of problems that can be solved without the aid of numerical methods are considered in this book. The two problem types are as follows:

Problem Type 1: Determine the moment reactions of a rigid body in which its angular velocity components are prescribed (known). The known angular velocity components are substituted into the right-hand side of Eq. (9.3-15) or (9.3-17) to obtain the moment reactions.

Problem Type 2: Determine moment reactions and angular acceleration components just after the rigid body is released from rest (the angular velocity components are zero). With the angular velocity components being equal to zero, Eqs. (9.3-15) and (9.3-17) become linear algebraic equations in terms of the unknown moment reactions and angular velocity components.

Examples

9.3-1 The heavy uniform rectangular plate shown, of mass 25 slugs with sides 3 ft and 6 ft, rotates about the point O. Determine the plate's angular acceleration when $\theta = 25°$.

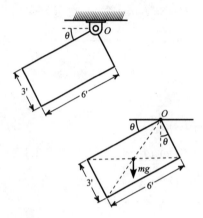

Example 9.3-1

Solution

The plate's angular acceleration is determined from Eq. (9.3-7a). The calculation of the plate's mass moment of inertia is described in Section 5.4. We get

$$I_O = \tfrac{1}{3}m(a^2 + b^2) = \tfrac{25}{3}(3^2 + 6^2) = 375 \text{ slug·ft}^2.$$

The resultant moment about the point O is produced by the gravitational force, which is located in the center of the plate. The resultant moment is

$$M_O = \frac{mg}{2}(6\cos\theta - 3\sin\theta) = 1678 \text{ 16·ft.}$$

The angular acceleration is

$$\alpha = M_O/I_O = 1678/375 = 4.47 \text{ rad/s}^2.$$

9.3-2 Let's compare the accelerations of a block and a uniform disk; each has mass m and each is subjected to a force F, as shown. The block slides on a smooth surface and the disk rolls without slipping. The radius of the disk is R.

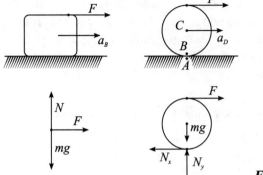

Example 9.3-2

Solution

(a) Let's first look at the block. From the free body diagram of the block, $F = ma_B$, so the acceleration of the block is $a_B = F/m$.

(b) Next, let's look at the disk. From the free body diagram of the disk, we get the three equations (summing forces and summing moments about the disk's mass center)

$$F - N_x = ma_D,$$
$$N_y - mg = 0,$$
$$-N_x R - FR = I_C \alpha$$

in which the moment of inertia of the disk is $I_C = \frac{1}{2}mR^2$ (see Section 5.4). These three equations are expressed in terms of the four unknowns N_x, a_D, N_y, and α, leaving us one equation short. The missing equation relates the acceleration a_D of the wheel's center C and the angular acceleration α of the wheel. This relationship is found by recognizing that the disk is rolling. To examine rolling, look at the points A and B, which are two points that are very close to the contact point. The **general rolling conditions** are

$$
\boxed{
\begin{aligned}
v_{Ax} &= v_{Bx}, \\
v_{Ay} &= v_{By}, \\
a_{Ax} &= a_{Bx}.
\end{aligned}
}
\tag{9.3-18}
$$

Notice that $a_{Ay} = a_{By}$ is **not** a condition. Indeed, $a_{Ay} \neq a_{By}$. In this example, the point A is not moving, so, from the general rolling conditions, Eqs. (9.3-18), we get $v_{Bx} = 0$, $v_{By} = 0$, and $a_{Bx} = 0$.

The points B and C are both fixed to the disk, so the velocity and acceleration of the point C are

$$\mathbf{v}_C = \mathbf{v}_B + \mathbf{v}_{C/B} = \mathbf{v}_B + \boldsymbol{\omega} \times \mathbf{r}_{C/B} = 0 + \omega \mathbf{k} \times R\mathbf{j} = -\omega R\mathbf{i},$$

$$\mathbf{a}_C = \mathbf{a}_B + \mathbf{a}_{C/B} = \mathbf{a}_B + \boldsymbol{\alpha} \times \mathbf{r}_{C/B} + \boldsymbol{\omega} \times (\boldsymbol{\omega} \times \mathbf{r}_{C/B})$$
$$= a_{By}\mathbf{j} + \alpha \mathbf{k} \times R\mathbf{j} - \omega \mathbf{k} \times (\omega \mathbf{k} \times R\mathbf{j})$$
$$= -\alpha R\mathbf{i} + (a_{By} - \omega^2 R)\mathbf{j} = a_C\mathbf{i},$$

so

$$a_C = -\alpha R,$$
$$a_{By} = \omega^2 R.$$

In the above, we have found a relationship between the angular velocity of a rolling disk and the velocity of its center, and we have found a relationship between the angular acceleration of a rolling disk and the acceleration of its center, given by

$$\boxed{v_{C/B} = -\omega R, \qquad a_{C/B} = -\alpha R.}$$
(9.3-19a,b)

Also, notice that the acceleration of the point B is $\omega^2 R$, which is a centrifugal effect.

Returning to the problem at hand, from the four equations that describe the motion of the disk, we now get

$$N_x = -\tfrac{1}{3}F \qquad a_D = \frac{4F}{3m}, \qquad N_y = mg, \qquad \alpha = -\frac{4F}{3mR}.$$

Comparing the acceleration of the disk, $a_D = 4F/3m$, with the acceleration of the block (recall that $a_B = F/m$), we find that the disk accelerates 33% faster than the block. This is attributed to the horizontal reaction force N_x, which increases the horizontal resultant force acting on the disk by 33%.

9.3-3   A ladder of mass 10 kg slides down a smooth wall and across a smooth floor. Determine the ladder's angular acceleration, immediately after it has been released from rest in the position shown. Determine the accelerations of the points A and B on the ladder at this instant too.

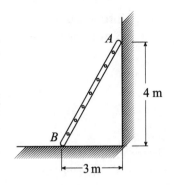

Solution

The points A and B are both on the ladder and $\boldsymbol{\omega} = \mathbf{0}$, so, from Eq. (9.2-3),

$$\mathbf{a}_{A/B} = \boldsymbol{\alpha} \times \mathbf{r}_{A/B},$$

so

$$\mathbf{a}_A = \mathbf{a}_B + \mathbf{a}_{A/B} = \mathbf{a}_B + \boldsymbol{\alpha} \times \mathbf{r}_{A/B},$$

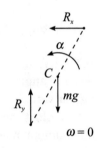

Example 9.3-3

in which

$$\mathbf{a}_A = a_A \mathbf{j}, \qquad \mathbf{a}_B = a_B \mathbf{i}, \qquad \boldsymbol{\alpha} = \alpha \mathbf{k}, \qquad \mathbf{r}_{A/B} = 3\mathbf{i} + 4\mathbf{j}.$$

We get the two equations

$$a_A = 3\alpha, \qquad a_B = 4\alpha.$$

The acceleration of the ladder's mass center is now expressed in terms of α as

$$\mathbf{a}_C = \mathbf{a}_B + \tfrac{1}{2}\mathbf{a}_{A/B} = 4\alpha \mathbf{i} + \tfrac{1}{2}(-4\alpha \mathbf{i} + 3\alpha \mathbf{j})$$

$$= \tfrac{1}{2}\alpha(4\mathbf{i} + 3\mathbf{j}) \, \text{m/s}^2.$$

Summing forces, summing moments about the mass center, and from Eq. (9.3-7a), we get the three equations

$$-R_x = ma_{Cx} = 10(2\alpha),$$
$$R_y - mg = ma_{Cy} = 10(1.5\alpha),$$
$$2R_x - 1.5R_y = I_C\alpha.$$

which are expressed in terms of the three unknowns R_x, R_y, and α. The moment of inertia of the ladder is $I_C = \frac{1}{12}mL^2 = \frac{1}{12}10(5^2) = 20.83 \text{ kg·m}^2$ (see Section 5.3). The solution is

$$\alpha = -1.77 \text{ rad/s}^2, \qquad R_x = 35.32 \text{ N}, \qquad R_y = 71.61 \text{ N}.$$

9.3-4 A motorized mechanism consists of a collar A that rotates at a constant angular velocity of ω_3 and a disk that rotates at a constant angular velocity of ω_1, as shown. The mass of the disk is m and its radius is R. Neglect the masses of the other components, and express the force and moment reactions at the fixed point O in terms of the prescribed

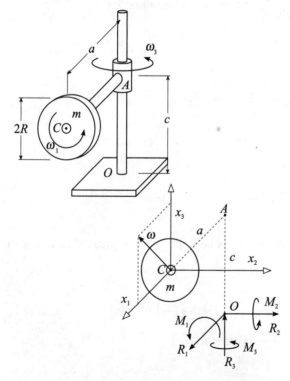

Example 9.3-4

angular velocity components (ω_1 and ω_3), the distances a and c, and the disk's mass m and radius R.

Solution

A set of body-fixed coordinates is set up with its origin located at the body's mass center C, as shown in the free body diagram. The angular velocity vector is $\boldsymbol{\omega} = \omega_1 \mathbf{i}_1 + \omega_3 \mathbf{i}_3$ and the angular acceleration vector is $\boldsymbol{\alpha} = \mathbf{0}$. From Eq. (9.2-3), the acceleration vector of the mass center C is $\mathbf{a}_C = \omega_3 \mathbf{i}_3 \times (\omega_3 \mathbf{i}_3 \times a\mathbf{i}_1) = -a\omega_3^2 \mathbf{i}_1$. From the free body diagram and Eqs. (6.2-5) and (9.3-7), we sum forces and sum moments about the point C, to get

$$R_1 = ma_1 = m(-a\omega_3^2) = -ma\omega_3^2,$$
$$R_2 = ma_2 = m(0) = 0,$$
$$R_3 - mg = ma_3 = m(0) = 0,$$
$$M_1 + cR_2 = I_{11}\alpha_1 + (I_{33} - I_{22})\omega_2\omega_3 = I_{11}(0) + (I_{33} - I_{22})(0)\omega_3 = 0,$$
$$M_2 - cR_1 + aR_3 = I_{22}\alpha_2 + (I_{11} - I_{33})\omega_3\omega_1 = I_{22}(0) + (I_{11} - I_{33})\omega_3\omega_1$$
$$= (I_{11} - I_{33})\omega_3\omega_1,$$
$$M_3 - aR_2 = I_{33}\alpha_3 + (I_{22} - I_{11})\omega_1\omega_2 = I_{33}(0) + (I_{22} - I_{11})\omega_1(0) = 0.$$

Since $I_{11} = \frac{1}{2}mR^2$ and $I_{22} = I_{33} = \frac{1}{4}mR^2$ (see Section 5.4), the solution is

$$R_1 = -ma\omega_3^2, \qquad R_2 = 0, \qquad R_3 = mg,$$
$$M_1 = 0, \qquad M_2 = -mac\omega_3^2 - amg + \tfrac{1}{4}mR^2\omega_3\omega_1, \qquad M_3 = 0.$$

9.3-5   A uniform rectangular plate is attached to a spherical joint at a point O, as shown in the figure overleaf. The plate is released from rest in the position shown. Determine the reactions at the joint just after the plate is released. Express your answers in terms of the plate's mass m, and its side lengths a and b.

Solution

Since $\boldsymbol{\omega} = \mathbf{0}$, from Eq. (9.2-3) the acceleration of the plate's mass center is

$$\mathbf{a}_C = \boldsymbol{\alpha} \times \mathbf{r}_C = (\alpha_1 \mathbf{i}_1 + \alpha_2 \mathbf{i}_2 + \alpha_3 \mathbf{i}_3) \times (\tfrac{1}{2}a\mathbf{i}_1 + \tfrac{1}{2}b\mathbf{i}_2)$$
$$= -\tfrac{1}{2}\alpha_3 b\mathbf{i}_1 + \tfrac{1}{2}\alpha_3 a\mathbf{i}_2 + \tfrac{1}{2}(\alpha_1 b - \alpha_2 a)\mathbf{i}_3.$$

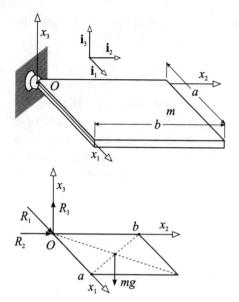

Example 9.3-5

Summing forces and summing moments about the point O yields

$$R_1 = ma_{C1} = m(-\tfrac{1}{2}\alpha_3 b),$$
$$R_2 = ma_{C2} = m(\tfrac{1}{2}\alpha_3 a),$$
$$R_3 - mg = ma_{C3} = m[\tfrac{1}{2}(\alpha_1 b - \alpha_2 a)],$$
$$-\tfrac{1}{2}bmg = I_{11}\alpha_1 + I_{12}\alpha_2 = \tfrac{1}{3}mb^2\alpha_1 - \tfrac{1}{4}mab\alpha_2,$$
$$\tfrac{1}{2}amg = I_{12}\alpha_1 + I_{22}\alpha_2 = -\tfrac{1}{4}mab\alpha_1 + \tfrac{1}{3}ma^2\alpha_2,$$
$$0 = I_{33}\alpha_3 = \tfrac{1}{3}m(a^2 + b^2)\alpha_3.$$

The solution is

$$\alpha_1 = -\frac{6g}{7b}, \qquad \alpha_2 = \frac{6g}{7a}, \qquad \alpha_3 = 0, \qquad R_1 = 0, \qquad R_2 = 0, \qquad R_3 = \tfrac{1}{7}mg.$$

Notice that $\alpha_3 = 0$, that the angular acceleration **axis** is directed from the point A to the point B, and that the **dynamic** reaction R_3 is about 15% of the **static** reaction $R_3 = mg$ (had the plate been held in the position shown statically).

Problems

9.3-1 A thin bar of mass 10 slug and length 1.4 ft rotates about a point O, and was released from rest when $\theta = 90°$. Determine the rod's angular acceleration when $\theta = 45°$, as shown.

9.3-2 Determine the acceleration of a block of mass m_1 suspended from a uniform disk of mass m_2 and radius R, as shown. If the block is released from rest, determine the amount of time required for the block to travel h units downward. How does the downward acceleration differ from the situation in which the block is replaced with a downward force of magnitude mg.

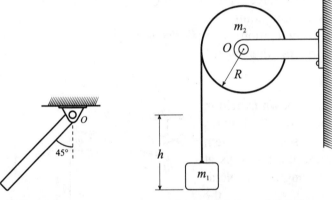

Problem 9.3-1   **Problems 9.3-2 and 9.3-3**

9.3-3 Determine the reactions at the point O in Problem 9.3-2. How do these **dynamic** reactions differ from the **static** reactions associated with the situation in which the disk is locked into position?

9.3-4 Two blocks of masses m_1 and m_2, respectively, are suspended from a pulley of mass m, as shown. Determine the accelerations a_1 and a_2 of the masses, and the reactions at the point C, each expressed in terms of the radii R_1 and R_2, and the pulley's radius of gyration r_g. *Note:* The **radius of gyration** of a rigid body is defined as the square root of the ratio of the body's moment of inertia and its mass, given by $r_g = \sqrt{I_C/m}$. The radius of gyration is the radius

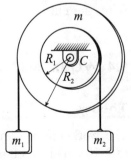

Problem 9.3-4

of a ring having the same mass and moment of inertia as the rigid body. As a matter of interest, the largest body having mass m and moment of inertia I_C is precisely a ring of radius r_g.

| **9.3-5** | A motorized system rotates a uniform disk of 0.10 ft radius and mass 1.25 slug at the constant rate of $\omega_1 = 240$ rad/s, while an external torque $T = 0.08$ lb·ft is applied, as shown. The connection at the point O carries force reactions R_1, R_2, and R_3, and moment reactions M_1 and M_3. Determine the force and moment reactions at the point O, and the angular acceleration α_2. Neglect the masses of the other members.

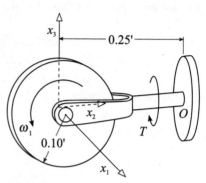

Problem 9.3-5

| **9.3-6** | The linkage shown is held in place with two thin wires. Both wires are cut, which releases the system from rest. Determine the reactions at the point O, and the angular accelerations of the members just after the wires have been cut. Assume that each slender member is uniform and has a mass of $m = 1.5$ kg and a length of $d = 0.4$ m, that the spring constant is $k = 150$ N/m, and that it is compressed $x = 0.1$ m before the wires are cut.

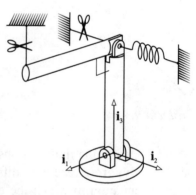

Problem 9.3-6

9.4 WORK AND KINETIC ENERGY

Theory

The work–energy equation for a system of particles was given by Eq. (8.2-10). This section modifies the expressions for work and kinetic energy for the case of a rigid body.

The work done by the forces acting on a rigid body can be divided into the work done that is associated with the translation of the rigid body and the work done that is associated with the rotation of the rigid body. Likewise, the kinetic energy can be divided into two components: one associated with the translations and the other associated with the rotations. Let's first examine the expression for work.

WORK

We begin by considering a rigid body that is rotating about a fixed point. Its work is exclusively **rotational**. From Eq. (9.2-1), the work done on a rigid body by the resultant external force is

$$U_{1-2} = \int \mathbf{F} \cdot d\mathbf{r} = \int \mathbf{F} \cdot \mathbf{v}\, dt = \int \mathbf{F} \cdot (\boldsymbol{\omega} \times \mathbf{r})\, dt$$

$$= \int (\mathbf{r} \times \mathbf{F}) \cdot \boldsymbol{\omega}\, dt = \int \mathbf{M}_O \cdot \boldsymbol{\omega}\, dt \qquad (9.4\text{-}1)$$

in which we have used vector identities of the type $\mathbf{A} \cdot (\mathbf{B} \times \mathbf{C}) = \mathbf{B} \cdot (\mathbf{C} \times \mathbf{A}) = \mathbf{C} \cdot (\mathbf{A} \times \mathbf{B})$. Thus, the work done on a rigid body that is fixed about a point O is

$$\boxed{U_{1-2} = \int \mathbf{M}_O \cdot \boldsymbol{\omega}\, dt.} \qquad (9.4\text{-}2)$$

In the case of planar motion, we can assume that the motion is in the x–y plane, and then let $\boldsymbol{\omega} = \omega \mathbf{k}$ and $\mathbf{M}_O = M_O \mathbf{k}$, in which case the work done by the force $\mathbf{F}$ is

$$\boxed{U_{1-2} = \int M_O \omega\, dt = \int M_O\, d\theta.} \qquad (9.4\text{-}3)$$

Of course, not all rigid bodies rotate about a fixed point. Examples of such cases are rolling wheels, wobbling footballs, spinning satellites, and the limbs of your body. In these cases, the work expression has two parts: a **translational** part and a **rotational** part. From Eq. (9.2-1),

$$U_{1-2} = \int \mathbf{F} \cdot d\mathbf{r} = \int \mathbf{F} \cdot \mathbf{v}\, dt = \int \mathbf{F} \cdot (\mathbf{v}_C + \boldsymbol{\omega} \times \mathbf{r}')\, dt$$

$$= \int \mathbf{F} \cdot \mathbf{v}_C\, dt + \int \mathbf{F} \cdot (\boldsymbol{\omega} \times \mathbf{r}')\, dt = \int \mathbf{F} \cdot \mathbf{v}_C\, dt + \int (\mathbf{r}' \times \mathbf{F}) \cdot \boldsymbol{\omega}\, dt,$$

$$(9.4\text{-}4)$$

in which the point C is the mass center. Equation (9.4-4) can be rewritten as the sum of the translational and rotational components of work, as

$$U_{1-2} = U_{T1-2} + U_{R1-2}, \tag{9.4-5a}$$

$$U_{T1-2} = \int \mathbf{F} \cdot \mathbf{v}_C \, dt \tag{9.4-5b}$$

$$U_{R1-2} = \int \mathbf{M}_C \cdot \boldsymbol{\omega} \, dt \tag{9.4-5c}$$

Notice that the translational component of work depends on $\mathbf{v}_C$ and that the rotational component of work depends on $\boldsymbol{\omega}$. In the case of planar motion, we can assume that the motion is in the x–y plane, and then let $\boldsymbol{\omega} = \omega\mathbf{k}$ and $\mathbf{M}_C = M_C\mathbf{k}$, in which case the rotational component of the work done by the force $\mathbf{F}$ is

$$U_{R1-2} = \int M_C\omega \, dt = \int M_C \, d\theta. \tag{9.4-6}$$

The expression for U_{T1-2} in the case of planar motion is the same as the expression for U_{T1-2} in the general case of three-dimensional motion, and is given by Eq. (9.4-2b).

KINETIC ENERGY

Let's now examine the expression for kinetic energy for the case of a rigid body. We begin by considering a rigid body that is rotating about a fixed point. Its kinetic energy is exclusively **rotational**. From Eqs. (8.1-4b) and (9.2-1), the kinetic energy is given by

$$T = \sum_{i=1}^{n} \tfrac{1}{2} m_i \mathbf{v}_i \cdot \mathbf{v}_i = \tfrac{1}{2} \int \mathbf{v}_i \cdot \mathbf{v}_i \, dm = \tfrac{1}{2} \int (\boldsymbol{\omega} \times \mathbf{r}) \cdot (\boldsymbol{\omega} \times \mathbf{r}) \, dm$$

$$= \tfrac{1}{2} \int \boldsymbol{\omega} \cdot [\mathbf{r} \times (\boldsymbol{\omega} \times \mathbf{r})] \, dm = \tfrac{1}{2} \boldsymbol{\omega} \cdot \int \mathbf{r} \times (\boldsymbol{\omega} \times \mathbf{r}) \, dm = \tfrac{1}{2} \boldsymbol{\omega} \cdot \mathbf{H}_O, \tag{9.4-7}$$

Thus,

$$T = \tfrac{1}{2} \boldsymbol{\omega} \cdot \mathbf{H}_O. \tag{9.4-8}$$

In the case of planar motion, Eq. (9.4-7) yields

$$T = \sum_{i=1}^{N} \tfrac{1}{2} m_i v_i^2 = \tfrac{1}{2} \int v^2 \, dm = \tfrac{1}{2} \int (\omega r)^2 \, dm = \tfrac{1}{2} \omega^2 \int r^2 \, dm,$$

so

$$\boxed{T = \tfrac{1}{2} I_0 \omega^2.} \tag{9.4-9}$$

Of course, not all rigid bodies rotate about a fixed point. In such cases, the kinetic energy has two parts: a **translational** part and a **rotational** part. The rotational part is identical in form to the rotational part obtained above, and the translational part is identical in form to the kinetic energy of a single particle. To show this, we start again with Eqs. (8.1-4b) and (9.2-1), and get

$$T = \sum_{i=1}^{n} m_i \mathbf{v} \cdot \mathbf{v}_i = \int \mathbf{v}_i \cdot \mathbf{v}_i \, dm = \int (\mathbf{v}_C + \boldsymbol{\omega} \times \mathbf{r}') \cdot (\mathbf{v}_C + \boldsymbol{\omega} \times \mathbf{r}') \, dm$$

$$= \tfrac{1}{2} \int [\mathbf{v}_C \cdot \mathbf{v}_C + 2\mathbf{v}_C \cdot (\boldsymbol{\omega} \times \mathbf{r}')\mathbf{v}_C + (\boldsymbol{\omega} \times \mathbf{r}') \cdot (\boldsymbol{\omega} \times \mathbf{r}')] \, dm$$

$$= \tfrac{1}{2} \mathbf{v}_C \cdot \mathbf{v}_C \int dm + \mathbf{v}_C \cdot \left(\boldsymbol{\omega} \times \int \mathbf{r}' \, dm \right) + \int (\boldsymbol{\omega} \times \mathbf{r}') \cdot (\boldsymbol{\omega} \times \mathbf{r}') \, dm$$

$$= \tfrac{1}{2} m(\mathbf{v}_C \cdot \mathbf{v}_C) + \tfrac{1}{2} \int \boldsymbol{\omega} \cdot [\mathbf{r}' \times (\boldsymbol{\omega} \times \mathbf{r}')] \, dm = \tfrac{1}{2} m v^2 + \tfrac{1}{2} \boldsymbol{\omega} \cdot \mathbf{H}_C,$$

$$\tag{9.4-10}$$

in which it has been recognized that $\int \mathbf{r}' \, dm = 0$ and that $v^2 = \mathbf{v}_C \cdot \mathbf{v}_C$. Thus,

$$\boxed{T = \tfrac{1}{2} m v^2 + \tfrac{1}{2} \boldsymbol{\omega} \cdot \mathbf{H}_C.} \tag{9.4-11}$$

As before, the rotational component of the kinetic energy, in the planar case, reduces to $\tfrac{1}{2} I_C \omega^2$. Thus, the kinetic energy of a general rigid body undergoing planar motion is

$$\boxed{T = \tfrac{1}{2} m v^2 + \tfrac{1}{2} I_C \omega^2.} \tag{9.4-12}$$

Examples

9.4-1   A uniform bar and a very long spring as shown (not drawn to scale) are released from rest (at $\theta = 0$) with the spring initially un-stretched. Determine (a) the angle θ at which the bar reaches its maximum angular velocity, (b) the maximum angular velocity of the bar, and (c) the largest angle θ that the bar reaches. Express your answers in terms of the bar's mass m, the spring constant k, and the bar's length L.

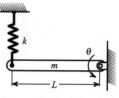

Example 9.4-1

Solution

The bar is subjected to reaction forces that do no work, a gravitational force that does conservative work V_g, and a spring force that does conservative work V_s. No non-conservative forces do work on the system, so the system is conservative. From Eqs. (8.2-10) and (9.4-9), the initial energy E_1 is equal to the energy E_2 at the angle θ, so

$$E_1 = T_1 + V_{g1} + V_{s1} = 0 + 0 + 0,$$

$$E_2 = T_2 + V_{g2} + V_{s2} = \frac{1}{2}I_O\omega^2 - mg\frac{L}{2}\sin\theta + \frac{1}{2}k(L\sin\theta)^2.$$

Thus,

$$\frac{1}{2}I_O\omega^2 = mg\frac{L}{2}\sin\theta - \frac{1}{2}k(L\sin\theta)^2 = \frac{mgL}{2}\sin\theta\left(1 - \frac{kL}{mg}\sin\theta\right).$$

We want to find the value of θ (or $\sin\theta$) where ω is a maximum. An easy way to do this is to solve for ω, take the derivative of both sides with respect to $\sin\theta$ (i.e. substitute $u = \sin\theta$ and take d/du of both sides). Next set the left side equal to zero and solve for $\sin\theta$. We see that the maximum angular velocity is achieved when $\sin\theta = \frac{1}{2}mg/kL$, that is, when

$$\theta = \sin^{-1}\left(\frac{1}{2}\frac{mg}{kL}\right),$$

and that the maximum angular velocity is

$$\omega_{max} = \frac{g}{L}\Big/\sqrt{\frac{k}{3m}}.$$

The maximum angle is reached when $\omega = 0$, which is when $\sin \theta = mg/kL$, that is, when

$$\theta = \sin^{-1} \left(\frac{mg}{kL} \right).$$

9.4-2 Compare the velocities of the block and the disk shown when they reach the bottoms of the inclines. Both the block and the disk have mass m and both start from rest at the tops of the inclines. The radius of the disk is R. Assume that the block slides without friction and that the disk rolls without slipping.

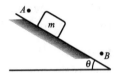

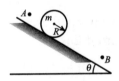

Example 9.4-2

Solution
Neither the block nor the disk are subjected to non-conservative forces that do work, so energy is conserved for each. In the case of the block,

$$E_A = T_A + V_A = 0 + mgh = mgh,$$
$$E_B = T_B + V_B = \tfrac{1}{2} m v_{\text{block}}^2,$$

so

$$v_{\text{block}} = \sqrt{2gh}.$$

In the case of the disk, from Eq. (9.4-12),

$$E_A = T_A + V_A = 0 + mgh = mgh,$$
$$E_B = T_B + V_B = \tfrac{1}{2} m v_{\text{disk}}^2 + \tfrac{1}{2} I_O \omega_{\text{disk}}^2 + 0 = \tfrac{1}{2} m v_{\text{disk}}^2 + \tfrac{1}{2} \left(\frac{mR^2}{2} \right) \left(\frac{v_{\text{disk}}}{R} \right)^2$$
$$= \tfrac{3}{4} m v_{\text{disk}}^2,$$

so

$$v_{\text{disk}} = \sqrt{\tfrac{4}{3} gh}.$$

The disk reaches the bottom of the incline at $\sqrt{\frac{4}{3}gh/2gh} = \sqrt{\frac{2}{3}} = 81.6\%$ of the speed of the block.

9.4-3   A uniform plate of mass m is pinned along an axis that lies in the x_2–x_3 plane, as shown, and is released from rest. Gravity acts downward in the $-x_3$ direction. Determine the maximum angular velocity of the plate.

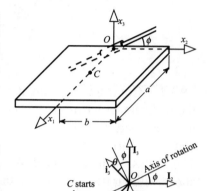

Solution

The plate and the forces to which it is subject represent a conservative system. Assuming that the plate rotates an angle θ around the axis of rotation, from Eqs. (8.2-10) and (9.4-8), we get

Example 9.4-3

$$E_1 = T_1 + V_1 = 0 + 0 = 0,$$
$$E_2 = T_2 + V_2 = \tfrac{1}{2}\mathbf{H}_O \cdot \boldsymbol{\omega} - mgz_C,$$

in which

$$T_2 = \tfrac{1}{2}\mathbf{H}_O \cdot \boldsymbol{\omega} = \tfrac{1}{2}(H_{O2}\omega_2 + H_{O3}\omega_3),$$
$$= \tfrac{1}{2}[(I_{22}\omega_2 + I_{23}\omega_3)\omega_2 + (I_{23}\omega_2 + I_{33}\omega_3)\omega_3] = \tfrac{1}{2}I_{22}\omega_2^2 + \tfrac{1}{2}I_{33}\omega_3^2$$
$$= \tfrac{1}{2}(\tfrac{1}{3}ma^2)(\omega\cos\phi)^2 + \tfrac{1}{2}[\tfrac{1}{12}m(a^2 + b^2) + m(\tfrac{1}{2}a)^2](\omega\sin\phi)^2$$
$$= \tfrac{1}{24}m\omega^2(4a^2 + b^2\sin^2\phi),$$

and, from the figure,

$$z_C = \tfrac{1}{2}a\cos\phi\sin\theta.$$

So, from conservation of energy.

$$0 = T_2 + V_2 = \tfrac{1}{24}m\omega^2(4a^2 + b^2\sin^2\phi) - mg(\tfrac{1}{2}a)\cos\phi\sin\theta.$$

It follows that, as we vary the angle of rotation θ, the angular velocity is at a maximum when $\sin\theta = 1$, at which point

$$\omega_{max} = \sqrt{\frac{12ga\cos\phi}{4a^2 + b^2\sin^2\phi}}.$$

Notice that when $b = 0$ and $\phi = 0$, the plate degenerates to a slender pendulum and the maximum angular velocity is $\omega_{max} = \sqrt{3g/a}$ (see Problem 9.4-1).

Problems

9.4-1   A uniform slender bar of mass m and length l is released from rest, as shown. Determine the bar's angular velocity when it reaches the vertical state.

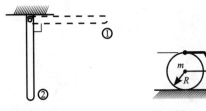

Problem 9.4-1   **Problem 9.4-2**

9.4-2   A uniform disk of mass m and radius R is released from rest. The spring is initially stretched an amount $s_0 = \frac{1}{4}R$. Determine the maximum angular velocity of the disk and the maximum distance of travel of the disk. Express the solutions in terms of m, R, and the spring constant k. Assume that the disk rolls without slipping.

9.4-3   Two identical uniform slender bars are pinned to each other and to a spring, as shown. Each slender bar has a mass of $4\,kg$ and a length of $6\sqrt{5} = 13.42\,m$. The spring constant is $100\,N/m$ and the unstretched length of the spring is $4\,m$. Assume that the system is released from rest in the position shown, and determine the maximum compression of the spring and the maximum angular velocity of the system.

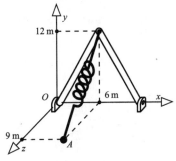

Problem 9.4-3

CHAPTER 10

Introduction to Deformation and Stress

Chapters 10–16 treat **mechanics of materials** problems. These are statics problems in which the interest lies in predicting such quantities as internal forces and moments, internal stresses and strains, and deformations. Knowledge of certain parameters that characterize material behavior is required in order to perform these predictions. The development begins by defining two material parameters: the modulus of elasticity and the modulus of rigidity. Structural members undergoing longitudinal deformations, torsional deformations, and bending deformations are examined in Chapters 11, 12, and 13. Stress and strain are treated in a more general framework in Chapter 14. This prepares the reader for the study of structural members that undergo combinations of longitudinal, torsional, and bending deformations in Chapter 15. Finally, Chapter 16 treats buckling, pressure vessels, composite beams, and temperature effects.

Recall from Chapters 2–5 that certain types of statics problems were **indeterminate**. A problem was said to be indeterminate when too many unknown reactions acted on the system. The problem was, in fact, indeterminate only in the sense that the unknown reactions that kept the system in static equilibrium depended on the deformation of the members in the system, and the tools for solving statics problems involving deformation had not yet been developed.

Perhaps the first thing to appreciate when examining deformation is that it depends on **material parameters**. When you pull on a spring with a force F, the spring will deform an amount $x = F/k$, where k is the spring constant. The spring constant k depends on material parameters, although k itself is not a material parameter. The spring could be made from a particular metal alloy, but not all springs made from that particular alloy have the same spring constant. The constant k is a **structural parameter**; it depends on material parameters, whatever they might be, in addition to geometric parameters such as the size and shape of the spring.

A material parameter is a parameter that depends on the type of material and, more importantly, does *not* depend on geometry.

Before discussing material parameters in detail, note that it has long been observed, when deformations in a structure are relatively small, that the relationship between force and deformation is linear. Just as in a conventional spring, twice the force induces twice the deformation, one third of the force induces one third of the deformation, and so on. This is called **linear elastic** behavior. A fundamental question asked in mechanics of materials problems is how to develop these linear relationships and how to sort out the material parameters from the geometric parameters. Without answering these questions, structural systems cannot be analyzed.

The three most basic material parameters are the **modulus of elasticity** E, the **shear modulus** G, and **Poisson's ratio** v. Simple tests that motivate how the modulus of elasticity and the shear modulus were developed are now described. Poisson's ratio is considered in Chapter 14.

10.1 MODULUS OF ELASTICITY

Theory

Let's perform what is called a **tension test**. We pull on a specimen of material and measure the deformation (see Fig. 10.1-1). The ASTM (American Society of Testing and Materials) standard specimen is 0.5 inch wide and has a gage length of 2 inches. The specimen is a round bar. As with a conventional spring, the applied force is linearly proportional to the deformation. In order to generate a material parameter for the specimen, we need to somehow account for the geometry. Longer specimens deform more, so we need to **divide out** the length of the specimen. Furthermore, larger cross-sectional areas require larger forces to generate the same deformation, so we need to divide out the cross-sectional area too. Thus, if the applied force is F, the deformation is u, and the proportionality constant is k, then, from the linear equation $F = ku$, we get

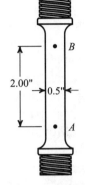

$$\frac{F}{A} = \left(\frac{kL}{A}\right)\frac{u}{L}, \tag{10.1-1}$$

where L is the length of the specimen and A is its cross-sectional area. Equation (10.1-1) is rewritten as

Figure 10.1-1

$$\boxed{\sigma = E\varepsilon,} \tag{10.1-2}$$

in which

$$\sigma = \frac{F}{A}, \qquad \varepsilon = \frac{u}{L}, \qquad E = \frac{kL}{A}.$$

Equation (10.1-2) is called **Hooke's Law**. The deformation per unit length ε is called the **normal strain**, and the force per unit area σ is called the **normal stress**. The constant of proportionality E is called the **modulus of elasticity**. The modulus of elasticity E is a **material parameter**. Notice that ε is unitless, which means that E has the same dimensions as σ. Equation (10.1-2) is a stress–strain relationship that is valid when the deformation is relatively small. Figure 10.1-2 shows a more general stress–strain relationship that is obtained from the tensile

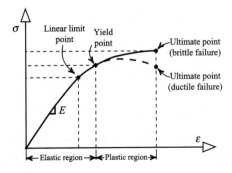

Figure 10.1-2

test. Notice that as the strain increases, the relationship between stress and strain becomes nonlinear.

In practice, the forces acting over a structure change in time. A force is applied for some time, then removed, applied again, and so on. These are called **cycles**. Referring to Fig. 10.1-2, the **stress–strain point** moves up the curve as an applied force is increased. There are several points along this curve that are **critical**.

Moving up the curve from the origin, the first critical point reached is the point at which the stress–strain relation ceases to be linear. Linear means that a change in strain ε produces a proportional change in stress σ. Below this point, the stress–strain relationship is linear, and above this point, the stress–strain relationship is nonlinear. This is called the **linear limit point**.

Continuing up the curve, the next critical point reached is called the **yield point**. Beyond this stress–strain point, the material undergoes internal molecular changes such that the stress–strain point will **not** follow the curve back down to the origin when the applied force is removed. The material will not return to its original shape. Points on the curve below the yield point are said to be in the **elastic** region of the material, and points on the curve that are above the yield point are said to be in the **plastic** region of the material.

The final critical stress–strain point is the **ultimate point**. This is the point where the material fails. In **brittle** materials, the failure is a rapid process in which the material cracks. In **ductile** materials, the failure is a slower process in which the material loses its stiffness. Ductile failure exhibits **necking**. (See Fig. 10.1-3.)

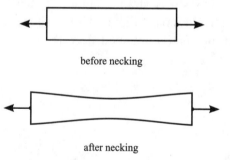

before necking

after necking

Figure 10.1-3

Example

10.1-1 A uniform bar of length 14 inches having a square cross-sectional area is designed to withstand a force of 10 000 lb before reaching the yield point. Determine the smallest cross-sectional area of the bar and the displacement of the bar as a function of *x*.

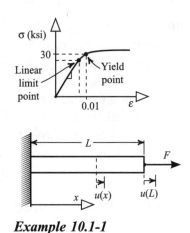

Solution

A free body diagram of a section of the bar reveals that the internal force in the bar is equal to the applied force of 10 000 lb. Denoting the bar's cross-sectional width by *a*, and referring to the stress–strain curve (where $1\,\text{ksi} = 1000\,\text{psi} = 1000\,\text{lb/in}^2$), the yield stress is $30\,000 = 10\,000/a^2\,\text{psi}$, so $a = 0.578$ inches. Also from the stress–strain curve, the strain in the bar is $\varepsilon = 0.01$. From Eq. (10.1-2), the deformation in the bar is proportional to the gage length *x*. The displacement is $u(x) = 0.01x$. The displacement at the end of the bar is $u = 0.01(14) = 0.14$ inches.

Example 10.1-1

Problems

10.1-1 A circular bar of length 3′ 4″ is subjected to a compressive force of $F = 5000$ lb as shown. Determine the smallest radius before which the bar reaches its linear limit point, and the smallest radius before which the bar reaches its yield point. Also, compare the associated displacements at the tip of the bar.

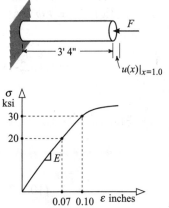

Problem 10.1-1

10.1-2 A rectangular bar of length 5′ and cross-sectional area of 0.2″ × 0.3″ is subjected to an applied force F as shown. Determine the maximum applied force before the bar reaches its linear limit point, and determine the maximum applied force before the bar reaches its yield point. Use the stress–strain curve given in Problem 10.1-1.

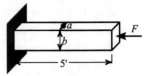

Problem 10.1-2

10.2 SHEAR MODULUS

Let's now perform what is called a **torsion test**. The specimen in this case is a thin tube. The thin tube is subjected to a torsional moment (see Fig. 10.2-1). As in a conventional torsional spring, the applied moment M is linearly proportional to the angular (twisting) deformation ϕ over the length L of the tube. In order to generate a material parameter for the specimen, we need to somehow account for the geometry. Longer specimens twist more, so we must **divide out** the length of the specimen. Furthermore, a larger cross-sectional

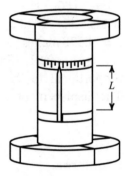

Figure 10.2-1

area A requires a larger moment to generate the same deformation, so we must divide out the cross-sectional area too. In the torsion test, the external moment produces **shearing** forces F in the plane of the cross section. The specimen undergoes a shearing angle γ (see Fig. 10.2-2). Thus, if the applied moment is $M = RF$, in which R denotes the radius of the specimen, then it follows from the linear equation $M = k\phi$ that

$$\frac{RF}{AR} = \left(\frac{kL}{AR}\right)\frac{\phi}{L} = \left(\frac{k}{AL}\right)\gamma, \tag{10.2-1}$$

since for small angles the arc length $\phi L = R\gamma$ (see Fig. 10.2-2). Equation (10.2-1) is rewritten as

$$\boxed{\tau = G\gamma,} \tag{10.2-2}$$

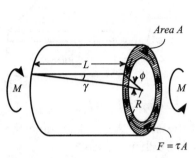

Figure 10.2-2

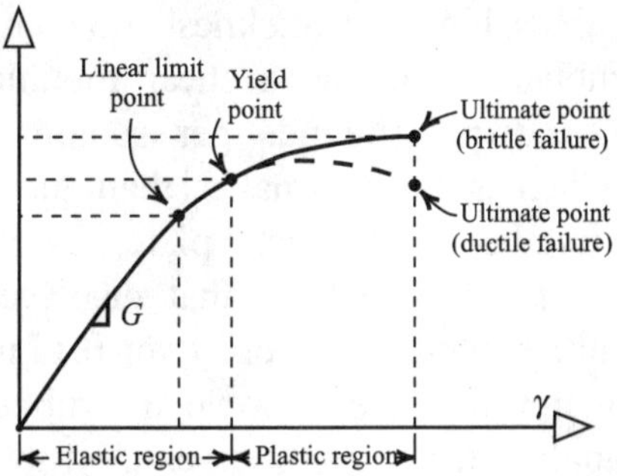

Figure 10.2-3

in which

$$\tau = \frac{F}{A}, \qquad G = \frac{k}{AL}.$$

The angle γ perpendicular to the plane of the cross section is called the **shear strain**, and the force per unit area τ acting in the plane of the cross section is called the **shear stress**. The constant of proportionality G is called the **shear modulus**. The shear modulus G is a **material parameter**. Equation (10.2-2) is a shear stress–strain relationship that is valid when the deformation is relatively small. Figure 10.2-3 shows a more general stress–strain curve that is obtained from the torsion test. The shear stress–strain curve and the normal stress–strain curve are similar in appearance. **Like the normal stress–strain curve, the shear stress–strain curve has a linear limit point, a yield point, and an ultimate point.**

The normal stress–strain curve and the shear stress–strain curve are important in design. **Structures can be designed to operate within certain regions of the stress–strain curves.** Depending on the design strategy, a structure can be designed so that the deformations are always within the linear region of the material, within the elastic region of the material, or simply within the region in which the material does not fail. The designer, again depending on the strategy, can **design for strength**, attempting to make the structure as strong as possible, or can **design for stiffness**, attempting to make the structure as stiff as possible.

Examples

10.2-1 A thin steel plate is subjected to an applied force F as shown in the figure overleaf. The steel plate is 5 inches × 5 inches and is glued to a

rubber block of thickness 1.5 inch. The
rubber block has a shear modulus of
$G = 30$ psi (pounds per square inch),
which is much smaller than the shear
modulus of the steel plate. Determine
the applied force F that displaces the
rubber block no more than 0.12 inches

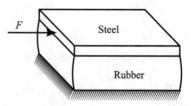

Example 10.2-1

at the interface between the rubber and the steel. Also, determine the
applied force that stresses the rubber no more than 0.3 ksi.

Solution

The applied force F induces a shear stress in the rubber of $\tau = F/25$ ksi
(again 1 ksi $= 1000$ psi or one kilopound per square inch). The shear
strain is $\gamma = \tau/G = F/((25)(30)) = F/750$. The rubber displaces by an
amount $u = 1.5\gamma = 1.5F/750 = 0.002F$ inches. The applied force that
displaces the rubber block no more than 0.12 inches is
$F = 0.12/0.0002 = 60$ lb. The applied force that stresses the rubber
no more than 0.3 ksi is $F = 25\tau = 25(0.3) = 7.5$ klb.

10.2-2 A thin hollow shaft is subjected to a maximum applied torque
of 14 N·m as shown. The ratio of the diameters of the shaft has been
constrained to $D_1/D_2 = 2$. (i) Designing for strength, determine the
diameter D_1 so that the shear stress in the shaft does not exceed
$100\,\text{kN/m}^2$. (ii) Designing for stiffness, determine the diameter D_1 so
that the twisting angle at the tip of the shaft does not exceed 0.1 rad.
Let $L_1 = 0.6$ m, $L_2 = 0.5$ m, and the thickness $t = 0.5$ cm. Let
$G = 7000\,\text{kN/m}^2$.

Solution

(i) The maximum shear stress acts in the small-diameter section of the
shaft (why?). From Eq. (10.2-1), the internal moment acting on the
small-diameter section of the shaft is

$$M = FR = \tau_{max}AR = \tau_{max}(2\pi R_2 t)R_2 = (2\pi t\tau_{max})R_2^2,$$

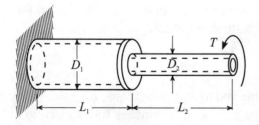

Example 10.2-2

so

$$R_2 = \sqrt{\frac{M}{2\pi t \tau_{\max}}} = \sqrt{\frac{14}{2(3.14)(0.005)(100 \times 10^3)}} = 0.06676 \text{ m}.$$

The large diameter of the shaft is then $D_1 = 4R_2 = 0.267$ m.
   (ii) From Eq. (10.2-1), the twisting angle of the shaft is

$$\phi = \frac{ML_1}{A_1 R_1^2 G} + \frac{ML_2}{A_2 R_2^2 G} = \frac{ML_1}{(2\pi R_1 t)R_1^2 G} + \frac{ML_2}{(2\pi R_2 t)R_2^2 G}$$

$$= \frac{M}{(2\pi R_2^3 t)G}\left(\frac{L_1}{8} + L_2\right)$$

$$= \frac{14}{2(3.14)(0.005)(7000 \times 10^3)}\left(\frac{0.6}{8} + 0.5\right)\frac{1}{R_2^3} = 3.66 \times 10^{-5}\frac{1}{R_2^3}$$

$$= 0.1,$$

so

$$R_2 = 0.07 \text{ m}.$$

The large diameter of the shaft is then $D_1 = 4R_2 = 0.286$ m.

Problems

| 10.2-1 | A circular peg of radius $R = 0.04$ m is subjected to a force $F = 100$ N as shown. Determine the average shear stress in the bolt. The shear modulus of the peg is $G = 20$ kN/m^2. Let $d = 0.1$ m.

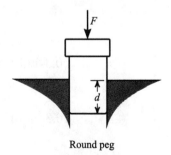

Round peg

Problem 10.2-1

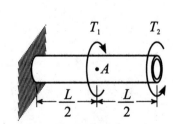

Problem 10.2-2

| 10.2-2 | A thin hollow shaft is subjected to maximum applied torques of $T_1 = 25$ N·m and $T_2 = 15$ N·m as shown. (i) Designing for strength, first

determine the diameter D so that the shear stress in the shaft does not exceed $120\,\text{kN/m}^2$. (ii) Designing for stiffness, determine the diameter D so that the twisting angle at the tip of the shaft does not exceed $0.12\,\text{rad}$. The length of the shaft is $L = 1.2\,\text{m}$, the thickness is $t = 0.005\,\text{m}$, $G = 7500\,\text{kN/m}^2$, and T_1 is applied in the center of the shaft.

10.3 POINT STRESS AND POINT STRAIN

Theory

The stresses and strains that were defined in Section 10.2 were **averages** over finite areas and finite lengths. The normal stress in the tensile test was a force per unit area over a cross-sectional area, and the normal strain was a displacement over a length. Likewise, the shear stress in the torsion test was a force per unit area over the tube's cross-sectional area, and the shear strain was an angular displacement over the thickness of the tube. In practice, however, the stresses and strains vary from point to point.

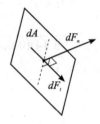

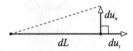

Figure 10.3-1

Thus, by looking at differential forces over differential areas and by looking at differential displacements over differential lengths, the **normal stress at a point** σ, the **normal strain at a point** ϵ, the **shear stress at a point** τ, and the **shear strain at a point** γ are defined as (see Fig. 10.3-1)

$$\sigma = \frac{dF_n}{dA}, \qquad \varepsilon = \frac{du_t}{dL}, \qquad \text{(10.3-1a,b)}$$

$$\tau = \frac{dF_t}{dA}, \qquad \gamma = \frac{du_n}{dx}, \qquad \text{(10.3-2a,b)}$$

where n indicates **normal** to the differential surface dA or normal to the differential length dL, and t indicates **tangential** to dA or tangential to dL. At this point, the reader may appreciate why a rectangular bar was used in the tension test, and why a thin tube was used in the torsion test.

The bar used in the tension test had the attractive property that its normal stresses and normal strains were **uniform** within the cross section and over the length of the specimen. Likewise, the thin tube used in the torsion test had the attractive property that its shear stresses and shear strains were **uniform** within the cross section and over the length of the specimen. Thus, the **average properties** measured in these tests were precisely the same as the **point properties**.

Example

10.3-1   A uniform bar of length L, cross-sectional area A, and weight density ρg hangs under its own weight as shown. Determine the displacement at the free end of the bar. Let $L = 2$ ft, $A = 0.04$ ft^2, $\rho g = 8$ lb/ft^3, and $E = 192$ lb/ft^2.

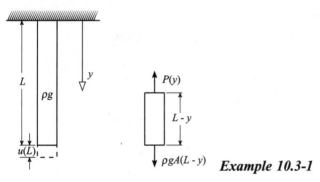

Example 10.3-1

Solution
From the free body diagram, the internal force in the bar at y is $P = \rho g A(L - y)$. The internal force is a linear function of y; its largest value is $\rho g A L$ at the top of the bar, and it reaches zero at the bottom of the bar. The stress is uniform over the cross section, so, from Eq. (10.1-2), the normal stress in the bar is $\sigma = P/A = \rho g(L - y)$ and the normal strain is $\varepsilon = \sigma/E = P/AE = \rho g(L - y)/E$. Since the strain varies with y, the displacement of the bar at y is determined by integrating Eq. (10.3-1b), to get

$$u = u(0) + \int_0^y \frac{\rho g(L - y)}{E} \, dy = \frac{\rho g}{E} y(L - \tfrac{1}{2}y).$$

The displacement at the tip of the bar is $u = \rho L^2/2E = 8(2^2)/[2(192)] = 0.0833$ ft $= 1$ inch. Also, notice that the displacement of the bar is independent of the bar's cross-sectional area.

Problem

10.3-1 A non-uniform bar of length $x = L$ is subjected to an applied force F, as shown. The cross-sectional area varies linearly with x and is of the form $A = A_0(2L - x)/3$. Determine the displacement at the free end of the bar. Let $L = 2$ ft, $A_0 = 0.04$ ft, $E = 192$ lb/ft^2, and $F = 1$ lb.

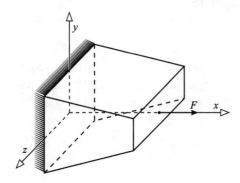

Problem 10.3-1

CHAPTER 11

Axial Deformation and Stress

11.1 GENERAL DEVELOPMENT

Theory

A structural member that is subjected to external forces along its long axis is referred to as a **bar**. A differential element of a bar is shown in Fig. 11.1-1. It is assumed that the normal stresses over any y–z cross section are uniform. Thus, the internal forces over any y–z cross section are given by

$$P(x) = \int \sigma \, dy \, dz = \sigma A,$$

so

$$\boxed{\sigma = \frac{P}{A},}$$

(11.1-1)

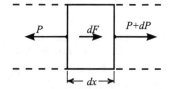

Figure 11.1-1

where A is the cross-sectional area. The differential element is in static equilibrium if the sum of the forces in the x direction is zero, so

$$(P + dP) - P + dF = 0.$$

Dividing by dx yields the equilibrium equation

$$\frac{dP}{dx} = -f, \qquad (11.1\text{-}2)$$

where $f = dF/dx$ is the external force per unit length. Assuming that the deformations are within the linear region of the stress–strain curve, the normal stresses are related to the normal strains by Eq. (10.1-2); that is,

$$\sigma = E\varepsilon, \qquad (11.1\text{-}3)$$

in which the modulus of elasticity E is assumed to be uniform over the cross section (not a function of y or z). The normal strain is related to the displacement of the bar by Eq. (10.3-1b); that is,

$$\varepsilon = \frac{du}{dx}. \qquad (11.1\text{-}4)$$

The internal force is now related to the displacement by substituting Eqs. (11.1-3) and (11.1-4) into (11.1-1) to get

$$P = AE\frac{du}{dx}. \qquad (11.1\text{-}5)$$

Integrating Eq. (11.1-5) with respect to x yields

$$u(x) = u(0) + \int_0^x \frac{P}{AE}\,dx. \qquad (11.1\text{-}6)$$

The internal force $P(x)$ in Eq. (11.1-6) is determined from the external force per unit length by integrating Eq. (11.1-2) to get

$$P(x) = P(0) - \int_0^x f\,dx. \qquad (11.1\text{-}7)$$

In a given problem, a bar is subjected to external forces. The bar, by the way in which it is secured at its boundaries, is also subjected to boundary conditions (see Fig. 11.1-2). The boundary conditions together with Eqs.

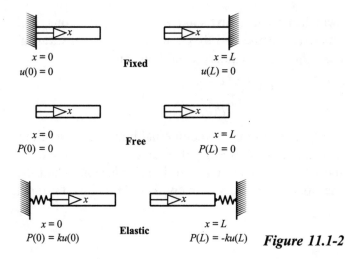

Figure 11.1-2

(11.1-6) and (11.1-7) are used to solve the given problem. Once the displacement and internal force of the bar have been found, Eq. (11.1-1) can be used to determine the stress in the bar.

When the internal force is specified at one or more boundaries, the internal force $P(x)$ can be determined from Eq. (11.1-7) without using Eq. (11.1-6). A bar of this kind is termed **determinate**, referring to the fact that the internal force does **not** depend on the bar's elasticity. On the other hand, if Eq. (11.1-6) is needed in order to determine the internal forces in the bar, then the bar is termed **indeterminate**, in which case the internal forces depend on the bar's elasticity.

Examples

| 11.1-1 | Two metal disks are stacked on top of each other and subjected to a force F as shown. Determine the displacement of the disks as a function of x. Let $D_1 = 0.1\,\text{m}$, $D_2 = 0.2\,\text{m}$, $h_1 = h_2 = 0.1\,\text{m}$, $E = 7500\,\text{kN/m}^2$, and $F = 2\,\text{kN}$.

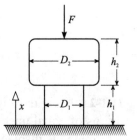

Example 11.1-1

Solution
Consider Eqs. (11.1-6) and (11.1-7) over the height of the lower disk ($0 < x < h_1$) and over the height of the upper disk ($h_1 < x < h_1 + h_2$). The applied force per unit length is zero over the height of both disks; that is, $f = 0$. The boundary conditions are divided into boundary conditions for the lower

and upper disks, which are called **external** boundary conditions, and boundary conditions between the two disks, which are called **internal** boundary conditions. The four boundary conditions are

External: $u(0) = 0$, $P(h_1 + h_2) = -F$,

Internal: $u(h_1 - \epsilon) = u(h_1 + \epsilon)$, $P(h_1 - \epsilon) = P(h_1 + \epsilon)$,

where ϵ is an arbitrarily small number. The internal boundary conditions at the interface between the blocks specify that the displacement and the internal force of the lower block are equal to the displacement and the internal force of the upper block, respectively. From Eqs. (11.1-6) and (11.1-7), we get

$$0 < x < h_1: \quad P = P(0) = P(h_1 - \epsilon) = P(h_1 + \epsilon),$$

$$u = 0 + \int_0^x \frac{P}{A_1 E} \, dx = \frac{P(0)x}{A_1 E},$$

$$u(h_1 - \epsilon) = \frac{P(0)h_1}{A_1 E} = u(h_1 + \epsilon);$$

$$h_1 < x < h_1 + h_2: \quad P = P(h_1 + \epsilon) = P(h_1 + h_2) = -F,$$

$$u = u(h_1 + \epsilon) + \int_{h_1 + \epsilon}^x \frac{P}{A_2 E} \, dx = -\frac{Fh_1}{A_1 E} - \frac{F(x - h_1)}{A_2 E},$$

$$u(h_1 + h_2) = -\frac{F}{E}\left(\frac{h_1}{A_1} + \frac{h_2}{A_2}\right) = -\frac{4F}{\pi E}\left(\frac{h_1}{D_1^2} + \frac{h_2}{D_2^2}\right)$$

$$= -\frac{4(2000)}{\pi(7 \times 10^6)}\left(\frac{0.1}{0.1^2} + \frac{0.1}{0.2^2}\right) = -0.0045 \text{ m}$$

$$= -4.5 \text{ mm}.$$

11.1-2 A rigid weight W is supported as shown by two concentric springs of equal unstretched length h having spring constants k_1 and k_2. Determine the displacement of the weight measured from the unstretched length of the springs. Let $k_1 = 20 \text{ klb/ft}$, $k_2 = 40 \text{ klb/ft}$, and $W = 2 \text{ klb}$.

Solution

The displacements of the springs are equal to each other at the interface between the springs and the weight. From the free body diagram of the

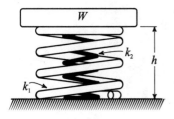

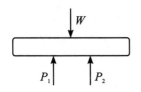

Example 11.1-2

block, $W = P_1 + P_2$, in which P_1 denotes the force acting on the outer spring and P_2 denotes the force acting on the inner spring. Thus,

$$W = P_1 + P_2 = k_1 u + k_2 u = (k_1 + k_2)u,$$

so

$$u = \frac{W}{k_1 + k_2} = \frac{2}{20 + 40} = 0.033 \text{ ft} = 0.4 \text{ inches.}$$

Notice that the solution is independent of the unstretched length h of the springs.

11.1-3 Two uniform bars support a weight W as shown in the figure overleaf. Determine the displacements Δx and Δy of the point A. Let $A_1 = A_2 = 8 \text{ mm}^2$, $E_1 = E_2 = 40 \times 10^6 \text{ N/m}^2$, and $W = 25 \text{ N}$.

Solution
Denote the length of the left bar by L_1, and the length of the right bar by L_2. Notice that $\sin\theta = L_1/25$ and $\cos\theta = 12/L_1$, from which it is determined that $L_1 = 15 \text{ cm}$. It follows that $\sin\theta = \frac{3}{5}$, $\cos\theta = \frac{4}{5}$, and $L_2 = 20 \text{ cm}$. From the free body diagram,

$$0 = -\tfrac{3}{5}P_1 + \tfrac{4}{5}P_2, \qquad 0 = \tfrac{4}{5}P_1 + \tfrac{3}{5}P_2 - W,$$

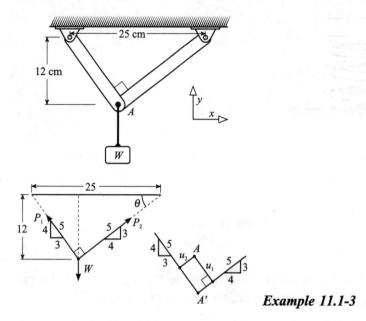

Example 11.1-3

so $P_1 = 20\,\text{N}$ and $P_2 = 15\,\text{N}$. From Eq. (11.1-6),

$$u_1 = \frac{P_1 L_1}{A_1 E} = \frac{20(0.15)}{8 \times 10^{-6}(40 \times 10^6)} = 0.01875\,\text{m} = 1.875\,\text{cm},$$

$$u_2 = \frac{P_2 L_2}{A_2 E} = \frac{15(0.20)}{8 \times 10^{-6}(40 \times 10^6)} = 0.01875\,\text{m} = 1.875\,\text{cm}.$$

These displacements are along the lengths of the bars. The point A moves to the point A', and by inspecting **the displacement compatibility** at the point A', we get

$$u_x = \tfrac{3}{5}u_1 - \tfrac{4}{5}u_2 = -0.375\,\text{cm},$$

$$u_y = -\tfrac{4}{5}u_1 - \tfrac{3}{5}u_2 = -2.625\,\text{cm}.$$

11.1-4 Determine the displacement at $x = 0$ of the bar shown. Let $A = 0.1\,\text{m}^2$, $a = 1.5\,\text{m}$, $E = 25\,\text{kN/m}^2$, $k = 750\,\text{N/m}$, and $F = 150\,\text{N}$.

Solution

The boundary condition at the left end is $P(0) = F$, and the force per unit length over the length of the bar is zero, so, from Eq. (11.1-7),

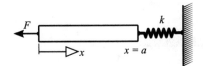

Example 11.1-4

$P(x) = P(0) = F$. The boundary condition at the right end is $P(a) = -ku(a)$. From Eq. (11.1-6),

$$u = u(0) + \int_0^x \frac{P}{AE}\, dx = u(0) + \frac{F}{AE}x,$$

so

$$u(a) = u(0) + \frac{Fa}{AE} = -\frac{F}{k}.$$

It follows that

$$u(0) = -F\left(\frac{a}{AE} + \frac{1}{k}\right) = -150\left[\frac{1.5}{0.1(25\,000)} + \frac{1}{750}\right] = -0.29 \text{ m} = -29 \text{ cm.}$$

Problems

11.1-1 A dense conical disk hangs from a ceiling as shown. The cross-sectional area of the disk is a quadratic function of x, given by

$$A = \frac{\pi}{4}\left(d_1 + \frac{d_2 - d_1}{h}x\right)^2.$$

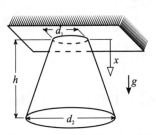

Problem 11.1-1

Determine the displacement of the disk at $x = h$. Let $d_1 = 0.1$ m, $d_2 = 0.2$ m, $h = 0.2$ m, and $E = 75$ kN/m². The mass per unit length is $\rho = 1000$ kg/m³. Where is the maximum normal stress located and what is its value?

11.1-2 A force F acts on a rigid platform that is supported as shown in the figure overleaf by two concentric posts of height h having elastic moduli E_1 and E_2. Determine the displacement of the rigid platform. Let $E_1 = 5000$ lb/ft², $E_2 = 1000$ lb/ft², $R_1 = 0.2$ ft, $R_2 = 0.4$ ft, $h = 0.6$ ft, and $F = 200$ lb.

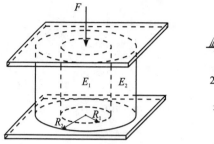

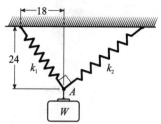

Problem 11.1-2   **Problem 11.1-3**

11.1-3 Two uniform springs support a weight W as shown. Determine the displacements Δx and Δy of the point A. Let $k_1 = 1600\,\text{N/m}$, $k_2 = 1200\,\text{N/m}$, and $W = 25\,\text{N}$.

11.1-4 A force F acts on the circular bar shown. Determine the displacement of the bar at $x = 6$ inches. Let $F = 150\,\text{lb}$ and $E = 25\,\text{ksi}$.

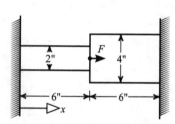

Problem 11.1-4

11.2 IMPULSE FUNCTIONS

Theory

As described in Section 11.1, the internal forces are determined from the external force per unit length using Eq. (11.1-7). **The external force per unit length represents any type of external force: a distributed force, a point force, or combinations of these.** A **distributed** force is a force that is spread out over the length of the bar. A **point** force is a force that is idealized as being concentrated at a point. The force per unit length associated with the point force is idealized as an infinitely large force per unit length over an infinitely small region of concentration. The immediate question arises how to mathematically represent the point force as a

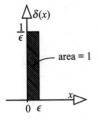

Figure 11.2-1

force per unit length in Eq. (11.1-7). Toward this end, the **impulse function** is defined as

$$\delta(x) = \begin{cases} \dfrac{1}{\epsilon}, & 0 < x < \epsilon, \\ 0, & x < 0 \text{ or } x > \epsilon, \end{cases} \qquad (11.2\text{-}1)$$

in which ϵ denotes an arbitrarily small number (see Fig. 11.2-1). Notice that the area under the curve of the impulse function is equal to 1, that is

$$\int_0^\epsilon \delta(x)\, dx = 1. \qquad (11.2\text{-}2)$$

A point force F located at $x = a$ is represented as a force per unit length as

$$f = F\delta(x - a). \qquad (11.2\text{-}3)$$

Let's now examine the internal forces in the bar surrounding the point force F. Substituting Eq. (11.2-3) into Eq. (11.1-7) using Eq. (11.2-2) yields

$$P(a + \epsilon) = P(a) - \int_a^{a+\epsilon} F\delta(x - a)\, dx$$

$$= P(a) - F \int_a^{a+\epsilon} \delta(x - a)\, dx = P(a) - F. \qquad (11.2\text{-}4)$$

Notice that **the internal force just to the right of the point force is less than the internal force just to the left of the point force by an amount F**.

Examples

11.2-1 A uniform rectangular bar of cross-sectional area A is subjected to point forces F_1 and F_2 as shown. Determine the displacement of the tip of the bar. Let $A = 1.44\,\text{in}^2$, $F_1 = 2\,\text{klb}$, $F_2 = 0.4\,\text{klb}$, and $E = 1200\,\text{ksi}$.

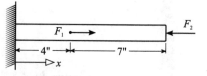

Example 11.2-1

Solution

The point forces are represented as a force per unit length as

$$f = F_1\delta(x-4) - F_2\delta(x-11).$$

From Eq. (11.1-7),

$$P = P(0) - \int_0^x f\,dx = \begin{cases} P(0), & x < 4 \text{ inches,} \\ P(0) - F_1, & 0 < x < 11 \text{ inches,} \\ P(0) - F_1 + F_2 & \text{at } x = 11 + \epsilon \text{ inches.} \end{cases}$$

The boundary conditions are

$$u(0) = 0, \qquad P(11 + \epsilon) = 0.$$

From the second boundary condition,

$$0 = P(0) - F_1 + F_2,$$

so

$$P(0) = F_1 + F_2 = 2 - 0.4 = 1.6 \text{ klb.}$$

From the first boundary condition and Eq. (11.1-6),

$$u = u(0) + \int_0^x \frac{P}{AE}\,dx$$

$$= \begin{cases} \dfrac{(F_1 + F_2)x}{AE} = 0.296x \text{ inches,} & x < 4 \text{ inches,} \\[2mm] \dfrac{(F_1 + F_2)(4)}{AE} + \dfrac{F_2(x-4)}{AE} = 0.231(20 - x) \text{ inches,} & \\[2mm] & 4 < x < 11 \text{ inches.} \end{cases}$$

In the calculation above P is discontinuous over $0 < x < 11$ and, in order to perform the integration, notice that the integral was divided into the two integrals over $0 < x < 4$ and $4 < x < 11$.

The displacement at the right end is $u(11) = 2.08$ inches.

Notice that the point force F_2 was treated as an applied force per unit length. Another way to treat the right point force is in the boundary condition. If the right point force is treated in the right boundary condition, then it should be removed from the expression for f and added to the right boundary condition to get

$$f = F_1 \delta(x - 4) \quad \text{and} \quad P(11 + \epsilon) = -F_2.$$

11.2-2 A force pulls on the bar–spring system shown. Determine the displacement of the bar at the right end (at $x = L$). Let $A = 0.1\,\text{m}^2$, $L = 2\,\text{m}$, $E = 2 \times 10^6\,\text{N/m}^2$, $k = 10^5\,\text{N/m}$, and $F = 500\,\text{N}$.

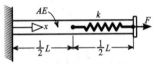

Example 11.2-2

Solution

Let's treat the spring force at $x = \frac{1}{2}L$ as a force per unit length and let's treat the spring force at $x = L$ and the applied force as part of the boundary conditions. The force per unit length and the boundary conditions are

$$f = k[u(L) - u(\tfrac{1}{2}L)]\delta(x - \tfrac{1}{2}L),$$

and

$$u(0) = 0, \quad P(L) = F - k[u(L) - u(\tfrac{1}{2}L)].$$

From Eqs. (11.1-6) and (11.1-7) and the left boundary condition,

$$P = P(0) - \int_0^x f\, dx = \begin{cases} P(0), & 0 < x < \tfrac{1}{2}L, \\ P(0) - k[u(L) - u(\tfrac{1}{2}L)], & \tfrac{1}{2}L < x < L, \end{cases}$$

$$u = u(0) + \int_0^x \frac{P}{AE}\, dx$$

$$= \frac{1}{AE} \begin{cases} P(0)x, & 0 < x < \tfrac{1}{2}L, \\ P(0)(\tfrac{1}{2}L) + \{P(0) - k[u(L) - u(\tfrac{1}{2}L)]\}(x - \tfrac{1}{2}L), & \tfrac{1}{2}L < x < L. \end{cases}$$

From the right boundary condition,

$$P(0) - k[u(L) - u(\tfrac{1}{2}L)] = F - k[u(L) - u(\tfrac{1}{2}L)],$$

so

$$P(0) = F.$$

The displacement at the right end is then

$$u(L) = \frac{1}{AE}\{F(\tfrac{1}{2}L) + \{F - k[u(L) - u(\tfrac{1}{2}L)]\}(\tfrac{1}{2}L)\}$$

$$= \frac{1}{AE}\left\{FL - \tfrac{1}{2}kL\left[u(L) - \frac{FL}{2AE}\right]\right\};$$

so

$$u(L) = \frac{FL}{AE}\left(1 + \frac{kL}{4AE}\right)\Big/\left(1 + \frac{kL}{2AE}\right)$$

$$= \frac{500(2)}{0.1(2 \times 10^6)}\left[1 + \frac{10^5(2)}{4(0.1)(2 \times 10^6)}\right]\Big/\left[1 + \frac{10^5(2)}{2(0.1)(2 \times 10^6)}\right]$$

$$= 4.17 \times 10^{-3} \text{ m} = 4.17 \text{ mm}.$$

Notice in the expression for $u(L)$ that the non-dimensional parameter kL/AE indicates the relative stiffness of the spring and that the effect of the spring on the tip displacement becomes negligible when kL/AE is very small **and** when kL/AE is very large.

Problems

11.2-1   A bar of length L and cross-sectional area A is subjected to two point forces $F = F_1 = F_2$, as shown. Determine the displacement as a function of x. At what points along the bar is the normal strain zero, the displacement zero, the normal strain a maximum, and the displacement a maximum? Let $A = 20 \text{ in}^2$, $L = 3 \text{ ft}$, $E = 20 \text{ ksi}$ and $F = 400 \text{ lb}$.

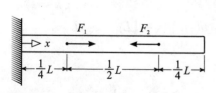

Problem 11.2-1

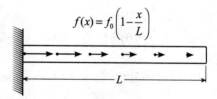

Problem 11.2-2

11.2-2 A bar of length L and cross-sectional area A is subjected to a linearly varying distributed force

$$f = f_0\left[1 - \left(\frac{x}{L}\right)\right],$$

as shown. Determine the displacement of the right end of the bar. Let $A = 0.15\,\text{m}^2$, $L = 1.5\,\text{m}$, $E = 200\,\text{kN/m}^2$, and $f_0 = 400\,\text{N/m}$.

11.2-3 Two identical bars are connected together by a spring k, as shown. The left bar is subjected to a point force F and the right bar is subjected to a uniformly distributed force per unit length f_0. Determine the displacement of the left end of the spring. Let $A = 0.1\,\text{m}^2$, $L = 0.8\,\text{m}$, $E = 250\,\text{kN/m}^2$, $F = 500\,\text{N}$, $f_0 = 400\,\text{N/m}$ and $k = 16\,\text{kN/m}$.

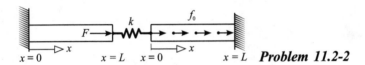

Problem 11.2-2

CHAPTER 12

Torsional Deformation and Stress

12.1 CIRCULAR SHAFTS

A structural member that is subjected to external moments about its long axis is referred to as a **shaft**. A differential element of a circular shaft is shown in Fig. 12.1-1. It is assumed that the external moment or torque T at x produces internal moments M that, in turn, are created by circumferentially directed shear stresses τ over the r–θ cross section. (See Fig. 12.1-2.) Mathematically, the shear stresses τ are functions of x and r and not functions of θ. The internal moment is related to the shear stresses by

$$M = \int r\tau \, dA. \tag{12.1-1}$$

The differential element is in static equilibrium if the sum of the moments about the x axis is zero, that is

$$(M + dM) - M + dT = 0.$$

Dividing by the differential length dx yields

$$\frac{dM}{dx} = -f, \tag{12.1-2}$$

261

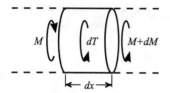

Figure 12.1-1

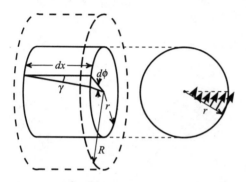

Figure 12.1-2

where $f = dT/dx$ is the external moment per unit length about the x axis. Assuming that the deformations are within the linear region of the stress–strain curve, the shear stresses are related to the shear strains by Eq. (10.2-2); repeated here,

$$\tau = G\gamma, \qquad (12.1\text{-}3)$$

in which the shear modulus G is assumed to be uniform over the cross section (that is, it is not a function of r and θ). Referring to Fig. 12.1-2, the shear strain is related to the twisting angle ϕ by equating the arc lengths,

$$\gamma = r\frac{d\phi}{dx}, \qquad (12.1\text{-}4)$$

in which the shaft's twisting angle ϕ is a function of x (and not a function of y or z) under the assumption that the cross section of the shaft twists like a rigid body; radial lines remain straight. From the above, the internal moment over the r–θ cross section is

$$M = \int r\tau\, dA = \int Gr^2\frac{d\phi}{dx}\, dA = GJ\frac{d\phi}{dx}, \qquad (12.1\text{-}5)$$

in which

$$J = \int r^2 \, dA, \qquad (12.1\text{-}6)$$

where J is the **polar moment of inertia** of the cross section. Substituting Eq. (12.1-5) into Eqs. (12.1-4) and (12.1-3) yields the shear stress

$$\boxed{\tau = \frac{M}{J} r,} \qquad (12.1\text{-}7)$$

which varies linearly with r. Integrating Eq. (12.1-5) with respect to x yields

$$\boxed{\phi(x) = \phi(0) + \int_0^x \frac{M}{JG} \, dx.} \qquad (12.1\text{-}8)$$

The internal moment M in Eq. (12.1-8) is determined by integrating Eq. (12.1-2), to get

$$\boxed{M(x) = M(0) - \int_0^x f \, dx.} \qquad (12.1\text{-}9)$$

In a given problem, a shaft is subjected to external moments. The shaft, by the way in which it is secured at its boundaries, is also subjected to boundary conditions (see Fig. 12.1-3). The boundary conditions together with Eqs. (12.1-8) and (12.1-9) are used to determine the internal moments and the twisting angles. The shear stresses are then determined from Eq. (12.1-7).

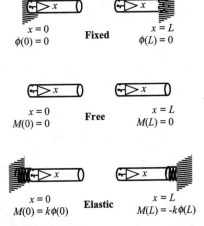

Figure 12.1-3

Notice that the internal moment $M(x)$ can be determined from Eq. (12.1-9) without using Eq. (12.1-8), provided one or more boundary conditions specifies the internal moment. This kind of shaft is referred to as **determinate**, indicating that the internal moment is independent of the shaft's elasticity. When determination of the internal moment requires Eq. (12.1-8), the shaft is referred to as **indeterminate**.

Examples

| 12.1-1 | Two metal shafts are stacked on top of each other and subjected to an applied torque T as shown. Determine the twisting angle of the shaft as a function of x. Let $D_1 = 0.1$ m, $D_2 = 0.2$ m, $h_1 = h_2 = 0.1$ m, $G = 7500$ kN/m^2, and $T = 200$ N·m.

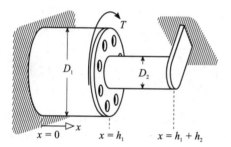

$x = 0$ $x = h_1$ $x = h_1 + h_2$ *Example 12.1-1*

Solution

Consider Eqs. (12.1-8) and (12.1-9) over the length of the left shaft $(0 < x < h_1)$ and over the length of the right shaft $(h_1 < x < h_1 + h_2)$. The applied torque per unit length is zero over the length of both shafts; that is, $f = 0$. The boundary conditions are divided into boundary conditions for the left and right shafts, which are called **external** boundary conditions, and boundary conditions between the two shafts, which are called **internal** boundary conditions. The four boundary conditions are

External: $\phi(0) = 0$, $\phi(h_1 + h_2) = 0$,

Internal: $\phi(h_1 - \epsilon) = \phi(h_1 + \epsilon)$, $M(h_1 - \epsilon) = M(h_1 + \epsilon) - T$,

where ϵ is an arbitrarily small number. The external boundary conditions at the interface specify the twisting angle of the left shaft to be equal to the twisting angle of the right shaft, and the internal twisting moment of the left shaft to be equal to the internal twisting moment of the right shaft minus the applied torque. From Eqs. (12.1-8) and (12.1-9), we get

$$0 < x < h_1: \qquad M = M(0) = M(h_1 - \epsilon) = M(h_1 + \epsilon) - T,$$

$$\phi = 0 + \int_0^x \frac{M}{J_1 G} \, dx = \frac{M(0)x}{J_1 G},$$

$$\phi(h_1 - \epsilon) = \frac{M(0)h_1}{J_1 G} = \phi(h_1 + \epsilon);$$

$$h_1 < x < h_1 + h_2 : \qquad M = M(h_1 + \epsilon) = M(h_1 + h_2) = M(0) + T$$

$$\phi = \phi(h_1 + \epsilon) + \int_{h_1 + \epsilon}^x \frac{M}{J_2 G} \, dx = \frac{M(0)h_1}{J_1 G} + \frac{[M(0) + T](x - h_1)}{J_2 G}$$

$$0 = \phi(h_1 + h_2) = \frac{M(0)}{G} \left(\frac{h_1}{J_1} + \frac{h_2}{J_2} \right) + \frac{Th_2}{J_2 G};$$

so

$$M(0) = -\frac{T}{1 + \left(\dfrac{h_1}{h_2} \right)\left(\dfrac{J_2}{J_1} \right)}.$$

The twisting angle as a function of x is

$$\phi = \begin{cases} -\dfrac{T}{1 + \left(\dfrac{h_1}{h_2} \right)\left(\dfrac{J_2}{J_1} \right)} \dfrac{x}{J_1 G}, & 0 < x < h_1, \\[4ex] \dfrac{T}{1 + \left(\dfrac{h_1}{h_2} \right)\left(\dfrac{J_2}{J_1} \right)} \dfrac{h_1}{h_2} \left(\dfrac{x - h_1 - h_2}{J_1 G} \right), & h_1 < x < h_1 + h_2, \end{cases}$$

$$= \begin{cases} -0.160x \text{ rad}, & 0 < x < 0.1, \\ 0.160(x - 0.2) \text{ rad}, & 0.1 < x < 0.2. \end{cases}$$

12.1-2 An applied torque T twists a rigid platform that is attached to two concentric torsional springs having spring constants k_1 and k_2, as shown in the figure overleaf. Determine the twisting angle of the platform measured from the unstretched angle of the springs. Let $k_1 = 20$ klb·ft/rad, $k_2 = 40$ klb·ft/rad, and $T = 2$ klb·ft.

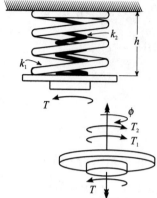

Example 12.1-2

Solution

The twisting angles of the springs are equal to each other at the interface between the springs and the platform. From the free body diagram of the platform, $T = T_1 + T_2$, in which T_1 denotes the torque acting on the outer spring and T_2 denotes the torque acting on the inner spring. Thus,

$$T = T_1 + T_2 = k_1\phi + k_2\phi = (k_1 + k_2)\phi,$$

so

$$\phi = \frac{T}{k_1 + k_2} = \frac{2}{20 + 40} = 0.033 \text{ rad} = 1.91°.$$

Also, make certain not to confuse a torsional spring constant with a linear spring constant.

12.1-3 A shaft is subjected to a uniformly distributed torque f as shown. The diameter of the shaft in the conical section is $D = D_1 + (D_2 - D_1)x/L_1$. Determine the twisting angle of the shaft as a function of x. Let $D_1 = 0.04$ m, $D_2 = 0.02$ m, $L_1 = L_2 = 0.05$ m, $f = 300$ N, and $G = 50 \times 10^6$ N/m².

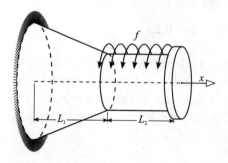

Example 12.1-3

Solution

The boundary conditions, the torque per unit length, and the polar moment of inertia are

$$\phi(0) = 0, \qquad M(L_1 + L_2) = 0,$$

$$f = \begin{cases} 0, & 0 < x < L_1, \\ 300, & L_1 < x < L_1 + L_2, \end{cases}$$

$$J = \tfrac{1}{2}\pi R^4 = \tfrac{1}{32}\pi D^4 = \begin{cases} \tfrac{1}{32}\pi D^4, & 0 < x < L_1, \\ \tfrac{1}{32}\pi D_2^4, & L_1 < x < L_1 + L_2. \end{cases}$$

From Eq. (12.1-9) and the second boundary condition, the internal twisting moment is

$$M = \begin{cases} M(0), & 0 < x < L_1, \\ M(0) - 300(x - L_1), & L_1 < x < L_1 + L_2, \end{cases}$$

from which

$$0 = M(0) - 300 L_2, \qquad \text{so} \qquad M(0) = 300 L_2,$$

and

$$M = \begin{cases} 300 L_2, & 0 < x < L_1, \\ 300(L_1 + L_2 - x), & L_1 < x < L_1 + L_2. \end{cases}$$

From the first boundary condition and Eq. (12.1-8), the twisting angle is

$$\phi = \begin{cases} \displaystyle\int_0^x \frac{300 L_2}{\tfrac{1}{32}\pi D^4 G}\,dx, & 0 < x < L_1, \\[4mm] \displaystyle\int_0^x \frac{300(L_1 + L_2 - x)}{\tfrac{1}{32}\pi D_2^4 G}\,dx, & L_1 < x < L_1 + L_2, \end{cases}$$

$$= \begin{cases} \displaystyle\frac{2400 L_1 L_2}{\pi G(D_1 - D_2)}\left(\frac{1}{D^3} - \frac{1}{D_1^3}\right), & 0 < x < L_1, \\[4mm] \displaystyle\frac{2400 L_1 L_2}{\pi G(D_1 - D_2)}\left(\frac{1}{D_2^3} - \frac{1}{D_1^3}\right) + \frac{9600}{\pi G D_2^4}(x - L_1)(\tfrac{3}{2}L_1 + L_2 - \tfrac{1}{2}x), \\[4mm] \hspace{6cm} L_1 < x < L_1 + L_2. \end{cases}$$

At $x = L_1 + L_2$, the twisting angle is

$$\phi(L_1 + L_2) = \frac{2400L_1L_2}{\pi GD_1^3 D_2^3}(D_1^2 + D_2^2 + D_1D_2) + \frac{9600}{\pi GD_2^4}L_2(L_1 + \tfrac{1}{2}L_2)$$

$$= \frac{2400(0.05)^2}{\pi(50 \times 10^6)(0.04)^3 0.02^3}[0.04^2 + 0.02^2 + (0.04)0.02]$$

$$+ \frac{9600}{\pi(50 \times 10^6)0.02^4}0.05(0.05 + \tfrac{1}{2}0.05)$$

$$= 1.64 \text{ rad} = 94.0°.$$

Problems

12.1-1 A solid shaft of diameter D and a torsional spring are fixed to each other and subjected to an applied torque T, as shown. Determine the twisting angle of the system as a function of x. Let $D = 0.1$ m, $L = 0.2$ m, $G = 7500$ kN/m^2, $k = 400$ N·m/rad and $T = 200$ N·m.

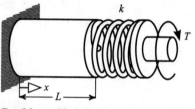

Problem 12.1-1

12.1-2 A pair of applied forces F twist a rigid platform that is attached to two concentric shafts having shear moduli of G_1 and G_2, as shown. Determine the twisting angle of the platform. Let $G_1 = 100$ klb/ft^2, $G_2 = 200$ klb/ft^2, $d = 1$ ft, $h = 1$ ft, $F = 1$ klb, $R_1 = 0.3$ ft, and $R_2 = 0.6$ ft.

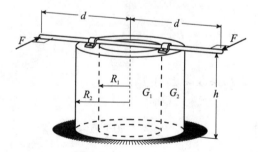

Problem 12.1-2

12.1-3 A uniform solid shaft having diameter D is subjected to a distributed twisting moment per unit length of $f = f_0(1 - x/L)x/L$ N, as shown. Determine the twisting angle of the free end. Also, determine the

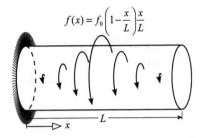

$$f(x) = f_0\left(1 - \frac{x}{L}\right)\frac{x}{L}$$

Problem 12.1-3

location d where a point moment T can be placed and produce the same twisting angle at the free end as the distributed moment. Assume that the total moment that is applied is the same in both cases; that is, let the point moment be equal to $T = \int f\, dx = \frac{1}{6} f_0 L$. Let $D = 0.02$ m, $L = 0.05$ m, $f_0 = 1800$ N, and $G = 50 \times 10^6$ N/m^2.

12.2 POINT MOMENTS

The internal moments are determined from the external moment per unit length using Eq. (12.1-8). The external moment per unit length represents any type of external moment, a distributed moment, a point moment, or combinations of these. A distributed moment is a moment that is spread out over the length of the shaft. A point moment is a moment that is concentrated at a point. The moment per unit length associated with the point moment is infinitely large over an infinitesimal region of concentration. The point moment T located at $x = a$ is represented as a moment per unit length using the impulse function, defined earlier in Eq. (11.2-1). We write

$$f = T\delta(x - a), \qquad (12.2\text{-}1)$$

Let's examine the internal moments surrounding the point moment T. Substituting Eq. (12.2-1) into Eq. (12.1-9) yields

$$M(a + \epsilon) = M(a) - \int_a^{a+\epsilon} T\delta(x - a)\, dx$$
$$= M(a) - T. \qquad (12.2\text{-}2)$$

Notice that **the internal moment just to the right of the point moment is less than the internal moment just to the left of the point moment by an amount T.**

Examples

12.2-1 A solid circular shaft of radius R is subjected to a pair of point moments T as shown in (a). Determine the twisting angle as a function of x. Let $R = 1$ inch, $T = 2$ klb·in, $G = 1200$ klb/in^2, and $L = 8$ inches.

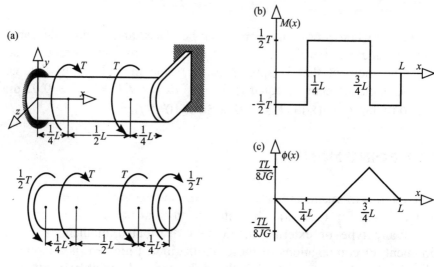

Example 12.2-1

Solution

The point moments are represented as a moment per unit length as $f = -T\delta(x - \frac{1}{4}L) + T\delta(x - \frac{3}{4}L)$, and the area polar moment of inertia is $J = \frac{1}{2}\pi R^4 = 1.57$ in^4. From Eq. (12.1-9),

$$M = M(0) - \int_0^x f \, dx = \begin{cases} M(0), & x < \frac{1}{4}L, \\ M(0) + T, & \frac{1}{4}L < x < \frac{3}{4}L, \\ M(0), & x > \frac{3}{4}L \end{cases}$$

[see the graph in (b)]. The boundary conditions are

$$\phi(0) = 0, \qquad \phi(L) = 0.$$

From Eq. (12.1-8) and the first boundary condition,

$$\phi = \phi(0) + \int_0^x \frac{M}{JG}\, dx$$

$$= \begin{cases} \dfrac{M(0)x}{JG}, & x < \frac{1}{4}L, \\[2ex] \dfrac{\frac{1}{4}M(0)L + [M(0) + T](x - \frac{1}{4}L)}{JG}, & \frac{1}{4}L < x < \frac{3}{4}L, \\[2ex] \dfrac{\frac{1}{2}TL + M(0)x}{JG}, & \frac{3}{4}L < x < L \end{cases}$$

[see the graph in (b)]. From the second boundary condition,

$$0 = \phi(L) = \frac{\frac{1}{2}TL + M(0)L}{JG},$$

so $M(0) = -\frac{1}{2}T$, in which case

$$\phi = \begin{cases} \dfrac{-Tx}{2JG}, & x < \frac{1}{4}L, \\[2ex] \dfrac{T(-L + 2x)}{4JG}, & \frac{1}{4}L < x < \frac{3}{4}L, \\[2ex] \dfrac{T(L - x)}{2JG}, & \frac{3}{4}L < x < L, \end{cases}$$

$$= \begin{cases} 5.31 \times 10^{-4}x \text{ rad}, & x < \frac{1}{4}L, \\[1ex] 2.65 \times 10^{-4}(-8.00 + 2x) \text{ rad}, & \frac{1}{4}L < x < \frac{3}{4}L, \\[1ex] 5.31 \times 10^{-4}(8.00 - x) \text{ rad}, & \frac{3}{4}L < x < L. \end{cases}$$

12.2-2 A circular shaft is subjected to a linearly varying distributed torque as shown. Determine the twisting angle of the shaft at the right

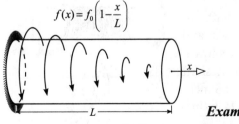

$$f(x) = f_0\left(1 - \frac{x}{L}\right)$$

Example 12.2-2

end (at $x = L$). Let $R = 0.1$ m, $L = 2$ m, $G = 2 \times 10^6$ N/m², and $f_0 = 500$ N.

Solution

The external moment per unit length and the boundary conditions are

$$f = f_0\left(1 - \frac{x}{L}\right), \qquad \phi(0) = 0, \qquad M(L) = 0.$$

From Eq. (12.1-9),

$$M(x) = M(0) - \int_0^x f\, dx = M(0) - f_0 x\left(1 - \frac{x}{2L}\right).$$

From the second boundary condition,

$$0 = M(L) = M(0) - \tfrac{1}{2}f_0 L,$$

so $M(0) = \tfrac{1}{2}f_0 L$, and then

$$M(x) = \frac{f_0 L}{2}\left[\left(\frac{x}{L}\right)^2 - 2\left(\frac{x}{L}\right) + 1\right] = \frac{f_0 L}{2}\left(1 - \frac{x}{L}\right)^2.$$

From Eq. (12.1-8) and the first boundary condition,

$$\phi(x) = \phi(0) + \int_0^x \frac{M}{JG}\, dx = \frac{f_0 L^2}{6JG}\left[1 - \left(1 - \frac{x}{L}\right)^3\right].$$

At $x = L$, $\qquad \phi(L) = \dfrac{f_0 L^2}{6JG} = 10.6$ rad.

12.2-3 Two identical circular shafts of radius R are connected by a spring k and subjected to a point torque T as shown. Determine the twist angles of each shaft as a function of x. Let $R = 0.3$ inch, $L = 6$ inches, $T = 8$ lb·in, $G = 10$ klb/in², , and $k = 20$ lb·in/rad.

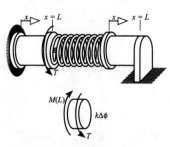

Example 12.2-3

Solution

The applied torque will be treated in the boundary condition rather than being represented as an applied torque per unit length. Therefore, the external force per unit length is zero for each shaft. The external and internal boundary conditions are

External : $\phi_1(0) = 0$, $\phi_2(L) = 0$,

Internal : $M_1(L) = k[\phi_2(0) - \phi_1(L)] + T$, $M_2(0) = k[\phi_2(0) - \phi_1(L)]$.

From Eq. (12.1-9),

$$M_1 = M_1(0) = M_1(L), \qquad M_2 = M_2(0) = M_1(L) = M_1(0),$$

and, from Eq. (12.1-8) and the first external boundary condition,

$$\phi_1 = \frac{M_1(0)x}{JG}, \qquad \phi_2 = \phi_2(0) + \frac{M_2(0)x}{JG}.$$

From the remaining boundary conditions, we get the three linear algebraic equations

$$M_1(0) = k\left[\phi_2(0) - \frac{M_1(0)L}{JG}\right] + T, \qquad M_2(0) = k\left[\phi_2(0) - \frac{M_1(0)L}{JG}\right],$$

$$0 = \phi_2(0) + \frac{M_2(0)L}{JG},$$

which are expressed in terms of the three unknowns $M_1(0)$, $M_2(0)$, and $\phi_2(0)$. Denoting the relative stiffness by $\beta = kL/JG$, we find that the solution is

$$M_1(0) = \frac{1+\beta}{1+2\beta}T, \qquad M_2(0) = -\frac{\beta}{1+2\beta}T, \qquad \phi_2(0) = \frac{\beta}{1+2\beta}\frac{TL}{JG}.$$

The twisting angles of each of the shafts are

$$\phi_1 = \frac{1+\beta}{1+2\beta}\frac{Tx}{JG} = 0.042x \text{ rad},$$

$$\phi_2 = \frac{\beta}{1+2\beta}\frac{T(L-x)}{JG} = 0.021(L-x) \text{ rad}.$$

Notice that when $\beta = 0$, $\phi_1 = Tx/JG$ and $\phi_2 = 0$, in which case the right shaft does not twist. When $\beta = \infty$, $\phi_1 = Tx/2JG$ and $\phi_2 = T(L - x)/2JG$, in which case the twisting is symmetric about the midpoint.

Problems

12.2-1 A circular shaft of radius R is subjected to two point torques T_1 and T_2 as shown. Determine the twisting angle as a function of x. Let $a = 0.2$ m, $b = 0.3$ m, $R = 0.04$ m, $G = 100$ kN/m², $T_1 = 50$ N·m, and $T_2 = 10$ N·m.

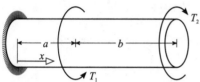

Problem 12.2-1

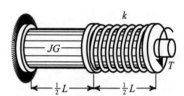

Problem 12.2-2

12.2-2 A circular shaft of radius R is stiffened by a spring k and subjected to a point torque T as shown. Determine the twisting angle of the shaft as a function of x. Let $L = 1.2$ m, $R = 0.04$ m, $G = 100$ kN/m², and $k = 200$ N·m/rad.

12.2-3 A circular shaft of radius R is stiffened by two identical springs having spring constants k and subjected to a pair of point torques T as shown. Determine the twisting angle of the shaft as a function of x. Let $L = 1.5$ ft, $R = 0.5$ inch, $G = 100$ klb/ft², $k = 1$ lb·ft, and $T = 4$ lb·ft.

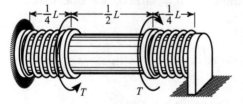

Problem 12.2-3

12.3 THIN IRREGULAR SHAFTS

The shear stresses τ in the r–ϕ plane act in the circumferential direction in the case of circular shafts. In the case of thin irregular shafts, the shear

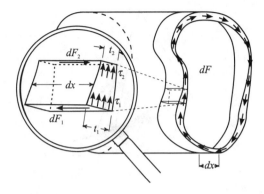

Figure 12.3-1

stresses in the r–θ plane act in directions that are tangent to the surface of the shaft. This changes the analysis of the shaft, specifically the development of Eqs. (12.1-1) and (12.1-3). The other equations in Section 12.1 remain valid for irregular shafts.

In static equilibrium, the forces on the two indicated surfaces in Fig. 12.3-1 balance, that is, $dF_1 = dF_2$, so

$$\tau_1 t_1 \, dx = \tau_2 t_2 \, dx,$$

where t is the thickness of the cross section. Also, notice that the shear stress in the r–θ cross section at the corner of surface 1 is equal to τ_1 and that the shear stress in the r–θ cross section at the corner of surface 2 is equal to τ_2. Thus, the product $q = \tau t$ is constant as we circle around the cross section. This product is called **shear flow**. We write

$$\boxed{q = \tau t = \text{constant.}} \qquad (12.3\text{-}1)$$

Figure 12.3-2(a) shows the differential element and (b) shows the differential deformation. As with circular shafts, the cross sections of the thin irregular shafts are assumed to be rigid. Referring to Fig. 12.3-2(a), by similarity of triangles, the differential length ds on the surface of the shaft is related to the differential angle $d\theta$ by

$$ds = r \, d\theta \left(\frac{r}{r_\perp}\right). \qquad (12.3\text{-}2)$$

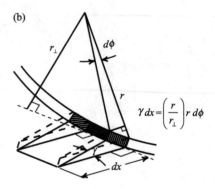

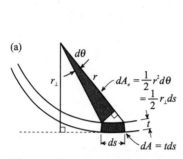

Figure 12.3-2

Similarly, referring to Fig. 12.3-2(b), by similarity of triangles, the infinitesimal twisting of the shaft produces a differential change in length on the surface of the thin irregular shaft $\gamma\,dx$ given by

$$\gamma\,dx = \left(\frac{r}{r_\perp}\right)r\,d\phi. \tag{12.3-3}$$

Substituting Eq. (12.3-2) into (12.3-3) yields

$$\frac{d\phi}{dx} = \gamma\frac{d\theta}{ds}. \tag{12.3-4}$$

Also, the area of the differential triangle generated by ds is

$$dA_e = \tfrac{1}{2}r^2\,d\theta = \tfrac{1}{2}r_\perp\,ds,$$

and the area of the differential element is $dA = t\,ds$.

The internal moment produced by the shear stresses can now be calculated. The differential internal moment is $dM = r_\perp \tau t\,ds$, so the internal moment produced by the internal stresses is

$$M = \int r_\perp \tau t\,ds = \int r_\perp q\,ds = q\int r_\perp\,ds = 2qA_e, \tag{12.3-5}$$

where A_e is the area enclosed by the cross section of the irregular shaft. Thus,

$$\boxed{q = \frac{M}{2A_e}.} \tag{12.3-6}$$

It also follows from Eq. (12.3-5) that

$$M = 2qA_e = 2q \int \frac{1}{2} r^2 \, d\theta = q \int r^2 \frac{1}{\gamma} \frac{d\phi}{dx} \, ds = q \int r^2 \frac{Gt}{\tau \, t} \, ds \, \frac{d\phi}{dx}$$

$$= G \int r^2 \, dA \, \frac{d\phi}{dx}.$$

Therefore,

$$M = GJ \frac{d\phi}{dx}, \tag{12.3-7}$$

where J is the polar moment of inertia of the thin irregular shaft. Notice that Eq. (12.1-5) associated with circular shafts is the same as Eq. (12.3-7) associated with thin irregular shafts.

As is the case in circular shafts, a thin irregular shaft is subjected to external moments and it is subject to boundary conditions. The boundary conditions together with Eqs. (12.1-8) and (12.1-9) are used to determine the internal moments and the twisting angles. The shear stresses are then determined from Eqs. (12.3-1) and (12.3-6).

Examples

12.3-1   A thin square tube of thickness t is subjected to a pair of twisting forces F (called a **couple**) as shown. Calculate the shear stresses in the cross section and the twisting angle at the tip of the tube (at $x = L$). Let $a = 5$ cm, $L = 20$ cm, $t = 5$ mm, $F = 10$ N, and $G = 600$ kN/m^2.

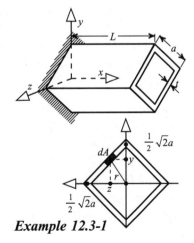

Example 12.3-1

Solution
The applied torque is treated in the boundary condition, so the external moment per unit length acting over the length of the tube is zero, that is, $f = 0$, and the boundary conditions are $\phi(0) = 0$ and $M(L) = aF$. From Eq.

(12.1-9), $M = M(0) = M(L) = aF$. From Eq. (12.3-6), the shear stresses
are

$$\tau = \frac{M}{2A_e t} = \frac{aF}{2a^2 t} = \frac{F}{2at} = 20 \text{ kN/m}^2.$$

Referring to the figure, the equation describing the cross section in the positive quadrant is $z = y - \frac{1}{2}\sqrt{3}a$ and $dA = \sqrt{2}t\,dy$, so the polar moment of inertia of the cross section is $J = \int r^2\,dA = 4\int_0^{\sqrt{2}a/2}(y^2 + z^2)\sqrt{2}t\,dy = \frac{4}{3}a^3 t = 8.33 \times 10^{-7} \text{ m}^4$.

From Eq. (12.1-8), the twisting angle is

$$\phi = \phi(0) + \int \frac{M}{JG}\,dx = \frac{aF}{\frac{4}{3}a^3 tG}x = \frac{3F}{4a^2 tG}x.$$

At the tip of the tube,

$$\phi(L) = \frac{3FL}{4a^2 tG} = 0.2 \text{ rad} = 11.5°.$$

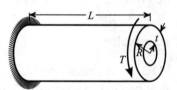

12.3-2 A thick circular tube of outer radius R and thickness t is subjected to a point torque T at its tip (at $x = L$), as shown. Calculate the shear stresses in the middle of the cross section in two ways: using Eq. (12.3-6) for thin irregular shafts and using Eq. (12.1-7) for circular shafts. Compare your results.

Example 12.3-2

Solution
The point torque is treated as a boundary condition, so the external moment per unit length is zero, that is, $f = 0$, and the boundary conditions are $\phi(0) = 0$ and $M(L) = T$. From Eq. (12.1-9), $M = M(0) = M(L) = T$. When using Eq. (12.3-6), the shear stresses are assumed to be uniform over the thickness of the tube. The first expression for the shear stress is

$$\tau = \frac{M}{2A_e t} = \frac{T}{2\pi R^2 t}.$$

The polar moment of inertia of the tube is

$$J = \int r^2 \, dA = \int_{R-t}^{R} r^2 \pi r \, dr = \tfrac{1}{2}\pi[R^4 - (R - t)^4].$$

When using Eq. (12.1-7), the shear stresses vary linearly through the cross section. In the middle of the cross section, the second expression for the shear stress is

$$\tau = \frac{M}{J} r = \frac{T}{\tfrac{1}{2}\pi[R^4 - (R - t)^4]}(R - \tfrac{1}{2}t)$$

$$= \frac{2T}{\pi} \frac{R - \tfrac{1}{2}t}{[R^2 + (R - t)^2][R + (R - t)][R - (R - t)]}$$

$$= \frac{T}{\pi t}\left[\frac{1}{R^2 + (R - t)^2}\right].$$

Comparing the two expressions for τ, first notice that when $t \ll R$, the term in large brackets is approximately equal to $1/(2R^2)$, from which it follows that the two expressions for the shear stress are approximately equal to each other. Letting $\alpha = t/R$, the term in large brackets can be expanded using a Taylor series approximation to get

$$\left[\quad \right] = \frac{1}{R^2} \frac{1}{[1 + (1 - \alpha)^2]} = \frac{1}{R^2} \frac{1}{\alpha^2 - 2\alpha + 2} = \frac{1}{R^2}(\tfrac{1}{2} + \tfrac{1}{2}\alpha + \tfrac{1}{4}\alpha^2 + \ldots).$$

By neglecting the first- and higher-order terms (letting $\alpha = 0$), the second expression reduces to the first expression, as indicated earlier. Neglecting only the α^2 term, the second expression for the shear stress becomes larger than the first expression for the shear stress by a factor α. The accuracy of the first expression is of the order of the ratio t/R.

Problems

12.3-1 A thin triangular tube of thickness t is subjected to a torque T as shown in the figure overleaf. Calculate the shear stress in the cross section and the twisting angle at the tip of the tube (at $x = L$). Let $a = 0.1$ m, $L = 1$ m, $t = 0.5$ cm, $T = 0.5$ N·m, and $G = 500$ kN/m². *Hint*: The polar moment of inertia must be calculated. Each of the three sides of the triangular cross section contributes equally to J. For

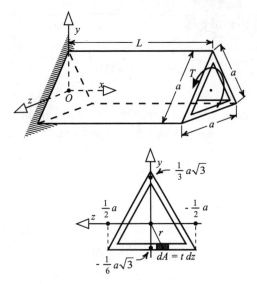

Problem 12.3-1

simplicity, compute the polar moment of the bottom leg of the triangular cross section and then multiply by three.

12.3-2 A thin channel of thickness t and length L is subjected to the torque T shown. The shaft is assumed to rotate about the point O as shown. Determine the shear stress in the shaft and the twisting angle at the tip (at $x = L$). Let $a = 0.1$ m, $L = 1$ m, $t = 0.5$ cm, $T = 0.5$ N·m, and $G = 500$ kN/m^2.

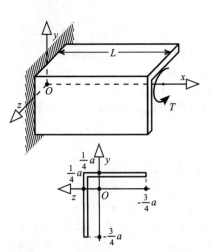

Problem 12.3-2

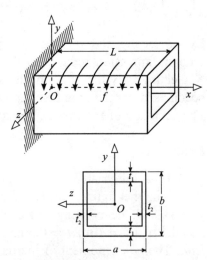

Problem 12.3-3

12.3-3   A rectangular tube is subjected to a uniformly distributed torque f as shown. The tube needs to be designed with minimal cross-sectional area such that the shear stresses do not exceed τ_{max}. Determine the thicknesses t_1 and t_2 of the rectangular tube. Let $a = 1$ inch, $b = 1.5$ inches, $L = 6$ inches, $f = 250$ lb·in/in, $G = 2.5$ ksi, and $\tau_{max} = 5$ ksi.

CHAPTER 13

Bending Deformation and Stress

13.1 NORMAL STRESS

Theory

A structural member that is subjected to external forces F and external moments B that are perpendicular to its long axis is referred to as a **beam**. It is assumed that the external force acts in the y direction and that the external moment acts about the z axis. The resulting displacement of the beam is in the y direction. A cross section of a beam is shown in Fig. 13.1-1. The normal stresses in the x direction are assumed to vary linearly with y and not to vary with z. The point at which the normal stress is zero along this line is called the **neutral point**. The neutral points along the x axis make up the **neutral axis** of the beam. Letting the x axis and the neutral axis coincide, the normal stresses in the x direction are linear functions of y of the form $\sigma = ky$, in which the proportionality constant k needs to be determined.

A differential element of the beam is shown in Fig. 13.1-2. The normal stresses are produced by the internal bending moment M. The sum of the internal forces acting on the cross section is zero, so

$$0 = \int \sigma \, dA - k \int y \, dA. \tag{13.1-1}$$

283

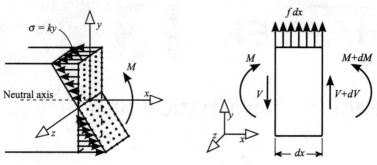

Figure 13.1-1   **Figure 13.1-2**

This equation reveals that **the beam's neutral axis is also the axis of the beam's area center**. Therefore, the beam's neutral axis is determined by locating the beam's area center.

The internal bending moment and the internal normal stresses are related by

$$M = -\int \sigma y \, dA = -\int ky^2 \, dA = -k \int y^2 \, dA. \qquad (13.1\text{-}2)$$

Substituting Eq. (13.1-2) into the form $\sigma = ky$ yields the normal stresses

$$\sigma = -\frac{M}{I_z} y, \qquad (13.1\text{-}3)$$

in which $I_z = \int y^2 \, dA$ denotes the beam's **area moment of inertia about the z axis**.

Examples

13.1-1 A hollow rectangular beam is subjected to the point force shown. Determine the normal stresses in the cross section of the beam. Let $a = 1.2 \, \text{cm}$, $b = 1.0 \, \text{cm}$, $t = 0.1 \, \text{cm}$, $F = 25 \, \text{N}$, and $L = 8 \, \text{cm}$. Where in the cross sections are the largest normal stresses located?

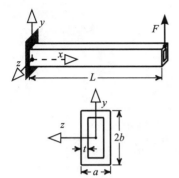

Example 13.1-1

Solution

The neutral axis of the beam is located in the center of the beam. The area moment of inertia I_z about the neutral axis is

$$I_z = \int y^2 \, dA = \int_{-b}^{b} y^2 (a \, dy) - \int_{-(b-t)}^{b-t} y^2 (a - 2t) \, dy$$

$$= \tfrac{2}{3}[ab^3 - (a - 2t)(b - t)^3] = 0.314 \, \text{cm}^4.$$

By cutting the beam at x, the free body diagram of the right section reveals that the internal bending moment is $M(x) = F(L - x)$. From Eq. (13.1-3),

$$\sigma = -\frac{3F(L - x)y}{2[ab^3 - (a - 2t)(b - t)^3]} = 79.6(8 - x)y \, \text{N/cm}^2.$$

The largest normal stresses in the cross section are located along a line segment in the z direction that is located at $y = b$. Among the cross sections, the largest normal stresses are located at $x = 0$. The largest normal stresses are along a line segment located at $y = b$ and $x = 0$, and are of magnitude $318.4 \, \text{N/cm}^2$.

13.1-2 A T-shaped beam is subjected to a point bending moment B as shown in the figure overleaf. Determine the normal stresses in the cross section of the beam. Let $a = 1.2 \, \text{inches}$, $t = 0.2 \, \text{inches}$, and $B = 200 \, \text{lb·in}$.

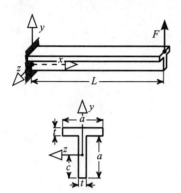

Example 13.1-2

Solution

The location c of the neutral axis measured from the bottom of the cross section is first computed. The location is

$$c = \frac{y_1 A_1 + y_2 A_2}{A_1 + A_2} = \frac{(b + \frac{1}{2}t)ta + \frac{1}{2}bta}{ta + ta}$$

$$= \frac{3}{4} + \frac{1}{4}\frac{t}{a} = 1 \text{ inch.}$$

The area moment of inertia about the neutral axis is

$$I_z = \int_{-c}^{a-c} y^2(t\,dy) + \int_{a-c}^{a-c+t} y^2 a\,dy$$

$$= \frac{1}{3}\{t[(a-c)^3 + c^3] + a[(a-c+t)^3 - (a-c)^3]\} = 0.096\,\text{in}^4.$$

Cutting the beam at x and drawing a free body diagram of the right piece yields $M(x) = B$. Thus, from Eq. (13.1-3),

$$\sigma = -\frac{3By}{t[(a-c)^3 + c^3] + a[(a-c+t)^3 - (a-c)^3]} = 2083.3y\,\text{lb/in}^2.$$

The maximum normal stresses in the cross sections act along the bottom surface of the beam at $y = -c$. The maximum normal stresses are $\sigma = -2.083.3\,\text{lb/in}^2$.

Problems

13.1-1 A solid rectangular beam is subjected to a point force F as shown. Determine the normal stresses in the cross section of the beam. Let $a = 1.2$ inch, $b = 1.0$ inch, $F = 250$ lb and $L = 20$ inch.

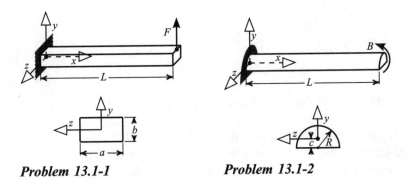

Problem 13.1-1   **Problem 13.1-2**

|13.1-2| A beam having a half-circular cross section is subjected to a point bending moment B as shown. Determine the normal stresses in the cross section of the beam. Let $R = 1.2$ inch, and $B = 200$ lb·in.

13.2 SHEAR FORCES AND BENDING MOMENTS

Theory

The differential element is assumed to deform in the y direction and to rotate an insignificant amount about the z axis. Referring to the free body diagram (Fig. 13.2-1), the sum of the forces in the y direction and the sum of the moments about the z axis are

$$0 = -V + (V + dV) + f\,dx, \qquad (13.2\text{-}1a)$$

$$0 = -M + (M + dM) + V\frac{dx}{2} + (V + dV)\frac{dx}{2}, \qquad (13.2\text{-}1b)$$

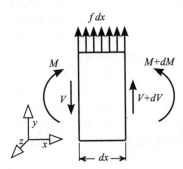

Figure 13.2-1

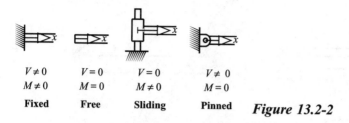

$V \neq 0$	$V = 0$	$V = 0$	$V \neq 0$
$M \neq 0$	$M = 0$	$M \neq 0$	$M = 0$
Fixed	Free	Sliding	Pinned

Figure 13.2-2

where f is the external force per unit length. Notice that by convention, the shear force on the right side of the differential element is taken to be positive upward and the shear force on the left side of the element is taken to be positive downward. The bending moment on the right side is positive counterclockwise and on the left side it is positive clockwise. Dividing Eqs. (13.2-1) by dx and neglecting infinitesimal quantities yields

$$\frac{dV}{dx} = -f, \tag{13.2-2a}$$

$$\frac{dM}{dx} = -V. \tag{13.2-2b}$$

Integrating Eqs. (13.2-2a,b) with respect to x yields

$$V(x) = V(0) - \int_0^x f\, dx, \tag{13.2-3a}$$

$$M(x) = M(0) - \int_0^x V\, dx. \tag{13.2-3b}$$

Equations (13.2-3a,b) are used to determine the internal shear forces and the internal bending moments along the x axis from the external forces and external moments. In order to complete the task of determining V and M as functions of x, notice that $V(0)$ and $M(0)$ are required. These are called boundary conditions (see Fig. 13.2-2). When two boundary conditions involving shear forces and bending moments are given, they can be used along with Eqs. (13.2-3a,b) to determine $V(0)$ and $M(0)$. These kinds of beams are called **determinant beams**.

Examples

13.2-1   A beam is subjected to a uniformly distributed load f as shown. Determine the shear force and the internal bending moment as functions of x. Let $f = 10\,\text{N/m}$ and $L = 2\,\text{m}$.

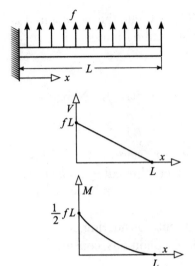

Example 13.2-1

Solution

Rather than cut the beam and determine the shear force and the internal bending moment from the associated free body diagram, they will be determined from Eqs. (13.2-3). At the right end of the beam, the internal bending moment is zero and the internal shear force is zero; that is, $M(L) = 0$ and $V(L) = 0$. Substituting the external force per unit length into Eq. (13.2-3a) yields

$$V(x) = V(0) - \int_0^x f\,dx = V(0) - fx.$$

By substituting the boundary condition $V(L) = 0$ into this equation, it is determined that $V(0) = fL$. The internal shear force is

$$V(x) = f(L - x) = 10(2 - x)\,\text{N}.$$

By substituting the internal shear force into Eq. (13.2-3b), the internal bending moment is obtained:

$$M(x) = M(0) - \int_0^x V(x)\,dx = M(0) - f(Lx - \tfrac{1}{2}x^2).$$

By substituting the boundary condition $M(L) = 0$ into this equation, it is determined that $M(0) = \frac{1}{2} fL^2$. The internal bending moment is

$$M(x) = f(\tfrac{1}{2}L^2 - Lx + \tfrac{1}{2}x^2) = \tfrac{1}{2}f(L-x)^2 = 5(2-x)^2 \text{ N·m}.$$

The internal shear force varies linearly with x, starting from 20 N at $x = 0$ and decreasing to 0 N at $x = L$. The internal bending moment is a quadratic function of x, starting from 20 N·m at $x = 0$ and decreasing to 0 N·m at $x = L$.

Note that this problem was solved without considering the beam's elasticity, because there were two boundary conditions constraining V and M.

13.2-2 A beam connected to a sliding joint at the left end and pinned at the right end is subjected to a point force F as shown. Determine the shear force and the internal bending moment as functions of x. Let $F = 200$ lb and $L = 2$ ft.

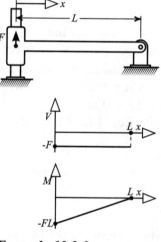

Example 13.2-2

Solution

Rather than cut the beam and determine the shear force and the internal bending moment from the associated free body diagram, they will be determined from Eq. (13.2-3). At the left end of the beam, the internal shear force is $V(0) = -F$. Recall that the shear force is taken to be positive downward on the left side of the differential element. At the right end of the beam, the internal bending moment is $M(L) = 0$. The external force per unit length throughout the beam is 0; that is, $f = 0$. Substituting the left boundary condition and $f = 0$ into Eq. (13.2-3a) yields

$$V(x) = V(0) - \int_0^x f\,dx = -F = -200 \text{ lb}.$$

By substituting the shear force into Eq. (13.2-3b), the internal bending moment is obtained as

$$M(x) = M(0) - \int_0^x V(x)\,dx = M(0) + Fx.$$

From the right boundary condition, $M(L) = 0 = M(0) + FL$, so $M(0) = -FL$, and the internal bending moment is then

$$M(x) = -F(L - x) = -200(2 - x)\,\text{lb·ft}$$

The shear force has a constant value of $-200\,\text{lb}$ throughout the beam, and the bending moment varies linearly with x, starting from $-400\,\text{lb·ft}$ at $x = 0$ and increasing to $0\,\text{lb·ft}$ at $x = L$.

Problems

13.2-1 A beam is subjected to a linearly varying distributed load $f = -f_0(L - x)\,\text{lb/ft}$ as shown. Using Eqs. (13.2-3), determine the shear force and the bending moment in the beam as functions of x. Let $f_0 = 1500\,\text{lb/ft}^2$ and $L = 8\,\text{ft}$.

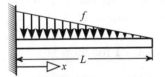

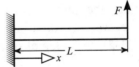

Problem 13.2-1   *Problem 13.2-2*

13.2-2 A beam is subjected to the point force F shown. Using Eqs. (13.2-3), determine the shear force and the bending moment of the beam as functions of x, Let $F = 300\,\text{N}$ and $L = 2.5\,\text{m}$.

13.3 POINT FORCES AND POINT MOMENTS

Theory

The external force per unit length can represent a distributed force, a point force, a point moment, or a combination of these. The external force per unit length can represent a point force using the unit impulse function that was defined earlier in Eq. (11.2-1) or using the unit impulse function that is defined below. The external force per unit length can represent a point moment using the **unit doublet function** that is also defined below. The unit impulse function and the unit doublet function are defined as

$$
\delta_1(x) = \begin{cases} \dfrac{1}{\epsilon^2}(\epsilon + x), & -\epsilon < x < 0, \\[2mm] \dfrac{1}{\epsilon^2}(\epsilon - x), & 0 < x < \epsilon, \\[2mm] 0, & x > \epsilon \text{ or } x < -\epsilon, \end{cases} \tag{13.3-1}
$$

$$
\delta_2(x) = \begin{cases} -\dfrac{1}{\epsilon^2} & -\epsilon < x < 0, \\[2mm] \dfrac{1}{\epsilon^2} & 0 < x < \epsilon, \\[2mm] 0, & x < -\epsilon \text{ or } x > \epsilon, \end{cases} \tag{13.3-2}
$$

in which ϵ denotes an arbitrarily small number (see Fig. 13.3-1). Notice that the area under the unit impulse function is equal to 1; that is, $\int_{-\epsilon}^{\epsilon} \delta_1(x)\, dx = 1$, as was found earlier in Eq. (11.2-2).

From Eq. (13.3-2), the integral of the unit doublet function is

$$
\int_{-\epsilon}^{x} \delta_2(x)\, dx = \begin{cases} -\dfrac{1}{\epsilon^2}(\epsilon + x), & -\epsilon < x < 0, \\[2mm] -\dfrac{1}{\epsilon^2}(\epsilon - x), & 0 < x < \epsilon. \end{cases}
$$

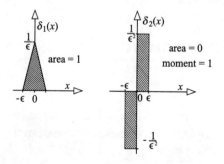

Figure 13.3-1

Thus,

$$\int_{-\epsilon}^{x} \delta_2 \, dx = -\delta_1(x).$$

(13.3-3)

The unit impulse function $\delta_1(x)$ and the unit impulse function $\delta(x)$ defined in Eq. (11.2-1) are interchangeable; both are equal to zero outside an infinitesimal interval, and the area under each function is equal to one; that is, $\int_{-\epsilon}^{\epsilon} \delta_1(x) \, dx = 1$, just as $\int_{0}^{\epsilon} \delta(x) \, dx = 1$. Also note that $\delta_1(x)$ has units of 1/length while $\delta_2(x)$ has units of $1/(\text{length})^2$

Returning to the unit doublet function, observe that the area under this function is equal to zero; that is,

$$\int_{-\epsilon}^{\epsilon} \delta_2(x) \, dx = 0.$$

(13.3-4)

The moment about $x = 0$ produced by the unit doublet is equal to 1; that is,

$$\int_{-\epsilon}^{\epsilon} \delta_2(x) x \, dx = 1.$$

(13.3-5)

The point force F located at $x = a$ is now represented as a force per unit length f_1 using the unit impulse function, and the point moment B located at $x = a$ is represented as a force per unit length f_2 using the unit doublet function, as

$$f_1 = F\delta_1(x - a), \qquad f_2 = B\delta_2(x - a).$$

(13.3-6a,b)

Now examine the internal shear forces and the internal bending moments surrounding the point force F and the point moment B. First consider the point force. Substituting Eq. (13.3-6a) into Eq. (13.2-3a) yields

$$V(a + \epsilon) = V(a) - \int_{a-\epsilon}^{a+\epsilon} F\delta_1(x - a) \, dx$$

$$= V(a) - F.$$

(13.3-7)

The internal shear force just to the right of the point force is less than the internal shear force just to the left of the point force by the amount F. Next consider the point moment. Substituting Eq. (13.3-6b) into Eq. (13.2-3a) and considering Eq. (13.3-3) yields

$$V(x) = V(a - \epsilon) - \int_{a-\epsilon}^{a+\epsilon} B\delta_2(x - a)\, dx$$

$$= V(a - \epsilon) + B\delta_1(x - a). \tag{13.3-8}$$

Evaluating Eq. (13.3-8) at $x = a + \epsilon$ yields

$$V(a + \epsilon) = V(a - \epsilon), \tag{13.3-9}$$

which states that **the shear force to the left of the point moment is equal to the shear force to the right of the point moment.** Substituting Eq. (13.3-8) into Eq. (13.2-3b) yields

$$M(a + \epsilon) = M(a - \epsilon) - \int_{a-\epsilon}^{a+\epsilon} V\, dx = M(a - \epsilon) - B, \tag{13.3-10}$$

which states that **the internal bending moment just to the right of the point moment is less than the internal bending moment just to the left of the point moment by the amount B.** These results are further clarified in the examples.

Examples

13.3-1 Point forces F act on a beam as shown. Determine the shear force and the internal bending moment in the beam as functions of x. Let $F = 100\,\text{N}$, $a = 0.2\,\text{m}$, and $L = 2\,\text{m}$.

Solution
The point forces are represented as a force per unit length using unit impulse functions as

$$f = -F\delta(x - \tfrac{1}{2}L + \tfrac{1}{2}a) - F\delta(x - \tfrac{1}{2}L - \tfrac{1}{2}a).$$

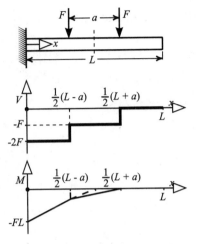

Example 13.3-1

The boundary conditions are $V(L) = 0$ and $M(L) = 0$. From Eq. (13.2-3a),

$$V(x) = V(0) - \int_0^x f\, dx = V(0) - \begin{cases} 0, & 0 < x < \tfrac{1}{2}L - \tfrac{1}{2}a, \\ -F, & \tfrac{1}{2}L - \tfrac{1}{2}a < x < \tfrac{1}{2}L + \tfrac{1}{2}a, \\ -2F, & \tfrac{1}{2}L + \tfrac{1}{2}a < x < L. \end{cases}$$

The boundary condition $V(L) = 0$ yields $0 = V(0) + 2F$, so $V(0) = -2F$ and

$$V(x) = \begin{cases} -2F, & 0 < x < \tfrac{1}{2}L - \tfrac{1}{2}a, \\ -F, & \tfrac{1}{2}L - \tfrac{1}{2}a < x < \tfrac{1}{2}L + \tfrac{1}{2}a, \\ 0, & \tfrac{1}{2}L + \tfrac{1}{2}a < x < L, \end{cases}$$

$$= \begin{cases} -200\,\text{N}, & 0 < x < \tfrac{1}{2}L - \tfrac{1}{2}a, \\ -100\,\text{N}, & \tfrac{1}{2}L - \tfrac{1}{2}a < x < \tfrac{1}{2}L + \tfrac{1}{2}a, \\ 0\,\text{N}, & \tfrac{1}{2}L + \tfrac{1}{2}a < x < L. \end{cases}$$

From Eq. (13.2-3b),

$$M(x) = M(0) - \int_0^x V(x)\, dx$$

$$= M(0) - \begin{cases} -2Fx, & 0 < x < \tfrac{1}{2}L - \tfrac{1}{2}a, \\ -F(x + \tfrac{1}{2}L - \tfrac{1}{2}a), & \tfrac{1}{2}L - \tfrac{1}{2}a < x < \tfrac{1}{2}L + \tfrac{1}{2}a, \\ -FL, & \tfrac{1}{2}L + \tfrac{1}{2}a < x < L. \end{cases}$$

Since $M(L) = 0, 0 = M(0) + FL$, so $M(0) = -FL$, and so

$$M(x) = \begin{cases} -F(L - 2x), & 0 < x < \frac{1}{2}L - \frac{1}{2}a, \\ -F(\frac{1}{2}L + \frac{1}{2}a - x), & \frac{1}{2}L - \frac{1}{2}a < x < \frac{1}{2}L + \frac{1}{2}a, \\ 0, & \frac{1}{2}L + \frac{1}{2}a < x < L, \end{cases}$$

$$= \begin{cases} -100(2 - 2x)\,\text{N·m}, & 0 < x < \frac{1}{2}L - \frac{1}{2}a, \\ -100(1.1 - x)\,\text{N·m}, & \frac{1}{2}L - \frac{1}{2}a < x < \frac{1}{2}L + \frac{1}{2}a, \\ 0\,\text{N·m}, & \frac{1}{2}L + \frac{1}{2}a < x < L. \end{cases}$$

The shear force is a constant value of $-200\,\text{N}$ to the left of the applied forces, is a constant value of $-100\,\text{N}$ between the applied forces, and is 0 to the right of the applied forces. The internal bending moment varies linearly to the left of the applied forces, starting from $-200\,\text{N·m}$ at the left end and increasing to $-20\,\text{N·m}$ at the left applied force. The internal bending moment varies linearly between the applied forces too. It starts from a value of $-20\,\text{N·m}$ at the left applied force and increases to $0\,\text{N·m}$ at the right applied force. The internal bending moment to the right of the applied forces is zero, as was the shear force.

Also notice that as a is allowed to approach zero, the solution approaches the solution associated with a single applied force of $2F$ located at the center of the beam.

13.3-2 A beam is subjected to a pair of point moments B as shown. Determine the shear force and the internal bending moment as functions of x. Let $B = 250\,\text{lb·ft}$, $a = 2\,\text{ft}$, and $L = 8\,\text{ft}$.

Solution

The pair of point moments is represented as an external force per unit length as

$$f = B\delta_2(x - \tfrac{1}{2}L + \tfrac{1}{2}a) + B\delta_2(x - \tfrac{1}{2}L - \tfrac{1}{2}a).$$

The beam is subject to the boundary conditions $M(0) = 0$ and $M(L) = 0$. From Eq. (13.2-3a),

$$V(x) = V(0) - \int_0^x f\,dx = V(0) + B\delta_1(x - \tfrac{1}{2}L + \tfrac{1}{2}a) + B\delta_1(x - \tfrac{1}{2}L - \tfrac{1}{2}a).$$

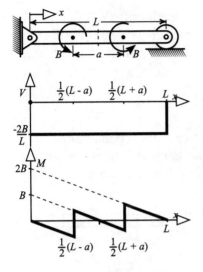

Example 13.3-2

From Eq. (13.2-3b),

$$M(x) = M(0) - \int_0^x V \, dx$$

$$= M(0) - V(0)x - \begin{cases} 0, & 0 < x < \tfrac{1}{2}L - \tfrac{1}{2}a, \\ B, & \tfrac{1}{2}L - \tfrac{1}{2}a < x < \tfrac{1}{2}L + \tfrac{1}{2}a, \\ 2B, & \tfrac{1}{2}L + \tfrac{1}{2}a < x < L. \end{cases}$$

From the boundary conditions $M(0) = 0$ and $M(L) = 0$, $0 = 0 - V(0)L - 2B$, so $V(0) = -2B/L$, and then $V(x) = -2B/L = -62.5$ lb·ft, and

$$M(x) = \begin{cases} B\left(0 - \dfrac{2x}{L}\right), & 0 < x < \tfrac{1}{2}L - \tfrac{1}{2}a, \\[2mm] B\left(1 - \dfrac{2x}{L}\right), & \tfrac{1}{2}L - \tfrac{1}{2}a < x < \tfrac{1}{2}L + \tfrac{1}{2}a, \\[2mm] B\left(2 - \dfrac{2x}{L}\right), & \tfrac{1}{2}L + \tfrac{1}{2}a < x < L, \end{cases}$$

$$= \begin{cases} 250\left(0 - \dfrac{2x}{L}\right) \text{ lb·ft}, & 0 < x < 3 \text{ ft}, \\[2mm] 250\left(1 - \dfrac{2x}{L}\right) \text{ lb·ft}, & 3 < x < 5 \text{ ft}, \\[2mm] 250\left(2 - \dfrac{2x}{L}\right) \text{ lb·ft}, & 5 < x < 8 \text{ ft}. \end{cases}$$

Notice that the shear force is independent of the location a. Likewise, the internal bending moment functions in each region are independent of a, although the ranges over which the functions exist (e.g., $[\frac{1}{2}L - \frac{1}{2}a, \frac{1}{2}L + \frac{1}{2}a]$) clearly depend on a.

Problems

13.3-1 A beam is subjected to a point force F and a point moment B as shown. Determine the shear force and the internal bending moment as functions of x. Let $F = 100\,\text{N}$, $B = 50\,\text{N·m}$, and $L = 1.5\,\text{m}$.

Problem 13.3-1   *Problem 13.3-2*

13.3-2 A beam is subjected to a pair of point forces F as shown. Determine the shear force and the internal bending moment as functions of x. Show that the shear forces and bending moments in this problem and in Problem 13.3-1 are the same. Let $F = 0$ in Problem 13.3-1 and let $F = B/a$, in which a tends to zero, in this problem. Let $F = 100\,\text{N}$, $a = 0.5\,\text{m}$, and $L = 1.5\,\text{m}$.

13.4 SHEAR STRESS

Theory

This section determines an equation describing the shear stresses in the cross section of the beam. As shown in Fig. 13.4-1, the differential element consists of the upper portion of a slice starting at y. The shear stresses in the beam produce an associated shear force over the cross section. Referring to Fig. 13.4-2, the resultant force in the horizontal direction is zero, so

$$0 = -\int_y^{\text{top}} \sigma(x)\, dA + \int_y^{\text{top}} \sigma(x + dx)\, dA + \tau t\, dx = \int_y^{\text{top}} d\sigma\, dA + \tau t\, dx$$

$$= \left(\int_y^{\text{top}} \frac{d\sigma}{dx}\, dA\right) dx + \tau t\, dx. \tag{13.4-1}$$

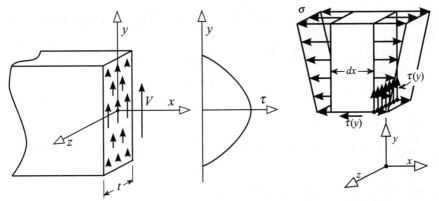

Figure 13.4-1   **Figure 13.4-2**

Thus, from Eq. (13.4-1),

$$\tau = -\frac{1}{t}\int_y^{top}\frac{d}{dx}\left(-\frac{My}{I_z}\right)dA = \frac{1}{I_zt}\frac{dM}{dx}\int_y^{top}y\,dA = \frac{VQ}{I_zt}, \qquad (13.4\text{-}2)$$

where

$$Q = \int_y^{top}y\,dA \qquad (13.4\text{-}3)$$

Notice in Eq. (13.4-3) that Q is zero when $y =$ top, and from Eq. (13.1-1) that Q is zero when $y =$ bottom. Thus, the shear stresses in the cross section are zero at the top and bottom surfaces. The shear stresses tend to be largest in the middle of the cross section.

Examples

13.4-1 A hollow rectangular beam is subjected to the point force P shown in the figure for Example 13.1-1. Determine the shear stresses in the cross section of the beam. Let $a = 1.2$ cm, $b = 1.0$ cm, $t = 0.1$ cm, $F = 25$ N, and $L = 8$ cm. Where in the cross sections are the largest shear stresses located?

Solution
The reader should refer to the solution of Example 13.1-1. The neutral axis is located in the center of the cross section, the area moment of inertia I_z about the neutral axis was shown to be $I_z =$

$\frac{2}{3}[ab^3 - (a - 2t)(b - t)^3] = 0.314\,\text{cm}^4$, and the shear force is $V(x) = P$.
From Eq. (13.4-3),

$$Q = \int_y^{top} y\, dA$$

$$= \begin{cases} \frac{1}{2}a[b^2 - y^2], & b - t < y < b, \\ t\,[(b - t)^2 - y^2] + \frac{1}{2}a[b^2 - (b - t)^2], & -(b - t) < y < b - t. \end{cases}$$

From Eq. (13.4-2), the shear stress is

$$\tau = \begin{cases} \dfrac{FQ}{I_z a}, & b - t < y < b, \\[2mm] \dfrac{FQ}{2I_z t}, & -(b - t) < y < b - t, \end{cases}$$

in which Q was expressed above as a function of y. Notice that the maximum shear stresses in the cross sections lie on a horizontal surface that cuts through the neutral axis (at $y = 0$). The maximum shear stresses are given by

$$\tau_{max} = \frac{3F[(b - t)^2(t - \frac{1}{2}a) + \frac{1}{2}ab^2]}{[ab^3 - (a - 2t)(b - t)^3]t} = 310.51\,\text{N/cm}^2.$$

13.4-2 A half-circular beam is subjected to a point bending moment B as shown in the figure for Problem 13.1-2. Determine the shear stresses in the cross section of the beam. Let $R = 1.2\,\text{in}$, and $B = 200\,\text{lb·in}$.

Solution
The location c of the neutral axis measured from the bottom of the cross section is first computed. Referring to the figure for Problem 13.1-2, the location is

$$c = \frac{1}{A}\int y\, dA = \frac{2}{\pi R^2}\int\int r\sin\theta\, r\, d\theta\, dr = \frac{2}{\pi R^2}\int_0^R r^2\, dr \int_0^\pi \sin\theta\, d\theta$$

$$= \frac{2}{\pi R^2}\frac{R^3}{3}2 = \frac{4}{3\pi}R = 0.509\,\text{inch}.$$

The area moment of inertia about the neutral axis is found using the parallel axis theorem:

$$I_z = I_{z0} - Ac^2 = \int\int (r\sin\theta)^2 r\,d\theta\,dr - Ac^2$$

$$= \int_0^R r^3\,dr \int_0^\pi \sin^2\theta\,d\theta - \frac{\pi R^2}{2}\left(\frac{4}{3\pi}R\right)^2$$

$$= \frac{R^4}{4}\frac{\pi}{2} - \frac{8R^4}{9\pi} = \pi R^4\left[\frac{1}{8} - 8\left(\frac{1}{3\pi}\right)^2\right] = 0.228\ \text{in}^4.$$

Cutting the beam at x and drawing a free body diagram of the right piece yields $V(x) = 0$ and $M(x) = B$. Thus, from Eq. (13.4-2),

$$\tau = \frac{VQ}{I_z t} = 0\ \text{lb/in}^2.$$

The shear stresses are zero throughout the cross sections.

Problems

13.4-1 A T-shaped beam is subjected to a point bending moment B as shown in the figure for Example 13.1-2. Determine the shear forces in the cross section of the beam. Let $a = 1.2$ inch, $t = 0.2$ inch, and $B = 200$ lb·in.

13.4-2 A solid rectangular beam is subjected to an applied force F as shown in the figure for Example 13.1-1. Determine the shear stresses in the cross section of the beam. Let $a = 1.2$ inch, $b = 1$ inch, and $F = 250$ lb.

13.5 DISPLACEMENTS AND SLOPES

Theory

As stated earlier, the internal shear forces and the internal bending moments in a beam can be determined from Eq. (13.2-3a,b) if two boundary conditions involving V and M are available. If this is not the case, then boundary conditions involving the slopes and the displace-

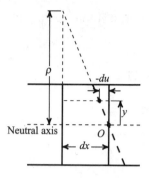

Figure 13.5-1

ments of the beam are needed in order to determine V and M. This latter case gives rise to **indeterminate beams** where the internal shear forces and bending moments produced in the beam depend on the beam's elasticity.

The normal stresses in the beam are related to the normal strains in the beam by

$$\varepsilon = \sigma/E, \tag{13.5-1}$$

and the normal strain $\varepsilon = du/dx$ is related to the beam's radius of curvature ρ by (see Fig. 13.5-1)

$$\varepsilon = -\frac{y}{\rho}. \tag{13.5-2}$$

Substituting Eqs. (13.5-1) and (13.5-2) into Eq. (13.1-3) yields

$$\frac{1}{\rho} = \frac{M}{EI_z}. \tag{13.5-3}$$

For small angles, the radius of curvature is related to the displacement w of the neutral axis by [see Eq. (7.5-5)]

$$\frac{d^2w}{dx^2} = \frac{1}{\rho}, \tag{13.5-4}$$

so, from Eq. (13.5-3),

$$\frac{d^2w}{dx^2} = \frac{M}{EI_z}. \tag{13.5-5}$$

Integrating Eq. (13.5-5) once yields the beam's slope as a function of x, since $dw/dx = \tan\theta \approx \theta$, and integrating once again yields the beam's displacement as a function of x:

$$\theta(x) = \theta(0) + \int_0^x \frac{M}{EI_z}\, dx, \qquad (13.5\text{-}6a)$$

$$w(x) = w(0) + \int_0^x \theta\, dx. \qquad (13.5\text{-}6b)$$

In summary, a beam is subjected to an external force per unit length. The beam is also subjected to conditions at its boundaries involving shear forces, bending moments, slopes, and displacements (see Fig. 13.5-2). Notice that at a given boundary point, the shear force V and the displacement w cannot be constrained simultaneously. Likewise, the bending moment M and the slope θ cannot be constrained simultaneously. From the external forces and the boundary conditions, the beam's internal shear forces, internal bending moments, slopes, and displacements can be determined throughout the beam. The internal

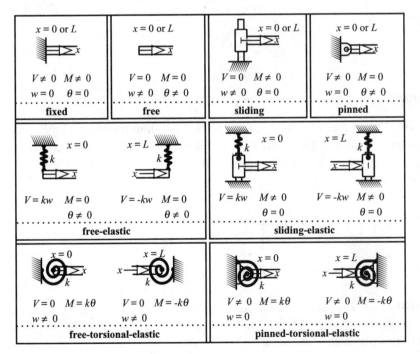

Figure 13.5-2

stresses throughout the beam can be determined too. The beam's internal shear forces, bending moments, slopes, and displacements are determined from the equations

$$V(x) = V(0) - \int_0^x f \, dx, \qquad (13.5\text{-}7a)$$

$$M(x) = M(0) - \int_0^x V \, dx, \qquad (13.5\text{-}7b)$$

$$\theta(x) = \theta(0) + \int_0^x \frac{M}{EI_z} \, dx, \qquad (13.5\text{-}7c)$$

$$w(x) = w(0) + \int_0^x \theta \, dx, \qquad (13.5\text{-}7d)$$

together with four boundary conditions. Once these have been solved, the internal stresses are given by

$$\sigma = -\frac{M}{I_z} y, \qquad (13.5\text{-}8a)$$

$$\tau = \frac{VQ}{I_z t}. \qquad (13.5\text{-}8b)$$

Examples

13.5-1 Determine the shear force, internal bending moment, slope, and displacement of the beam shown as functions of x. The beam has a square cross section with thickness a. Let $L = 1\,\text{m}$, $f_0 = 100\,\text{N/m}$, $a = 0.1\,\text{m}$, and $E = 15\,\text{MN/m}^2$.

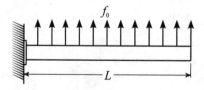

Example 13.5-1

Solution

The solution begins with the beam's external force per unit length and boundary conditions, which are

$$f_0 = 100, \quad w(0) = 0, \quad \theta(0) = 0, \quad M(L) = 0, \quad V(L) = 0.$$

Next, Eqs. (13.5-7a–d) are solved. Together, they are

$$V = V(0) - 100x,$$

$$M = M(0) - V(0)x + 100\frac{x^2}{2},$$

$$\theta = \theta(0) + \frac{1}{EI_z}\left[M(0)x - V(0)\frac{x^2}{2} + 100\frac{x^3}{6}\right],$$

$$w = w(0) + \theta(0)x + \frac{1}{EI_z}\left[M(0)\frac{x^2}{2} - V(0)\frac{x^3}{6} + 100\frac{x^4}{24}\right].$$

The boundary conditions yield

$$0 = V(0) - 100L,$$

$$0 = M(0) - V(0)L + 100\frac{L^2}{2},$$

so $V(0) = 100L$ and $M(0) = 50L^2$. The shear force, internal bending moment, slope, and displacement as functions of x are

$$V = 100L - 100x,$$

$$M = 50L^2 - 100Lx + 100\frac{x^2}{2},$$

$$\theta = \frac{1}{EI_z}\left(50L^2x - 100L\frac{x^2}{2} + 100\frac{x^3}{5}\right),$$

$$w = \frac{1}{EI_z}\left(50L^2\frac{x^2}{2} - 100L\frac{x^3}{6} + 100\frac{x^4}{24}\right).$$

Substituting the values of the parameters into these expressions yields the solutions

$$V = 100(1 - x)\,\text{N},$$

$$M = 50\frac{(1 - x^2)}{2} = 25(1 - x^2)\,\text{N·m},$$

$$\theta = \tfrac{1}{9}x(x^2 - 3x + 3)\,\text{rad},$$

$$w = \tfrac{1}{36}x^2(x^2 - 4x + 6)\,\text{m}.$$

Notice at $x = 0$ that $V(0) = 100\,\text{N}$ and $M(0) = 50\,\text{N·m}$. At $x = 1$, $\theta(1) = \tfrac{1}{9} = 0.111\,\text{rad}$ and $w(1) = \tfrac{1}{12} = 0.083\,\text{m}$. Also notice that the beam is determinate.

13.5-2 Determine the shear force, internal bending moment, slope, and displacement of the beam shown as functions of x. The beam has a solid, circular cross section with radius R. Let $L = 1$ ft, $F = 100$ lb, $R = 0.05$ ft, and $E = 5$ Mlb/ft^2

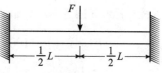

Example 13.5-2

Solution

The solution begins with the beam's external force per unit length and boundary conditions, which are

$$f = -F\delta(x - \tfrac{1}{2}L), \qquad w(0) = 0, \qquad \theta(0) = 0, \qquad w(L) = 0, \qquad \theta(L) = 0.$$

Next, Eqs. (13.5-7a–d) are solved. They are given by

$$V = V(0) - \begin{cases} 0, & x < \tfrac{1}{2}L, \\ -F, & x > \tfrac{1}{2}L, \end{cases}$$

$$= \begin{cases} V(0), & x < \tfrac{1}{2}L, \\ V(0) + F, & x > \tfrac{1}{2}L, \end{cases}$$

$$M = \begin{cases} M(0) - V(0)x, & x < \tfrac{1}{2}L, \\ M(0) - V(0)x - F(x - \tfrac{1}{2}L), & x > \tfrac{1}{2}L, \end{cases}$$

$$\theta = \theta(0) + \frac{1}{EI_z} \begin{cases} M(0)x - \tfrac{1}{2}V(0)x^2, & x < \tfrac{1}{2}L, \\ M(0)x - \tfrac{1}{2}V(0)x^2 - \tfrac{1}{2}F(x - \tfrac{1}{2}L)^2, & x > \tfrac{1}{2}L, \end{cases}$$

$$w = w(0) + \theta(0)x + \frac{1}{EI_z} \begin{cases} \tfrac{1}{2}M(0)x^2 - \tfrac{1}{6}V(0)x^3, & x < \tfrac{1}{2}L, \\ \tfrac{1}{2}M(0)x^2 - \tfrac{1}{6}V(0)x^3 - \tfrac{1}{6}F(x - \tfrac{1}{2}L)^3, & x > \tfrac{1}{2}L. \end{cases}$$

The boundary conditions yield

$$0 = M(0)L - \tfrac{1}{2}V(0)L^2 - \tfrac{1}{8}FL^2,$$

$$0 = \tfrac{1}{2}M(0)L^2 - \tfrac{1}{6}V(0)L^3 - \tfrac{1}{48}FL^3,$$

so $M(0) = -\frac{1}{8}FL$ and $V(0) = -\frac{1}{2}F$. The shear force, internal bending moment, slope, and displacement as functions of x are

$$V = \begin{cases} -\frac{1}{2}F, & x < \frac{1}{2}L, \\ \frac{1}{2}F, & x > \frac{1}{2}L, \end{cases}$$

$$M = \begin{cases} -\frac{1}{8}FL + \frac{1}{2}Fx, & x < \frac{1}{2}L, \\ -\frac{3}{8}FL + \frac{1}{2}Fx, & x > \frac{1}{2}L, \end{cases}$$

$$\theta = \frac{1}{EI_z} \begin{cases} \frac{1}{8}Fx(2x - L), & x < \frac{1}{2}L, \\ -\frac{1}{8}F(x - L)(2x - L), & x > \frac{1}{2}L, \end{cases}$$

$$w = \frac{1}{EI_z} \begin{cases} \frac{1}{48}Fx^2(4x - 3L), & x < \frac{1}{2}L, \\ F(x - L)^2(L - 4x), & x > \frac{1}{2}L. \end{cases}$$

The area moment of inertia is $I_z = \frac{1}{2}\pi R^4$. Substituting the values of the parameters yields

$$V = \begin{cases} -50 \text{ lb}, & x < \frac{1}{2}L, \\ 50 \text{ lb}, & x > \frac{1}{2}L \end{cases}$$

$$M = \begin{cases} -12.5 + 50x \text{ lb·ft}, & x < \frac{1}{2}L, \\ -37.5 + 50x \text{ lb·ft}, & x > \frac{1}{2}L, \end{cases}$$

$$\theta = \frac{1}{EI_z} \begin{cases} 12.5x(2x - 1) \text{ rad}, & x < \frac{1}{2}L, \\ -12.5(x - 1)(2x - 1) \text{ rad}, & x > \frac{1}{2}L, \end{cases}$$

$$w = \frac{1}{EI_z} \begin{cases} 0.0833x^2(4x - 3L) \text{ ft}, & x < \frac{1}{2}L, \\ 0.0833(x - L)^2(L - 4x) \text{ ft}, & x > \frac{1}{2}L. \end{cases}$$

Notice that $\theta(\frac{1}{2}) = 0$, $w(\frac{1}{2}) = 0.00177 \text{ ft} = 0.02 \text{ inch}$, and that the beam is indeterminate.

13.5-3 Determine the shear force, internal bending moment, slope, and displacement of the beam shown as functions of x. The beam has a square

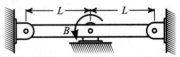

Example 13.5-3

cross section with thickness a. Let $L = 1\,\text{m}$, $B = 100\,\text{N·m}$, $a = 0.1\,\text{m}$, and $E = 15\,\text{MN/m}^2$.

Solution

The solution begins with the beam's external force per unit length and the boundary conditions, which are

$$f = 100\delta_2(x - \tfrac{1}{2}L) + P\delta_1(x - \tfrac{1}{2}L), \qquad w(0) = 0, \qquad M(0) = 0,$$
$$w(L) = 0, \qquad M(L) = 0, \qquad w(\tfrac{1}{2}L) = 0.$$

Notice that the external force per unit length includes the unknown reaction force P at $x = \tfrac{1}{2}L$. From Eqs. (13.5-7a–d) and the boundary conditions,

$$V = \begin{cases} V(0), & x < \tfrac{1}{2}L, \\ V(0) - P + B\delta_1(x - \tfrac{1}{2}L), & x > \tfrac{1}{2}L \end{cases}$$

$$M = \begin{cases} -V(0)x, & x < \tfrac{1}{2}L, \\ -V(0)x + P(x - \tfrac{1}{2}L) - B, & x > \tfrac{1}{2}L, \end{cases}$$

$$\theta = \theta(0) + \frac{1}{EI_z} \begin{cases} -\tfrac{1}{2}V(0)x^2, & x > \tfrac{1}{2}L, \\ -\tfrac{1}{2}V(0)x^2 + \tfrac{1}{2}P(x - \tfrac{1}{2}L)^2 - B(x - \tfrac{1}{2}L), & x > \tfrac{1}{2}L, \end{cases}$$

$$w = \theta(0)x + \frac{1}{EI_z} \begin{cases} -\tfrac{1}{6}V(0)x^3, & x > \tfrac{1}{2}L, \\ -\tfrac{1}{6}V(0)x^3 + \tfrac{1}{6}P(x - \tfrac{1}{2}L)^3 - \tfrac{1}{2}B(x - \tfrac{1}{2}L)^2, & x > \tfrac{1}{2}L. \end{cases}$$

The boundary conditions that have not yet been considered yield the three equations

$$0 = -V(0)L + \tfrac{1}{2}PL - B,$$

$$0 = \theta(0)L + \frac{1}{EI_z}[-\tfrac{1}{6}V(0)L^3 + \tfrac{1}{48}PL^3 - \tfrac{1}{8}BL^2],$$

$$0 = \tfrac{1}{2}\theta(0)L + \frac{1}{EI_z}[-\tfrac{1}{48}V(0)L^3],$$

expressed in terms of the three unknowns $V(0)$, P, and $\theta(0)$. The solution
is

$$V(0) = -\frac{B}{L}, \qquad P = 0, \qquad \theta(0) = -\frac{BL}{24EI_z}.$$

Substituting these quantities into the expressions for V, M, θ, and w
yields

$$V = -\frac{B}{L}\,\text{N},$$

$$M = B\begin{cases} \dfrac{x}{L}\,\text{N·m}, & x < \tfrac{1}{2}L, \\[2mm] \dfrac{x}{L} - 1\,\text{N·m}, & x > \tfrac{1}{2}L, \end{cases}$$

$$\theta = \frac{B}{EI_z}\begin{cases} -\dfrac{L}{24} + \dfrac{x^2}{2L}\,\text{rad}, & x > \tfrac{1}{2}L, \\[3mm] \dfrac{11L}{24} + \dfrac{x^2}{2L} - x\,\text{rad}, & x > \tfrac{1}{2}L, \end{cases}$$

$$w = \frac{B}{EI_z}\begin{cases} -\dfrac{Lx}{24} + \dfrac{x^3}{6L}\,\text{m}, & x > \tfrac{1}{2}L, \\[3mm] \dfrac{11Lx}{24} + \dfrac{x^3}{6L} - \dfrac{x^2}{2} - \dfrac{L^2}{8}\,\text{m}, & x > \tfrac{1}{2}L. \end{cases}$$

Notice that

$$\theta(0) = \theta(L) = -\frac{B}{24EI_z} = -\frac{1}{30} = -0.033\ \text{rad},$$

that

$$\theta(\tfrac{1}{2}L) = \frac{B}{12EI_z} = -2\theta(0),$$

and that

$$V(0) = V(L) = -\frac{B}{L} = -100\ \text{N}.$$

Problems

| 13.5-1 | Determine the shear force, internal bending moment, slope, and displacement of the beam shown as functions of x. The beam has a square cross section with thickness a. Let $L = 1\,\text{m}$, $f_0 = 100\,\text{N/m}$, $a = 0.1\,\text{m}$, and $E = 15\,\text{MN/m}^2$.

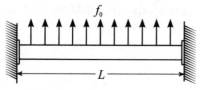

Problem 13.5-1   **Problem 13.5-2**

| 13.5-2 | Determine the shear force, internal bending moment, slope, and displacement of the beam shown as functions of x. The beam has a solid, circular cross section with radius R. Let $L = 1\,\text{m}$, $F = 150\,\text{N}$, $R = 0.05\,\text{m}$, and $E = 15\,\text{MN/m}^2$.

| 13.5-3 | Determine the shear force, internal bending moment, slope, and displacement of the beam shown as functions of x. The beam has a square cross section with thickness a. Let $L = 1\,\text{m}$, $B = 200\,\text{N·m}$, $a = 0.1\,\text{m}$, and $E = 15\,\text{MN/m}^2$.

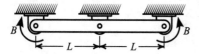

Problem 13.5-3

CHAPTER 14

General Stress–Strain Relationships

14.1 PLANE STRESS

Theory

Chapters 11–13 treated special situations. Bars undergoing longitudinal displacements, shafts undergoing twisting angles, and beams undergoing bending displacements were examined. In Chapters 15 and 16, more general problems will be treated. Problems that involve combinations of bending, twisting, and longitudinal deformations will be considered. Before doing this though, stress and strain need to be examined in a more general manner. This chapter examines **planar stresses on inclined surfaces**, and determines at which incline angles the stresses are the greatest. This is of central importance in structural analysis and design.

A general differential element is shown in Fig. 14.1-1. The lengths of the sides of the differential element are dx, dy, and dz. As shown, the x–z surfaces have normal and shear stresses σ_y and τ_{yx}, and the y–z surfaces have normal and shear stresses σ_x and τ_{xy}. The first subscript on τ indicates the axis of the normal to the surface and the second subscript indicates the direction of the stress. The stresses on the **positive surfaces** (the surfaces in view) act in the positive directions and the stresses on the

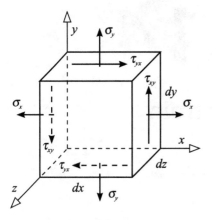

Figure 14.1-1

negative surfaces (the hidden surfaces) act in the negative directions. The resultant moment about the z axis is

$$0 = dx\, \tau_{xy}(dy\, dz) - dy\, \tau_{yx}(dx\, dz) = (\tau_{xy} - \tau_{yx})(dx\, dy\, dz),$$

from which it follows that

$$\tau_{xy} = \tau_{yx}.$$

Similarly, by taking moments about the x and y axis, it is concluded that

$$\tau_{xy} = \tau_{yx}, \qquad \tau_{yz} = \tau_{zy}, \qquad \tau_{zx} = \tau_{xz}. \tag{14.1-1}$$

Equation (14.1-1) states that the **shear stresses are symmetric**.

In this section, our attention is restricted to the situation in which no stresses act in the z direction. This situation is called **plane stress**. It follows from the symmetry of the shear stresses, Eq. (14.1-1), that no stresses act on the x–y surfaces in plane stress problems. Thus, it will be sufficient for our purposes to write the shear stress as τ and to drop its subscripts. Plane stress problems can arise in structures that are subjected to the appropriate loading conditions and in structures in which one length dimension is **large** compared with the other two (as in bars, shafts, and beams).

Now consider the wedge shown in Fig. 14.1-2. The normal and shear stresses on the inclined surface will be expressed in terms of the normal

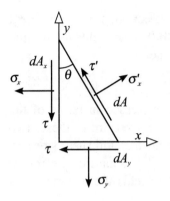

Figure 14.1-2

and shear stresses in the x and y directions. Summing forces in the x' and y' directions yields

$$0 = \sigma' \, dA - (\sigma_x \cos\theta + \tau \sin\theta) \, dA_x - (\sigma_y \sin\theta + \tau \cos\theta) \, dA_y,$$

$$0 = \tau' \, dA + (\sigma_x \sin\theta - \tau \cos\theta) \, dA_x - (\sigma_y \cos\theta - \tau \sin\theta) \, dA_y,$$

where

$$dA_x = dy \, dz, \quad dA_y = dx \, dz, \quad \frac{dA_x}{dA} = \cos\theta, \quad \frac{dA_y}{dA} = \sin\theta.$$

Dividing by dA yields

$$\sigma'_x = \sigma_x \cos^2\theta + \sigma_y \sin^2\theta + 2\tau \cos\theta \sin\theta, \qquad (14.1\text{-}2a)$$

$$\tau' = (\sigma_y - \sigma_x)\cos\theta \sin\theta + \tau(\cos^2\theta - \sin^2\theta). \qquad (14.1\text{-}2b)$$

Substituting the trigonometric identities $\sin 2\theta = 2\cos\theta \sin\theta$ and $\cos 2\theta = \cos^2\theta - \sin^2\theta$ into Eqs. (14.1-2a,b) yields the stresses on the inclined surface:

$$\boxed{\begin{aligned} \sigma' &= \tfrac{1}{2}(\sigma_x + \sigma_y) + \tfrac{1}{2}(\sigma_x - \sigma_y)\cos 2\theta + \tau \sin 2\theta, \qquad (14.1\text{-}3a)\\ \tau' &= -\tfrac{1}{2}(\sigma_x - \sigma_y)\sin 2\theta + \tau \cos 2\theta. \qquad (14.1\text{-}3b) \end{aligned}}$$

The stress σ' is the normal stress in the x' direction; that is, $\sigma' = \sigma'_x$. Notice that replacing θ in Eqs. (14.1-3a,b) with $\theta + \frac{1}{2}\pi$ yields the normal stress on the other side of the differential element:

$$\sigma'_y = \tfrac{1}{2}(\sigma_x + \sigma_y) - \tfrac{1}{2}(\sigma_x - \sigma_y)\cos 2\theta - \tau \sin 2\theta. \qquad (14.1\text{-}4)$$

In Eqs. (14.1-3a,b), the inclined stresses are expressed in terms of the incline angle. The incline angle can be removed from these two equations to create an equation that relates the normal and shear stresses. Subtracting $\sigma_{avg} = \frac{1}{2}(\sigma_x + \sigma_y)$ from Eq. (14.1-3a), squaring the result, squaring Eq. (14.1-3b), and adding the results yields the equation of a circle with radius R:

$$\boxed{(\sigma' - \sigma_{avg})^2 + \tau'^2 = R^2,} \qquad (14.1\text{-}5)$$

in which

$$\boxed{R = \sqrt{\left(\frac{\sigma_x - \sigma_y}{2}\right)^2 + \tau^2}.} \qquad (14.1\text{-}6)$$

This circle is called **Mohr's circle**, as shown in Fig. 14.1-3. Notice that the center of the circle is shifted to the right by an amount σ_{avg}. The

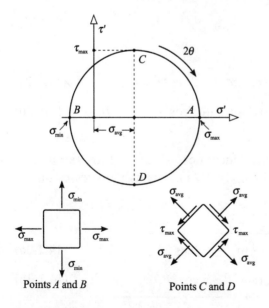

Points A and B Points C and D *Figure 14-1-3*

stress states (σ', τ') are points on the circle. Rotating **clockwise** on the circle by an amount 2θ corresponds to rotating the inclined surface **counterclockwise** by an amount θ. Points on the circle that are $180°$ apart correspond to surfaces that are $90°$ apart. Notice that the two points A and B on the circle that intersect the σ' axis correspond to the differential element that has maximum and minimum normal stresses and zero shear stresses. The two points C and D at the top and bottom of the circle correspond to the differential element that has the average normal stresses and maximum shear stresses. These two differential elements are depicted in Fig. 14.1-3. The normal and shear stresses on these differential elements are given by

$$\sigma_{\text{max,min}} = \tfrac{1}{2}(\sigma_x + \sigma_y) \pm \sqrt{\left(\frac{\sigma_x - \sigma_y}{2}\right)^2 + \tau^2}, \qquad \tau' = 0, \qquad (14.1\text{-}7a)$$

$$\sigma'_x = \sigma'_y = \tfrac{1}{2}(\sigma_x + \sigma_y), \qquad \tau_{\text{max,min}} = \pm\sqrt{\left(\frac{\sigma_x - \sigma_y}{2}\right)^2 + \tau^2}. \qquad (14.1\text{-}7b)$$

Structures tend to fail in the direction of maximum shear stress or maximum normal stress, depending on the **micromechanical properties** of the material. This book does not consider micromechanical properties, nor do most structural engineers. For most applications it is sufficient to prevent failure by requiring that the maximum normal stress and the maximum shear stress be below the ultimate level given in the associated stress–strain curve.

Examples

14.1-1   Given the **uniaxial** stress state shown, use Mohr's circle to determine the stress state (σ', τ') for the rotated element. Let $\theta = 30°$, $\sigma_x = 50$ ksi, $\sigma_y = 0$ ksi, and $\tau = 0$ ksi.

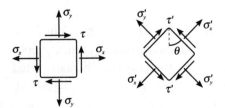

Examples 14.1-1 and 14.1-2 and Problems 14.1-1 and 14.1-2

Solution

The average normal stress is $\sigma_{avg} = 25$ and the radius of Mohr's circle is $R = 25$. Rotating the element counterclockwise 30° as shown corresponds to rotating clockwise 60° on Mohr's circle, starting from the original state of stress. The original state of stress is located at the circle's right intercept, so the rotated state of stress is located at $(\sigma', \tau') = (25 + 25\cos(60°)), -25\sin(60°)) = (37.5, -21.65)$ ksi. Rotating another 180° on Mohr's circle yields the stress $\sigma_y' = 12.5$ ksi.

14.1-2 Given the **biaxial** stress state shown, use Mohr's circle to determine the stress state (σ', τ') for the rotated element. Let $\theta = 45°$, $\sigma_x = 100$ ksi, $\sigma_y = 110$ ksi, and $\tau = 12$ ksi.

Solution

The average normal stress is $\sigma_{avg} = 105$ and the radius of Mohr's circle is $R = 13$. Rotating the element counterclockwise 45° corresponds to rotating clockwise 90° on Mohr's circle, starting from the original state of stress. The original state of stress is an angle β counterclockwise from the circle's right intercept, in which β is determined from the relationship $\tau = R\sin\beta$; that is, $\sin\beta = \frac{12}{13}$, so $\beta = 67.38°$ or $112.62°$. To check which value of β is correct, notice that $\sigma_x = \sigma_{avg} + R\cos\beta$, that is, $100 = 105 + 13\cos\beta$, so $\cos\beta = -\frac{5}{13}$. Thus $\beta = 112.62°$. The rotated state of stress is then $(\sigma', \tau') = (117.00, 5.00)$ ksi. Rotating another 180° on Mohr's circle yields the stress $\sigma_y' = 13$ ksi. Notice that the normal stresses in the rotated element are nearly identical and the shear stress is small in comparison, just as in the original element.

Problems

14.1-1 Given the planar stress state shown, use Mohr's circle to determine the stress state (σ', τ') for the element rotated clockwise 10°. Also, determine the maximum normal stress and the associated angle of rotation of the element to this stress state. Let $\sigma_x = -10$ ksi, $\sigma_y = 0$ ksi, and $\tau = 100$ ksi.

14.1-2 Given the planar stress state shown, use Mohr's circle to determine the stress state (σ', τ') for the element rotated counterclockwise 30°. Also determine the maximum shear stress and the associated angle of rotation of the element to this stress state. Let $\sigma_x = 50$ ksi, $\sigma_y = 50$ ksi, and $\tau = 50$ ksi.

14.2 PLANE STRAIN

Theory

Consider the differential element shown in Fig. 14.2-1, which is assumed to deform in the plane as shown. This is a state of **plane strain**. Plane strain problems arise under the appropriate loading conditions and when one length dimension is **small** compared with the other two (e.g., in-plane loaded thin plates). The question arises how normal and shear strains in the x and y directions are related to the normal and shear strains in the inclined directions. Since the strains are small, the normal strain and the shear strain in the inclined directions, ε' and γ', are determined by examining the effects of ε_x, ε_y, and γ on ε' and γ'. As shown in Fig. 14.2-1,

$$\varepsilon' \, ds = \varepsilon_x \, dx \cos \theta + \varepsilon_y \, dy \sin \theta + \tfrac{1}{2}\gamma \, dy \cos \theta + \tfrac{1}{2}\gamma \, dx \sin \theta. \qquad (14.2\text{-}1)$$

Also, $dx/ds = \cos \theta$ and $dy/ds = \sin \theta$, so dividing by ds yields

$$\varepsilon' = \varepsilon_x \cos^2 \theta + \varepsilon_y \sin^2 \theta + \gamma \sin \theta \cos \theta. \qquad (14.2\text{-}2)$$

Similarly, the shear strain in the inclined direction is determined by adding the individual angle changes:

$$\tfrac{1}{2}\gamma' \, ds = -\varepsilon_x \, dx \sin \theta + \varepsilon_y \, dy \cos \theta - \tfrac{1}{2}\gamma \, dy \sin \theta + \tfrac{1}{2}\gamma \, dx \cos \theta.$$

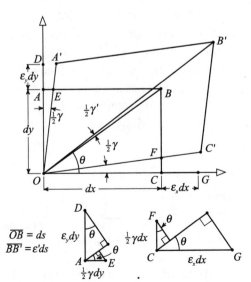

Figure 14.2-1

Dividing by ds yields

$$\tfrac{1}{2}\gamma' = -(\varepsilon_x - \varepsilon_y)\sin\theta\cos\theta + \tfrac{1}{2}\gamma(\cos^2\theta - \sin^2\theta). \qquad (14.2\text{-}3)$$

Substituting the trigonometric identities $\sin 2\theta = 2\sin\theta\cos\theta$ and $\cos 2\theta = \cos^2\theta - \sin^2\theta$ into Eqs. (14.2-2) and (14.2-3) yields

$$\varepsilon' = \tfrac{1}{2}(\varepsilon_x + \varepsilon_y) + \tfrac{1}{2}(\varepsilon_x - \varepsilon_y)\cos 2\theta + \tfrac{1}{2}\gamma\sin 2\theta, \qquad (14.2\text{-}4a)$$

$$\tfrac{1}{2}\gamma' = -\tfrac{1}{2}(\varepsilon_x - \varepsilon_y)\sin 2\theta + \tfrac{1}{2}\gamma\cos 2\theta. \qquad (14.2\text{-}4b)$$

Notice that these equations look like Eqs. (14.1-3a,b). The **plane strain state** $(\varepsilon', \tfrac{1}{2}\gamma')$ is analogous to the plane stress state (σ, τ). Indeed, **Mohr's circle for plane stress can also be used for plane strain.** In Fig. 14.1-3, simply replace σ with ε, and τ with $\tfrac{1}{2}\gamma$.

Examples

14.2-1 Given the **uniaxial** strain state shown, use Mohr's circle for strain to determine the strain state $(\varepsilon', \tfrac{1}{2}\gamma)$ for the rotated element. Let $\theta = 30°$, $\varepsilon_x = 0.05$, $\varepsilon_y = 0$, and $\gamma = 0$.

Example 14.2-1

Solution

The average normal strain is $\varepsilon_{avg} = 0.025$ and the radius of Mohr's circle is $R = 0.025$. Rotating the element counterclockwise 30° as shown corresponds to rotating clockwise 60° on Mohr's circle, starting from the original state of strain. The original state of strain is located at the circle's right intercept, so the rotated state of strain is located at the point B, as shown. The rotated state of strain is $(\varepsilon', \tfrac{1}{2}\gamma') = (0.0375, -0.0216)$. Rotating another 180° on Mohr's circle yields the strain $\varepsilon'_y = 0.011$.

14.2-2 Given the **biaxial** strain state shown, use Mohr's circle to determine the strain state $(\varepsilon', \tfrac{1}{2}\gamma')$ for the rotated element. Let $\theta = 45°$, $\varepsilon_x = 0.1$, $\varepsilon_y = 0.11$, and $\gamma = 0.024$.

Example 14.2-2

Solution

The average normal strain is $\sigma_{avg} = 0.105$ and the radius of Mohr's circle is $R = 0.013$. Rotating the element counterclockwise 45° corresponds to rotating clockwise 90° on Mohr's circle starting from the original state of strain. The original state of strain is an angle β counterclockwise from the circle's right intercept, in which β is determined from the relationship $\frac{1}{2}\gamma = R \sin \beta$, that is, $\sin \beta = 0.012/0.013 = \frac{12}{13}$, so $\beta = 67.38°$ or 112.62°. Notice by looking at ε_x that $\beta = 112.62°$. The rotated state of strain is $(\varepsilon', \frac{1}{2}\gamma') = (0.025 \quad 0.005)$. Rotating another 180° on Mohr's circle yields the strain, $\varepsilon'_y = 0.001$.

Problems

14.2-1 Given the planar strain state shown in Fig. 14.2-1, use Mohr's circle to determine the strain state $(\varepsilon', \frac{1}{2}\gamma')$ for the element rotated clockwise 10°. Also, determine the maximum normal strain and the associated angle of rotation of the element to this strain state. Let $\varepsilon_x = -0.01$, $\varepsilon_y = 0$, and $\gamma = 0.2$.

14.2-2 Given the planar strain state shown in Fig. 14.2-1, use Mohr's circle to determine the strain state $(\varepsilon', \frac{1}{2}\gamma')$ for the element rotated counterclockwise 30°. Also determine the maximum shear strain and the associated angle of rotation of the element to this strain state. Let $\varepsilon_x = 0.05$, $\varepsilon_y = 0.05$, and $\gamma = 0.1$.

14.3 POISSON'S RATIO

Theory

As was found in Chapter 10, the normal strain in the direction of the normal stress is linearly proportional to the normal stress when the deformation is sufficiently small. Such materials were said to exhibit linear elastic behavior. Another linear elastic behavior that a material undergoes is in the **lateral** direction of the normal stress. When the deformation is sufficiently small, a normal strain that is lateral to the normal stress is produced in proportion to the normal strain (see Fig. 14.3-1). In plane stress problems, the normal strains are related to the normal stresses by

$$\varepsilon_x = \frac{1}{E}\sigma_x - \frac{\nu}{E}\sigma_y, \tag{14.3-1a}$$

$$\varepsilon_y = \frac{1}{E}\sigma_y - \frac{\nu}{E}\sigma_x, \tag{14.3-1b}$$

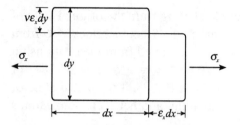

Figure 14.3-1

where v is called **Poisson's ratio**. Notice that a positive normal stress induces a negative normal strain in the lateral direction. Typical values of Poisson's ratio lie between 0 and 0.5.

The behavior of elastic structures is characterized through three material parameters: the modulus of elasticity E, the shear modulus G, and Poisson's ratio v. These three parameters are dependent; the shear modulus can be expressed in terms of E and v. To show this, consider the two differential elements shown in Fig. 14.3-2. The first differential element is in a pure state of shear stress and shear strain, and rotating this differential element 45° produces a differential element that is in a pure state of normal stress and normal strain. Based on Mohr's circle for plane stress, the normal stresses on the two sides of the rotated differential element are opposite and their magnitudes are equal to the magnitude of the shear stresses in the first differential element ($\sigma'_x = \tau$, $\sigma'_y = -\tau$). Based on Mohr's circle for plane strain, the relationship between the shear strain in the first differential element and the normal strain in the

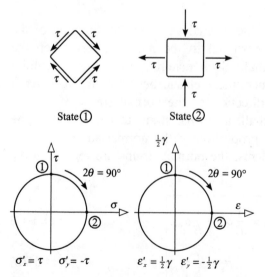

Figure 14.3-2

second differential element is $\varepsilon'_x = \frac{1}{2}\gamma$. Thus, from Eqs. (14.3-1a) and (10.2-2),

$$\tfrac{1}{2}\gamma = \varepsilon'_x = \frac{\sigma'_x}{E} - v\frac{\sigma'_y}{E} = \left(\frac{1+v}{E}\right)\tau = \frac{1}{2G}\tau,$$

so

$$G = \frac{E}{2(1+v)}. \qquad (14.3\text{-}2)$$

Notice that as the Poisson effect decreases, the ratio of the shear modulus to the elastic modulus increases. Materials that have a small Poisson effect are stiffer in shear relative to their normal elastic stiffness.

Examples

14.3-1 Determine the shear modulus of a material for which $E = 17.5$ Mlb/in^2 and $v = 0.2$.

Solution
Using Eq. (14.3-2), $G = 17.5/[2(1+0.2)] = 7.29$ Mlb/in^2.

14.3-2 A thin plate is stretched in its plane. Determine the maximum normal strain of the plate if $\sigma_x = 100$ ksi, $\sigma_y = 200$ ksi, and $\tau = 0$ ksi. Let $E = 1200$ ksi and $v = 0.2$.

Solution
Using Eq. (14.3-1), the normal strains are

$$\varepsilon_x = \frac{1}{12 \times 10^6}100 \times 10^3 - \frac{0.2}{12 \times 10^6}200 \times 10^3 = 0.005,$$

$$\varepsilon_y = \frac{1}{12 \times 10^6}200 \times 10^3 - \frac{0.2}{12 \times 10^6}100 \times 10^3 = \frac{19}{1200} = 0.01583$$

The shear strain is $\gamma = \tau/G = 0$. From Mohr's circle for strain, the average normal strain is 0.0104 and the radius of Mohr's circle is 0.05417. The maximum normal strain is 0.01583. The surface of the rotated element is rotated 180° from the original orientation of the element.

Problems

14.3-1 A block of rubber is compressed with a force F as shown. Determine the lateral displacement Δz of the rubber, in the z direction, at $(x, y, z) = (0, 0, 0)$. Let $a = 18$ inches, $b = 3$ inches, $c = 12$ inches, $F = 100$ lb, $E = 500$ psi, and $v = 0.2$. Notice that the longitudinal displacement Δx is larger than the lateral displacement Δz and that both are significantly larger than the out-of-plane displacement Δy.

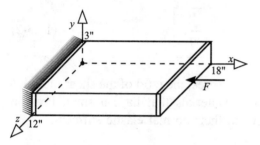

Problem 14.3-1

14.3-2 A thin plate is stretched in its plane as shown. Determine the maximum shear strain of the plate if $\sigma_x = 100$ ksi, $\sigma_y = 100$ ksi, and $\tau = 100$ ksi. Let $E = 1200$ ksi and $v = 0.2$.

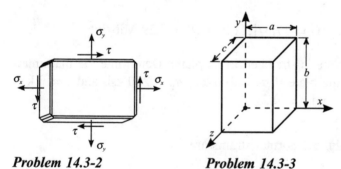

Problem 14.3-2 *Problem 14.3-3*

14.3-3 In a general three-dimensional stress state, Eq. (14.3-1) is extended to

$$\varepsilon_x = \frac{\sigma_x - v\sigma_y - v\sigma_z}{E}, \qquad \varepsilon_y = \frac{\sigma_y - v\sigma_x - v\sigma_z}{E}, \qquad \varepsilon_z = \frac{\sigma_z - v\sigma_x - v\sigma_y}{E}.$$

Assume that a block having length dimensions a, b, and c as shown is subjected to the stresses σ_x, σ_y, and σ_z. Determine the value of Poisson's ratio for which the block will **not** undergo a volume change. Assume that the strains are small ($\ll 1$).

CHAPTER 15

Combined Deformation and Stress

Theory

Consider **structural systems** composed of assemblages of long slender **members** connected together. Each member can undergo longitudinal deformation, torsional deformation, and bending deformation. The structural systems are subjected to external forces and external moments, expressed mathematically as external forces per unit length. This chapter presents an organized method of determining the internal forces, the internal moments, the deformations, and the internal stresses in these structural systems.

Begin by labeling the structural members as member 1, member 2,..., member n. A coordinate system is set up for each member. These coordinate systems are called **local coordinate systems**. In addition, one coordinate system of your choice is designated as a **global coordinate system** for the structural system.

A typical structural member is shown in Fig. 15.1-1. As shown, the member is subjected to external forces per unit length f_x, f_y, and f_z, and an external moment per unit length T_x. The external forces per unit length f_x, f_y, and f_z represent distributed or discrete forces in the x, y, and z directions, and they can contribute as point moments about the y and z axes. Distributed external moments per unit length about the y and z axes are not considered because they are rare. The external moment per unit length T_x represents either a distributed moment or a point moment about the x axis.

The internal forces and the internal moments in the cross section are denoted by P_x, P_y, P_z, M_x, M_y, and M_z, respectively. The member undergoes displacements u_x, u_y, and u_z, and slopes ϕ_x, ϕ_y, and ϕ_z.

323

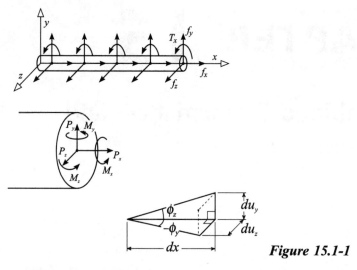

Figure 15.1-1

The external force per unit length f_x, the internal force P_x, and the displacement u_x are associated with longitudinal behavior. The external force per unit length f_y, the internal force P_y, the external bending moment B_z, the internal moment M_z, the displacement u_y, and the slope $\phi_z = du_y/dx$ are associated with bending behavior in the x–y plane. The external force per unit length f_z, the internal force P_z, the external bending moment B_y, the internal moment M_y, the displacement u_z, and the slope $\phi_y = -du_z/dx$ are associated with bending behavior in the x–z plane. The external moment per unit length T_x, the internal moment M_x, and the slope ϕ_x are associated with torsional behavior.

Equations (11.1-7), (12.1-8), and (13.2-3a,b) are collected together to express the internal forces and the internal moments in terms of the external forces per unit length and the external moment per unit length, as

$$P_x(x) = P_x(0) - \int_0^x f_x\,dx, \qquad (15.1\text{-}1a)$$

$$P_y(x) = P_y(0) - \int_0^x f_y\,dx, \qquad (15.1\text{-}1b)$$

$$P_z(x) = P_z(0) - \int_0^x f_z\,dx, \qquad (15.1\text{-}1c)$$

$$M_x(x) = M_x(0) - \int_0^x T_x\,dx, \qquad (15.1\text{-}1d)$$

$$M_y(x) = M_y(0) + \int_0^x P_z\,dx, \qquad (15.1\text{-}1e)$$

$$M_z(x) = M_z(0) - \int_0^x P_y\,dx \qquad (15.1\text{-}1f)$$

Equations (11.1-6), (12.1-7), and (13.5-7c,d) are collected together to express the displacements and slopes in terms of the internal forces and internal moments as

$$\phi_x(x) = \phi_x(0) + \int_0^x \frac{M_x}{JG}\, dx, \tag{15.1-2a}$$

$$\phi_y(x) = \phi_y(0) + \int_0^x \frac{M_y}{EI_y}\, dx, \tag{15.1-2b}$$

$$\phi_z(x) = \phi_z(0) + \int_0^x \frac{M_z}{EI_z}\, dx, \tag{15.1-2c}$$

$$u_x(x) = u_x(0) + \int_0^x \frac{P_x}{AE}\, dx, \tag{15.1-2d}$$

$$u_y(x) = u_y(0) + \int_0^x \phi_z\, dx, \tag{15.1-2e}$$

$$u_z(x) = u_z(0) - \int_0^x \phi_y\, dx, \tag{15.1-2f}$$

Equations (15.1-1) and (15.1-2) together with the **external** boundary conditions and the **internal** boundary conditions (between the members) are used to determine the internal forces, the internal moments, the displacements, and the slopes. From Eqs. (11.1-1), (12.1-7) or (12.3-6), and (13.1-3), the internal stresses in the cross section are given by

$$\sigma_x = \frac{P_x}{A} - \frac{M_z}{I_z}y + \frac{M_y}{I_y}z, \tag{15.1-3a}$$

$$\tau_{xy} = \frac{P_y Q_z}{I_z t_z} + \frac{M_x}{J}z, \tag{15.1-3b}$$

$$\tau_{xz} = \frac{P_z Q_y}{I_y t_y} + \frac{M_x}{J}y. \tag{15.1-3c}$$

By virtue of linear behavior, the combined state of stress given in Eqs. (15.1-3a–c) was determined by simply adding together the longitudinal, torsional, and bending effects.

Examples

$\boxed{\text{15.1-1}}$ A rectangular beam is subjected to an axial force F and a bending moment B as shown in the figure overleaf. Determine the normal stresses and the shear stresses in the y–z cross sections of the beam as functions of x, y, and z. Let $F = 100\,\text{lb}$, $B = 50\,\text{lb·in}$, $a = 1$ inch and $L = 12$ inch.

Solution

The structural system consists of one member: a beam that bends in the y direction and undergoes longitudinal deformations in the x direction. The associated boundary conditions and external forces and moments per unit length are

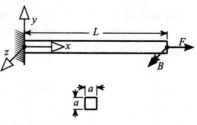

Example 15.1-1

$$u_x(0) = 0, \quad u_y(0) = 0, \quad \phi_z(0) = 0, \quad P_x(L) = F, \quad P_y(L) = 0,$$
$$M_z(L) = B, \quad f_x = 0, \quad f_y = 0.$$

Note that these are the only boundary conditions that come into play. Considering bending only in the y direction and longitudinal deformation, Eqs. (15.1-1a,b,f) and the external forces yield

$$P_x = P_x(0), \quad P_y = P_y(0), \quad M_z = M_z(0) - P_y(0)x.$$

From Eqs. (15.1-2c–e),

$$\phi_z = \frac{1}{EI_z}\left[M_z(0)x - P_y(0)\frac{x^2}{2}\right],$$

$$u_x = \frac{P_x(0)x}{AE}, \quad u_y = \frac{1}{EI_z}\left[M_z(0)\frac{x^2}{2} - P_y(0)\frac{x^3}{6}\right].$$

From the boundary conditions at $x = L$,

$$P_x(0) = F, \quad P_y(0) = 0, \quad M_z(0) - (0)L = M_z(0) = B.$$

The internal forces, internal moments, displacements, and slopes are given by the expressions

$$P_x = F, \quad P_y = 0, \quad M_z = B, \quad \phi_z = \frac{B}{EI_z}x,$$

$$u_x = \frac{F}{AE}x, \quad u_y = \frac{B}{2EI_z}x^2.$$

The moment of inertia about the neutral axis of the square cross section is $I_z = \frac{1}{12}a^4$. Substituting values yields

$$P_x = 100\,\text{lb}, \qquad P_y = 0\,\text{lb}, \qquad M_z = 50\,\text{lb·in}$$

$$\phi_z = \frac{50}{12\,000(1.0^4/12)}x = 0.05\,\text{rad}$$

$$u_x = \frac{100}{1.0^2(12\,000)}x = 0.0083x\,\text{inch}$$

$$u_y = \frac{50}{2(12\,000)(1.0^4/12)}x^2 = 0.025x^2\,\text{inch.}$$

Notice that the bending displacement is three times greater than the longitudinal displacement. Given that

$$Q_z = \int y\,dA = \int_y^{a/2} ya\,dy = a^3\left[\frac{1}{8} - \frac{1}{2}\left(\frac{y}{a}\right)^2\right],$$

the normal and shear stresses in the beam are determined from Eqs. (15.1-3a–c), to give

$$\sigma_x \frac{P_x}{A} - \frac{M_z}{I_z}y = \frac{F}{a^2} - \frac{12B}{a^4}y = 100 - 600y\,\text{psi},$$

$$\tau_{xy} = \frac{P_y Q_z}{I_z t_z} = \frac{Fa^3\left[\frac{1}{8} - \frac{1}{2}(y/a)^2\right]}{\left(\frac{1}{12}a^4\right)a} = 75 - 300y^2\,\text{psi},$$

$$\tau_{xz} = 0\,\text{psi.}$$

15.1-2   A round beam is subjected to a bending force and a torsional moment as shown. Determine the normal and shear stresses in the y–z cross sections of the beam as functions of x, y, and z. Let $F = 100\,\text{N}$, $B = 500\,\text{N·m}$, $R = 0.01\,\text{m}$, and $L = 0.8\,\text{m}$.

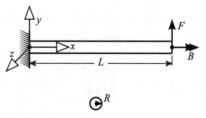

Example 15.1-2

Solution

The structural system consists of one member: a beam that bends in the y direction and twists about the x axis. The associated boundary conditions and external forces per unit length and external moments per unit length are

$$\phi_x(0) = 0, \qquad u_y(0) = 0, \qquad \phi_z(0) = 0, \qquad M_x(L) = B, \qquad P_y(L) = F,$$

$$M_z(L) = 0, \qquad T_x = 0, \qquad f_y = 0.$$

Considering bending only in the y direction and torsion, the external force per unit length and the external moment per unit length are substituted into Eqs. (15.1-1b,d,f) to give

$$P_y = P_y(0), \qquad M_x = M_x(0), \qquad M_z = M_z(0) - P_y(0)x.$$

From the boundary conditions at $x = L$,

$$P_y(0) = F, \qquad M_x(0) = B, \qquad M_z(0) - P_y(0)L = 0,$$

so $M_z(0) = FL$. The internal forces and moments are thus

$$P_y = F, \qquad M_x = B, \qquad M_z = F(L - x).$$

From Eqs. (15.1-2a,c,e) and the boundary conditions at $x = 0$, the slopes and displacements are

$$\phi_x = \frac{B}{JG}x, \qquad \phi_z = \frac{F}{EI_z}(Lx - \tfrac{1}{2}x^2), \qquad u_y = \frac{F}{EI_z}(\tfrac{1}{2}Lx^2 - \tfrac{1}{6}x^3).$$

The area moment of inertia of the round cross section about the neutral axis is

$$I_z = \int y^2 \, dA = \int_0^R \left(\int_0^{2\pi} r^2 \sin^2 \theta \, r \, d\theta \right) dr = \int_0^R r^3 \, dr \int_0^{2\pi} \sin^2 \theta \, d\theta = \tfrac{1}{4}\pi R^4,$$

The polar moment of the round cross section is

$$J = \int r^2 \, dA = \int_0^R r^2 (2\pi r \, dr) = \tfrac{1}{2}\pi R^4,$$

and

$$Q_z = \int_y^{top} y \, dA = 2 \int_{\theta_0}^{\pi/2} \left(\int_{y/\sin\theta}^{R} r\sin\theta \, r \, dr \right) d\theta$$

$$= 2 \int_{\theta_0}^{\pi/2} \sin\theta \left(\int_{y/\sin\theta}^{R} r^2 \, dr \right) d\theta$$

$$= \tfrac{2}{3} \int_{\theta_0}^{\pi/2} (R^3 \sin\theta - y^3 \operatorname{cosec}^2 \theta) \, d\theta = \tfrac{2}{3} [-R^3 \cos\theta + y^3 \cot\theta]_{\theta_0}^{\pi/2}$$

$$= \frac{2R^3}{3} \left[1 - \left(\frac{y}{R} \right)^2 \right]^{3/2},$$

in which $y = R\sin\theta_0$ and $\cos\theta_0 = \sqrt{1 - \sin^2\theta_0}$. From Eqs. (15.1-3a–c), and $t_z = 2(R/\cos\theta_0)$ the normal stresses and the shear stresses are

$$\sigma_x = -\frac{M_z}{I_z} y = -\frac{4F}{\pi R^4} y(L - x) = -12\,732 y(L - x)\,\text{MN/m}^2,$$

$$\tau_{xy} = \frac{P_y Q_z}{I_z t_z} = \frac{4F}{3\pi R^2} \left[1 - \left(\frac{y}{R} \right)^2 \right] = 424 \left[1 - \left(\frac{y}{R} \right)^2 \right] \text{kN/m}^2,$$

$$\tau_{xz} = 0.$$

15.1-3 A bent beam is subjected to the force F shown. Determine the normal and shear stresses in the round cross sections of the beam as functions of $x, y,$ and z. Let $F = 100\,\text{N}$, $R = 0.01\,\text{m}$, $L_1 = 0.8\,\text{m}$, $L_2 = 1.2\,\text{m}$, and $E = 12$ kN/m^2.

Solution
The structural system is composed of two structural members; each member is a beam that undergoes bending in the z direction (in local coordinates) and twisting about the x axes (in local coordinates). Notice that the z axes for both structural members are the same but the x and y coordinates are not the same for

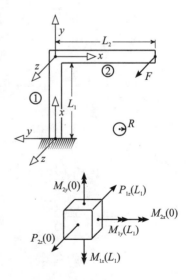

Example 15.1-3

each member. Quantities associated with a particular structural member are indicated by the first subscript. For example, P_{1x} is the internal force in the x direction of the first structural member and M_{2y} is the internal bending moment about the y axis of the second structural member.

The boundary conditions associated with bending in the z direction can involve P_z, M_y, ϕ_y, and u_z, and the boundary conditions associated with torsion involve M_x and ϕ_x. The boundary conditions, the external forces per unit length associated with bending, and the external moments per unit length associated with torsion are as follows:

External : $\phi_{1y}(0) = 0,$ $u_{1z}(0) = 0,$ $\phi_{1x}(0) = 0,$ $P_{2z}(L_2) = F,$
$M_{2y}(L_2) = 0,$ $M_{2x}(L_2) = 0;$

Internal : $P_{1z}(L_1) = P_{2z}(0),$ $M_{1y}(L_1) = -M_{2x}(0),$ $M_{1x}(L_1) = M_{2y}(0),$
$\phi_{1y}(L_1) = -\phi_{2x}(0),$ $u_{1z}(L_1) = u_{2z}(0),$ $\phi_{1x}(L_1) = \phi_{2y}(0),$
$f_{1z} = 0,$ $T_{1x} = 0,$ $f_{2z} = 0,$ $T_{2x} = 0.$

The internal boundary conditions above involving internal forces and internal moments were determined from the free body diagram of the corner element shown. Note that moment vectors are distinguished from force vectors in the diagram through the use of double arrows. From Eqs. (15.1-1c–e),

$$P_{1z} = P_{1z}(0), \quad M_{1x} = M_{1x}(0), \quad M_{1y} = M_{1y}(0) + P_{1z}(0)x,$$
$$P_{2z} = P_{2z}(0), \quad M_{2x} = M_{2x}(0), \quad M_{2y} = M_{2y}(0) + P_{2z}(0)x.$$

From Eqs. (15.1-2a,b,f) and three of the twelve boundary conditions,

$$\phi_{1x} = \frac{M_{1x}(0)}{JG}x, \quad \phi_{1y} = \frac{1}{EI_z}[M_{1y}(0)x + \tfrac{1}{2}P_{1z}(0)x^2],$$

$$u_{1z} = -\frac{1}{EI_y}[\tfrac{1}{2}M_{1y}(0)x^2 + \tfrac{1}{6}P_{1z}(0)x^3],$$

$$\phi_{2x} = \phi_{2x}(0) + \frac{M_{2x}(0)}{JG}x, \quad \phi_{2y} = \phi_{2y}(0) + \frac{1}{EI_z}[M_{2y}(0)x + \tfrac{1}{2}P_{2z}(0)x^2],$$

$$u_{2z} = u_{2z}(0) - \phi_{2y}(0)x - \frac{1}{EI_y}[\tfrac{1}{2}M_{2y}(0)x^2 + \tfrac{1}{6}P_{2z}(0)x^3].$$

The twelve expressions above are expressed in terms of nine unknowns. The nine unknowns are determined from the nine boundary conditions that have not yet been invoked. Considering these remaining boundary conditions yields the nine unknowns:

$$P_{2z}(0) = F,$$

$$M_{2y}(0) + P_{2z}(0)L_2 = 0, \qquad \text{so} \qquad M_{2y}(0) = -FL_2,$$

$$M_{2x}(0) = 0, \qquad P_{1z}(0) = F,$$

$$M_{1y}(0) + FL_1 = 0, \qquad \text{so} \qquad M_{1y}(0) = -FL_1,$$

$$M_{1x}(0) = -FL_2, \qquad \phi_{2x}(0) = \frac{FL_1^2}{2EI_y},$$

$$u_{2z}(0) = -\frac{1}{EI_y}\left(-\frac{FL_1^3}{2} + \frac{FL_1^3}{6}\right) = \frac{FL_1^3}{3EI_y}, \qquad \phi_{2y}(0) = -\frac{FL_1L_2}{JG}.$$

The forces, moments, slopes, and displacements in the first member are

$$P_{1z} = F, \qquad M_{1x} = -FL_2, \qquad M_{1y} = F(x - L_1),$$

$$\phi_{1y} = \frac{F}{2EI_y}x(x - 2L_1), \qquad \phi_{1x} = -\frac{FL_2}{JG}x, \qquad u_{1z} = \frac{F}{6EI_y}x^2(x - 3L_1).$$

The forces, moments, slopes, and displacements in the second member are

$$P_{2z} = F, \qquad M_{2x} = 0, \qquad M_{2y} = F(x - L_2),$$

$$\phi_{2y} = -\frac{FL_1L_2}{JG} + \frac{F}{2I_y}x(x - 2L_2), \qquad \phi_{2x} = \frac{FL_1^2}{2EI_y},$$

$$u_{2z} = \frac{FL_1^3}{3EI_y} + \frac{FL_1L_2}{JG} - \frac{F}{6EI_y}x^2(x - 3L_2).$$

From Eqs. (15.1-3a–c), the normal stresses and shear stresses in the first member are

$$\sigma_{1x} = \frac{4F}{\pi R^4}(x - L_1)z, \qquad \tau_{1xy} = -\frac{2FL_2}{\pi R^4}z,$$

$$\tau_{1xz} = \frac{4F}{3\pi R^2}\left[1 - \left(\frac{z}{R}\right)^2\right] - \frac{2FL_2}{\pi R^4}y.$$

The normal stresses and shear stresses in the second member are

$$\sigma_{2x} = \frac{4F}{\pi R^4}(x - L_2)z, \qquad \tau_{2xy} = 0, \qquad \tau_{2xz} = \frac{4F}{3\pi R^2}\left[1 - \left(\frac{z}{R}\right)^2\right].$$

15.1-4   A structural system consists of two identical members pinned together as shown. The system is subjected to a force F that is applied at the pin connection, and the angle between the members is β. The cross section of the members is square, with sides of length a. Determine expressions for the internal forces, internal moments, normal stresses, and shear stresses in terms of the system parameters L, E, a, and F.

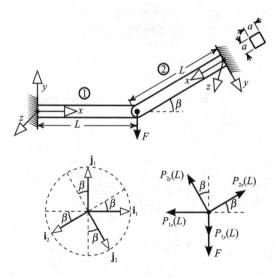

Example 15.1-4

Solution

The applied force induces in both members bending in the y direction and longitudinal deformation. The boundary conditions and the associated external forces are as follows:

External: $\phi_{1z}(0) = 0,$ $u_{1y}(0) = 0,$ $u_{1x}(0) = 0,$ $\phi_{2z}(0) = 0,$

$u_{2y}(0) = 0,$ $u_{2x}(0) = 0;$

Internal: $P_{2x}(L) = \cos\beta \, P_{1x}(L) + \sin\beta \, [P_{1y}(L) + F],$

$P_{2y}(L) = -\sin\beta \, P_{1x}(L) + \cos\beta \, [P_{1y}(L) + F],$

$M_{1z}(L) = 0,$ $M_{2z}(L) = 0,$

$u_{1x}(L) = -\cos\beta \, u_{2x}(L) + \sin\beta \, u_{2y}(L),$

$u_{1y}(L) = -\sin\beta \, u_{2x}(L) - \cos\beta \, u_{2y}(L),$

$f_{1x} = 0,$ $f_{1y} = 0,$ $f_{2x} = 0,$ $f_{2y} = 0.$

The internal boundary conditions involving internal forces were determined from the free body diagram of the pin connection as shown. The internal boundary conditions involving displacements were determined using the transformation (rotation) of the associated unit vectors as shown. From Eqs. (15.1-1a,b,f) and the given external forces per unit length,

$$P_{1x} = P_{1x}(0), \quad P_{1y} = P_{1y}(0), \quad M_{1z} = M_{1z}(0) - P_{1y}(0)x,$$
$$P_{2x} = P_{2x}(0), \quad P_{2y} = P_{2y}(0), \quad M_{2z} = M_{2z}(0) - P_{2y}(0)x,$$

From Eqs. (15.1-2c–e) and the external boundary conditions,

$$\phi_{1z} = \frac{1}{EI}[M_{1z}(0)x - \tfrac{1}{2}P_{1y}(0)x^2],$$

$$u_{1y} = \frac{1}{EI}[\tfrac{1}{2}M_{1z}(0)x^2 - \tfrac{1}{6}P_{1y}(0)x^3], \quad u_{1x} = \frac{P_{1x}(0)}{AE}x,$$

$$\phi_{2z} = \frac{1}{EI}[M_{2z}(0)x - \tfrac{1}{2}P_{2y}(0)x^2],$$

$$u_{2y} = \frac{1}{EI}[\tfrac{1}{2}M_{2z}(0)x^2 - \tfrac{1}{6}P_{2y}(0)x^3], \quad u_{2x} = \frac{P_{2x}(0)}{AE}x.$$

From the internal boundary conditions,

$$M_{1z}(0) = P_{1y}(0)L, \qquad M_{2z}(0) = P_{2y}(0)L,$$

$$P_{2x}(0) = \cos \beta \, P_{1x}(0) + \sin \beta \, P_{1y}(0) + \sin \beta \, F,$$

$$P_{2y}(0) = -\sin \beta \, P_{1x}(0) + \cos \beta \, P_{1y}(0) + \cos \beta \, F,$$

$$\frac{P_{1x}(0)L}{AE} = -\cos \beta \, \frac{P_{2x}(0)L}{AE} + \sin \beta \, \frac{P_{2y}(0)L^3}{3EI_z},$$

$$\frac{P_{1y}(0)L^3}{3EI_z} = -\sin \beta \, \frac{P_{2x}(0)L}{AE} - \cos \beta \, \frac{P_{2y}(0)L^3}{3EI_z}.$$

Substituting $P_{2x}(0)$ and $P_{2y}(0)$ into the last two of these equations yields the two equations

$$P_{1x}(0)(1 + c^2 + \gamma s^2) + P_{1y}(0)cs(1 - \gamma) = -cs(1 - \gamma)F,$$

$$P_{1x}(0)cs(1 - \gamma) + P_{1y}[\gamma(1 + c^2) + s^2] = -(\gamma c^2 + s^2)F,$$

which are expressed in terms of the two unknowns $P_{1x}(0)$ and $P_{1y}(0)$, and where $c = \cos \beta$, $s = \sin \beta$, and

$$\gamma = \frac{AEL^2}{3EI_z} = \left(\frac{2L}{a}\right)^2$$

is the relative stiffness (longitudinal stiffness divided by lateral stiffness) of a member. After some manipulation and reduction, the solution of these two equations is

$$P_{1x}(0) = \frac{-cs\gamma(1 - \gamma)F}{s^2(1 + \gamma^2) + 2\gamma(1 + c^2)}, \qquad P_{1y}(0) = \frac{-[s^2 + \gamma(1 + c^2)]F}{s^2(1 + \gamma^2) + 2\gamma(1 + c^2)}.$$

To check these results, notice that when $\beta = 0$, the force reactions at the boundaries reduce to

$$P_{1x}(0) = 0, \qquad P_{1y}(0) = -\tfrac{1}{2}F, \qquad P_{2x}(0) = 0, \qquad P_{2y}(0) = \tfrac{1}{2}F,$$

and when $\beta = \tfrac{1}{2}\pi$, the force reactions at the boundaries reduce to

$$P_{1x}(0) = 0, \qquad P_{1y}(0) = -\frac{F}{1 + \gamma}, \qquad P_{2x}(0) = 0, \qquad P_{2y}(0) = \frac{\gamma F}{1 + \gamma}.$$

From Eqs. (15.1-3), the general expressions for the normal stresses and the shear stresses in the first member become

$$\sigma_{1x} = \frac{P_{1x}(0)}{a^2} - \frac{12P_{1y}(0)}{a^4}(L-x)y, \quad \tau_{1xy} = \frac{6P_{1y}(0)}{a^2}\left[\frac{1}{4} - \left(\frac{y}{a}\right)^2\right], \quad \tau_{1xz} = 0.$$

The general expressions for the normal stresses and the shear stresses in the second member become

$$\sigma_{2x} = \frac{P_{2x}(0)}{a^2} - \frac{12P_{2y}(0)}{a^4}(L-x)y, \quad \tau_{2xy} = \frac{6P_{2y}(0)}{a^2}\left[\frac{1}{4} - \left(\frac{y}{a}\right)^2\right], \quad \tau_{2xz} = 0.$$

Problems

15.1-1 A beam is subjected to a longitudinal force F and a bending moment B as shown. Determine the normal stresses and the shear stresses in the member. Let $L = 5$ inch, $R = 0.2$ inch, $E = 120$ psi, $F = 150$ lb, and $B = 15$ lb·in.

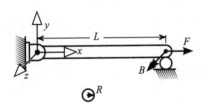

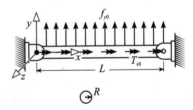

Problem 15.1-1   *Problem 15.1-2*

15.1-2 A beam is subjected to a uniformly distributed longitudinal force and a uniformly distributed torsional moment as shown. Determine the normal stresses and shear stresses in the beam. Let $L = 2.4$ m, $R = 10$ cm, $E = 120$ N/m², $f_{y0} = 15$ N/m and $T_{x0} = 15$ N·m/m.

15.1-3 A structural system consists of two identical beams that are pinned together at one end and welded to a wall at the other end as shown in the figure overleaf. Determine the maximum normal stress and the maximum shear stress in the y–z cross sections. Let $L = 5$ inch, $a = 0.2$ inch, $E = 120$ psi, $F = 150$ lb, and $\beta = 45°$. What are the values of the quantities sought when $\beta = 0°$ and when $\beta = 90°$?

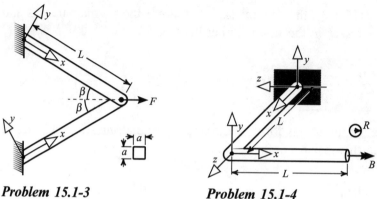

Problem 15.1-3   **Problem 15.1-4**

15.1-4 A structural system is subjected to a torsional moment as shown. Determine expressions for the internal forces and moments, the slopes and displacements, and the normal and shear stresses in terms of the system parameters G, L, E, R, and B.

CHAPTER 16

Special Topics

16.1 BUCKLING

Theory

Static displacements can be **discontinuous** functions of the applied forces. In other words, at some position a small increase in force can produce a large change in position. This situation is called **buckling**. Figure 16.1-1 illustrates this phenomenon. A spring of unstretched length a is attached to a rigid bar of equal length a. The rigid body is subjected to a spring moment M_s and an applied moment M_a. The system is in static equilibrium provided the sum of the moments is zero, that is, $0 = M_s + M_a$. The angle θ of the bar is positive clockwise measured from the bar's axis as shown.

First, assume that the applied moment M_a is zero. The equilibrium angles are located at the intercepts of the spring moment and the horizontal axis of the graph. The angles $\theta = 0°$ and $\theta = 90°$ are stable equilibrium angles and the angles $45°$ and $225°$ are unstable equilibrium angles. Notice that when $dM_s/d\theta < 0$ at any of these equilibrium points, the spring moment is a **restoring** moment and the equilibrium angles are **stable**.

Next, let the dashed line have an intercept equal to the negative of the applied moment and assume that the applied moment M_a is non-zero. The dashed line crosses the axis at $-M_a$. The equilibrium angles are located at the intersection points of the dashed line and the spring

337

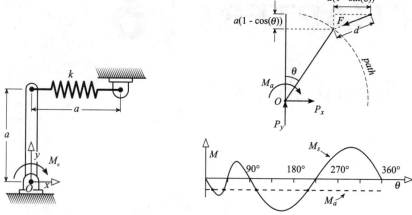

Figure 16.1-1

moment. The dashed line moves downward as the applied moment is increased. The equilibrium angle changes continuously as the applied moment is increased, until a point is reached at which there is a discontinuity (where M_a meets M_s at one point on the first dip); at this point, the slightest increase in the applied moment causes the equilibrium angle to move over from $0°$ to the neighborhood of $90°$. The finite change in a structure's position (or angle) due to an infinitesimal change in the applied force is called **buckling**.

Buckling is undesirable in many structural systems (but not all, like in switches), and can be considered a **failure mechanism**. In many systems, buckling accompanies prohibitively high stresses, resulting in failure. This section examines buckling in structural members that undergo combined longitudinal deformations and bending deformations.

Consider a structural member subjected to an applied force F at its end, as shown in Fig. 16.1-2. Whereas in Chapter 15, the longitudinal forces were assumed to have no effect on lateral displacements, the longitudinally applied force F shown is assumed to be sufficiently large to allow the possibility of inducing lateral displacements. The resulting internal force and internal moment at x are

$$P = F, \qquad M(x) = F[w(L) - w(x)]. \qquad (16.1\text{-}1)$$

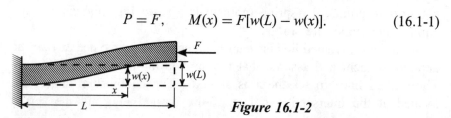

Figure 16.1-2

Substituting Eq. (13.5-5) into Eq. (16.1-1) yields the linear differential equation governing the lateral equilibrium positions of the structural member:

$$\frac{d^2w}{dx^2} + \beta^2 w = \beta^2 w(L), \qquad (16.1\text{-}2)$$

in which

$$\beta^2 = \frac{F}{EI_z}. \qquad (16.1\text{-}3)$$

Notice that $w = 0$ is one of the solutions to Eq. (16.1-2) and that non-zero solutions to Eq. (16.1-2) are possible too, depending on the applied force F.

Equation (16.1-2) is subject to boundary conditions that depend on the particular problem. The example problems provide further clarification and show how to solve the linear differential equation, Eq. (16.1-2).

Examples

16.1-1   A beam is subjected to the applied force F as shown. Determine the equilibrium positions (i.e., buckling modes) and the associated applied forces (i.e., buckling loads) of the beam.

Solution
The boundary conditions associated with bending are

$$w(0) = 0, \qquad w(L) = 0, \qquad M(0) = 0, \qquad M(L) = 0.$$

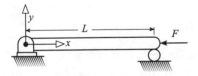

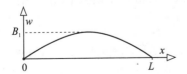

Example 16.1-1

Thus the right-hand side of Eq. (16.1-2) is zero. With the right-hand side equal to zero, Eq. (16.1-2) is called a **homogeneous** linear differential equation. By substitution, it is a simple matter to verify that the solution of Eq. (16.1-2) is of the general form

$$w = A \cos \beta x + B \sin \beta x,$$

where A and B are constants that are determined from the boundary conditions

$$w(0) = 0, \qquad M(0) = EI_z \frac{d^2 w(x)}{dx^2}\bigg|_{x=0}.$$

Substituting these boundary conditions into the general form of the solution yields the two equations

$$0 = A, \qquad 0 = -B\beta^2 \sin \beta L.$$

Notice that one of the solutions of this pair of equations is $A = B = 0$, so $w = 0$ is one of the solutions of the differential equation, as expected. A second possibility is that $B \neq 0$ and $\beta L = r\pi$ $(r = 1, 2, \ldots)$. Then the differential equation has the set of solutions [taking Eq. (16.1-3) into account]

$$w_r = B \sin \beta_r x, \qquad \beta_r = \frac{r\pi}{L} = \sqrt{\frac{P_r}{EI_z}}, \qquad r = 1, 2, \ldots.$$

These solutions are called **buckling modes**, and the corresponding applied forces

$$P_r = EI_z \left(\frac{r\pi}{L}\right)^2$$

are called **buckling loads**. As the applied force is increased from zero, the lateral displacement remains zero until the first buckling load $P_1 = \pi^2 EI_z/L^2$ is reached. At this point, the beam buckles to the first buckling mode.

16.1-2 A beam is subjected to an applied force F as shown. Determine the beam's buckling modes and buckling loads.

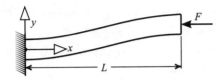

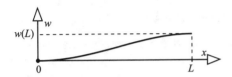

Example 16.1-2

Solution

The boundary conditions associated with bending are

$$w(0) = 0, \qquad \theta(0) = 0, \qquad M(L) = 0, \qquad V(L) = 0.$$

The right-hand side of Eq. (16.1-2) is $\beta^2 w(L)$. Since the right-hand side is non-zero, Eq. (16.1-2) is called a **non-homogeneous** linear differential equation. Linear differential equations are not treated systematically in this book. But it is useful to know that the general solutions are the sum of **homogeneous** solutions and **particular** solutions. By substitution, it is a simple matter to verify that one of the particular solutions of Eq. (16.1-2) is $w = w(L)$. The associated homogeneous solutions (letting the right-hand side equal zero) are of the general form

$$w = A \cos \beta x + B \sin \beta x,$$

where A and B are constants that are determined from the boundary conditions. Adding the particular solution to the homogeneous solutions yields a general form of the solution of the non-homogeneous equation:

$$w = A \cos \beta x + B \sin \beta x + w(L).$$

The boundary conditions $w(0) = 0$ and $\theta(0) = dw(x)/dx|_{x=0}$ and $w(L)$ are substituted into the general form of the non-homogeneous solution to yield the three equations

$$0 = A + w(L),$$
$$0 = \beta B,$$
$$w(L) = A \cos \beta L + B \sin \beta L + w(L).$$

These three equations are expressed in terms of the three unknowns A, B, and $w(L)$. One possible solution is $w = 0$, as expected. The second possibility yields the three unknowns $A = -w(L)$, $B = 0$, and $\cos \beta_r L = 1$, $r = 1, 2, \ldots$, and the associated solutions

$$w_r = w(L)(1 - \cos \beta_r x), \qquad \beta_r = \frac{(2r - 1)\pi}{2L} = \sqrt{\frac{P_r}{EI_z}}, \qquad r = 1, 2, \ldots.$$

These solutions are the buckling modes, and the corresponding applied forces

$$P_r = EI_z \left[\frac{(2r - 1)\pi}{2L}\right]^2$$

are the buckling loads. As the applied force is increased from zero, the lateral displacement remains zero until the first buckling load $P_1 = \pi^2 EI_z/4L^2$ is reached. At this point, the beam buckles to the first buckling mode.

Problems

16.1-1 Determine the buckling modes and the associated buckling loads of the beam shown.

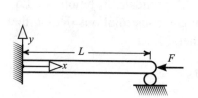

Problem 16.1-1   **Problem 16.1-2**

16.1-2 Determine the buckling modes and the associated buckling loads of the beam shown. Notice that $EI(dw^3/dx^3) = Kw$ at $x = 2$, and that the right side of Eq. 16.1-2 is $[\beta^2 - (\alpha/L)^2(1 - x/L)]w(4)$ in which $\beta^2 = F/EI_z$ and $\alpha^2 = kL^3/EI_z$.

16.2 THIN-WALLED PRESSURE VESSELS

Theory

This section examines thin-walled pressure vessels that are spherical and cylindrical.

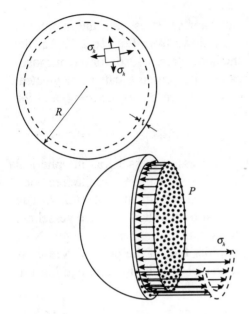

Figure 16.2-1

SPHERICAL PRESSURE VESSELS

The stresses in spherical thin-walled pressure vessels, because of spherical symmetry, can be determined with relative ease. One half of a pressure vessel is shown in Fig. 16.2-1. The pressure vessel is in static equilibrium provided

$$0 = -p(\pi R^2) + \sigma_h(2\pi Rt), \tag{16.2-1}$$

where p is the relative pressure of the fluid inside the vessel, R is the radius of the vessel, σ_h is the circumferential stress in the vessel, called the **hoop stress**, and t is the vessel's thickness. From Eq. (16.2-1),

$$\sigma_h = \frac{pR}{2t}. \tag{16.2-2}$$

By symmetry, the hoop stress is the same in all of the cross-sectional planes that pass through the origin of the sphere. All of the stresses on the outer surface of the sphere are zero. Thus, the state of stress on the outer surface is planar. On the inner surface, the stress consists of a normal pressure p and no shear stresses.

Let's now determine the maximum normal stress and the maximum shear stress in the pressure vessel. It follows from Eq. (16.2-2) that the pressure p is small compared with the hoop stress σ_h, so, for the purposes of determining the maximum stresses, the pressure can be neglected. Using Mohr's circle, the maximum normal stress is equal to σ_h, and the maximum shear stress is equal to zero.

CYLINDRICAL PRESSURE VESSELS

The stresses in cylindrical thin-walled pressure vessels, as in spherical vessels, can be determined with relative ease. First consider the free body diagram of half of the pressure vessel as shown in Fig. 16.2-2. The circumferential stresses in the cylindrical thin-walled pressure vessel are the same as the circumferential stresses in the spherical vessel. Next consider the free-body diagram of another half of the pressure vessel as shown in Fig. 16.2-2. This pressure vessel half is in static equilibrium provided

$$0 = -p(2Rb) + \sigma_l(2bt), \qquad (16.2\text{-}3)$$

where p is the relative pressure of the fluid inside the vessel, R is the radius of the vessel, σ_l is the **longitudinal stress**, and t is the vessel's thickness. From Eqs. (16.2-2) and (16.2-3),

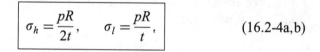

$$\sigma_h = \frac{pR}{2t}, \qquad \sigma_l = \frac{pR}{t}. \qquad (16.2\text{-}4a,b)$$

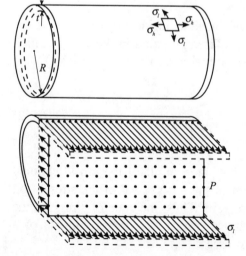

Figure 16.2-2

The stresses on the outer surface of the cylindrical vessel are zero. Thus, the state of stress on the outer surface is planar. On the inner surface, the stress is equal to the pressure p. Again, the pressure p is small compared with the hoop stress σ_h and the longitudinal stress σ_l, so the pressure can be neglected for the purpose of determining maximum stresses in the vessel. Using Mohr's circle, the maximum normal stress is equal to σ_l and the maximum shear stress is equal to

$$\tau_{max} = \frac{pR}{4t} = \tfrac{1}{4}\sigma_l. \tag{16.2-5}$$

Notice that the maximum normal stress in the cylindrical pressure vessel is twice the maximum normal stress in the spherical pressure vessel.

Examples

16.2-1   Determine the maximum permissible pressure in a spherical pressure vessel of radius $R = 3$ ft, and thickness $t = 1$ inch. The yield stress of the material is $\sigma_y = 15$ ksi.

Solution
From Eq. (16.2-2),

$$p = \frac{2t\sigma_y}{R} = \frac{2(\tfrac{1}{12})(15\,000)}{3} = 833 \text{ psi.}$$

16.2-2   A cylindrical pressure vessel of radius $R = 2$ inches is reinforced at its ends (so that the greatest stresses occur away from its ends). Determine the smallest thickness of the vessel that will allow a relative internal pressure of 1000 psi. The yield stress of the material is 20 ksi.

Solution
The maximum stress in the vessel is the longitudinal stress. From Eq. (16.2-4),

$$t_{min} = \frac{pR}{\sigma_y} = \frac{1000(2)}{20\,000} = 0.1 \text{ ft} = 1.2 \text{ inches.}$$

Problems

16.2-1 A cylindrical pressure vessel is made from two pieces that are welded together along an inclined line that is rotated 30° from the longitudinal axis. Determine the normal stress along the weld line. Let $p = 2000\,\text{psi}$, $R = 1.5\,\text{ft}$, and $t = 2\,\text{inch}$.

16.2-2 Toxic waste contained in an enormous cylindrical barrel was thought to have begun to undergone chemical reactions, causing the internal pressure to reach dangerous levels. The normal stresses σ' in the barrel were measured on each side of a square element at an angle of 30° relative to the longitudinal axis. Estimate the internal pressure if $R = 15\,\text{ft}$, $t = 2.5\,\text{inch}$, $\sigma'_x = 5\,\text{ksi}$, and $\sigma'_y = 7\,\text{ksi}$.

16.3 COMPOSITE BEAMS

Theory

A composite beam is composed of more than one material through its thickness (see Fig. 16.3-1). As in a simple beam, the strain in a composite beam is assumed to vary linearly with y; that is, $\varepsilon = ky$, in which k is presently unknown. For simplicity, assume that a composite beam is composed of two materials, with one sandwiched on top of the other. The results presented in this section can be extended to composite beams composed of more than two layered materials.

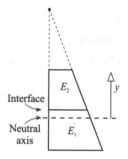

Figure 16.3-1

Let the bottom material have a modulus of elasticity E_1 and the top material have a modulus of elasticity E_2. The normal stresses are then

$$\sigma = \begin{cases} E_1 ky, & \text{bottom layer,} \\ E_2 ky, & \text{top layer,} \end{cases} \qquad (16.3\text{-}1)$$

as shown in Fig. 16.3-2. In a simple beam, the neutral axis is located at the beam's area center. This is not the case in a composite beam.

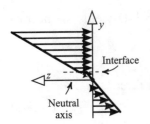

Interface

Neutral
axis

Figure 16.3-2

Let's now locate the neutral axis in a composite beam. Assuming that the resultant external force acting on the beam in the x direction is zero, it follows from Eq. (16.3-1) that

$$0 = \int \sigma \, dA = \int_{y_1}^{y^*} E_1 ky \, dA + \int_{y^*}^{y_2} E_2 ky \, dA = k\left(E_1 \int_{y_1}^{y^*} y \, dA + E_2 \int_{y^*}^{y_2} y \, dA \right),$$

where y_1 is the location of the bottom surface, y^* is the location of the interface, and y_2 is the location of the top surface. Dividing by k yields

$$0 = E_1 \int_{y_1}^{y^*} y \, dA + E_2 \int_{y^*}^{y_2} y \, dA. \qquad (16.3\text{-}2)$$

Equation (16.3-2) is used to locate the position of the neutral axis (detailed in Example 16.3-1). **Notice that for a composite beam, the location of the neutral axis depends on the relative elasticity of the materials, whereas for a simple beam, the location of the neutral axis does *not*.**

The moment produced by the normal stresses is now given by

$$M = -\int \sigma y \, dA = -k\left(\int_{y_1}^{y^*} E_1 y^2 \, dA + \int_{y^*}^{y_2} E_2 y^2 \, dA \right) = -k(E_1 I_1 + E_2 I_2),$$

$$(16.3\text{-}3)$$

in which

$$I_1 = \int_{y_1}^{y^*} y^2 \, dA, \qquad I_2 = \int_{y^*}^{y_2} y^2 \, dA. \qquad (16.3\text{-}4a,b)$$

Substituting for k from Eq. (16.3-3) into Eq. (16.3-1) yields the normal stresses in the cross section:

$$\sigma = \begin{cases} \dfrac{E_1}{E_1 I_1 + E_2 I_2} My, & \text{bottom layer,} \\[3mm] \dfrac{E_2}{E_1 I_1 + E_2 I_2} My, & \text{top layer,} \end{cases} \qquad (16.3\text{-}5)$$

Notice that the normal stresses in a composite beam depend explicitly on the elasticity of the materials, whereas the normal stresses in a simple beam do not.

Next consider the shear stresses. The shear stresses in a composite beam are determined by substituting Eq. (16.3-5) into Eq. (13.4-1). First assume that y is in the bottom layer, in which case

$$\tau = -\frac{1}{t}\int_y^{\text{top}} \frac{d\sigma}{dx} dA = -\frac{1}{t}\frac{dM}{dx}\frac{1}{E_1 I_1 + E_2 I_2}\left(E_1 \int_y^{y^*} y\, dA + E_2 \int_{y^*}^{y_2} y\, dA\right).$$

Thus, incorporating Eq. (13.2-2b),

$$\tau = \frac{V(E_1 Q_1 + E_2 Q_2)}{(E_1 I_1 + E_2 I_2)t}, \qquad (16.3\text{-}6)$$

in which

$$Q_1 = \int_y^{y^*} y\, dA, \qquad Q_2 = \int_{y^*}^{y_2} y\, dA. \qquad (16.3\text{-}7a,b)$$

The shear stress in Eq. (16.3-6) is contained in the bottom layer, that is, $y_1 < y < y^*$. The formula for the shear stress in the bottom layer, Eq. (16.3-6), is the same as the formula for the shear stress in the top layer, except that the quantities Q_1 and Q_2 change. Rather than Q_1 and Q_2 being given by Eq. (16.3-7a,b), they are given by

$$Q_1 = 0, \qquad Q_2 = \int_y^{y_2} y\, dA \qquad (16.3\text{-}8a,b)$$

for values of y contained in the top layer, that is, for $y^* < y < y_2$.

Equations (16.3-5) and (16.3-6) replace the expressions for the normal and shear stresses given in Eq. (13.5-8a,b) developed for simple beams. The other expressions developed for simple beams, Eqs. (13.5-7a–d), are valid for composite beams.

Example

16.3-1 A composite beam of length L is subjected to a force F as shown. The beam cross section is rectangular. Determine expressions for the normal stress and the shear stress in the beam. Assume that $E_1/E_2 \geqslant (b/a)^2$.

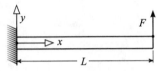

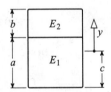

Example 16.3-1

Solution

The location of the neutral axis is determined first. Referring to the figure, $y_1 = -c$, $y^* = b - c$, $y_2 = a - c + b$, and $dA = t\,dy$ in Eq. (16.3-2), so

$$0 = E_1 \int_{-c}^{a-c} yt\,dy + E_2 \int_{a-c}^{a-c+b} yt\,dy$$

$$= -c(aE_1 + bE_2) + \tfrac{1}{2}(E_1 a^2 + E_2 b^2 + 2E_2 ab);$$

then

$$\frac{c}{a} = \frac{\dfrac{E_1}{E_2} + \dfrac{b}{a}\left(2 + \dfrac{b}{a}\right)}{2\left(\dfrac{E_1}{E_2} + \dfrac{b}{a}\right)}.$$

The location c of the neutral axis is measured from the bottom of the cross section. This expression for c/a is valid when c is in the bottom layer, that is, when $c/a \leqslant 1$. Substituting this condition into the expression above for c/a yields the condition $E_1/E_2 \geqslant (b/a)^2$. Notice that this condition is precisely the assumption that was made in the problem statement. Thus, the neutral axis is located in the bottom layer of the composite beam.

Next an expression for the normal stresses is determined. The internal shear force and bending moment in the beam are $V(x) = F$ and $M(x) = F(L - x)$. From Eqs. (16.3-4),

$$I_1 = \int_{-c}^{a-c} y^2 t \, dy = \tfrac{1}{3} at[a^2 - 3c(a - c)],$$

$$I_2 = \int_{a-c}^{a-c+b} y^2 t \, dy = \tfrac{1}{3} bt[b^2 + 3(a - c + b)(a - c)],$$

so, from Eq. (16.3-5),

$$\sigma = \begin{cases} \dfrac{3E_1 F}{E_1 at[a^2 - 3c(a - c)] + E_2 bt[b^2 + 3(a - c + b)(a - c)]}(L - x)y, \\[1em] \qquad\qquad\qquad\qquad\qquad\qquad\qquad\qquad\qquad\qquad \text{bottom layer} \\[1em] \dfrac{3E_2 F}{E_1 at[a^2 - 3c(a - c)] + E_2 bt[b^2 + 3(a - c + b)(a - c)]}(L - x)y, \\[1em] \qquad\qquad\qquad\qquad\qquad\qquad\qquad\qquad\qquad\qquad \text{top layer} \end{cases}$$

Finally an expression for the shear stresses is determined. From Eqs. (16.3-7) and (16.3-8), for y in the bottom layer,

$$Q_1 = \int_{y_1}^{y^*} y \, dA = \int_{y}^{a-c} yt \, dy = \tfrac{1}{2} t[(a - c)^2 - y^2],$$

$$Q_2 = \int_{a-c}^{a-c+b} yt \, dy = \tfrac{1}{2} tb[b + 2(a - c)],$$

and for y in the top layer,

$$Q_1 = 0, \qquad Q_2 = \int_{y}^{a-c+b} yt \, dy = \tfrac{1}{2} t[(a - c + b)^2 - y^2].$$

From Eq. (16.3-6), the shear stresses are

$$\tau = \frac{3F}{2t\{E_1a[a^2 - 3c(a-c)] + E_2b[b^2 + 3(a-c+b)(a-c)]\}}$$

$$\times \begin{cases} E_1[(a-c)^2 - y^2] + E_2b[b + 2(a-c)], & \text{bottom layer,} \\ E_2[(a-c+b)^2 - y^2], & \text{top layer.} \end{cases}$$

Notice that the shear stresses are zero on the bottom and top surfaces, as expected.

Problem

16.3-1 The composite beam in Example 16.3-1 now has a cross section composed of two rectangular beams stacked on top of each other as shown. Determine the normal stresses and the shear stresses in the beam. Let $L = 1\,\text{m}$, $F = 100\,\text{N}$, $a = 0.1\,\text{m}$, $E_1 = 10\,\text{MN/m}^2$, and $E_2 = 20\,\text{MN/m}^2$.

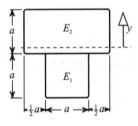

Problem 16.3-1

16.4 TEMPERATURE EFFECTS

Theory

Temperature changes that are small compared with the absolute temperature have long been known to cause deformations that are linearly proportional to the temperature change. The change in length of a given specimen is governed by

$$\Delta L = k\,\Delta T,$$

in which ΔL denotes the change in length, ΔT denotes the change in temperature, and k is a linear proportionality constant. Dividing by length yields

$$\boxed{\varepsilon_t = \alpha\,\Delta T,} \tag{16.4-1}$$

in which $\varepsilon_t = \Delta L/L$ is the **thermal strain**, and $\alpha = k/L$ is a **material parameter** called the **thermal coefficient of expansion**. The thermal strain as defined is an **average** strain over the length ΔL. The thermal strain at a **point** is also given by Eq. (16.4-1), in which the thermal coefficient of expansion α and the temperature change ΔT can now be functions of x. This section examines the longitudinal effect and the lateral effect caused by temperature changes.

LONGITUDINAL EFFECT

The strain in a material can be composed of a thermal strain and mechanical strain, so $\varepsilon = \varepsilon_m + \varepsilon_t$. Equation (11.1-4) is modified to yield

$$\frac{du}{dx} = \varepsilon_m + \varepsilon_t = \frac{P}{AE} + \alpha\,\Delta T. \tag{16.4-2}$$

Integrating Eq. (16.4-2) with respect to x yields

$$u(x) = u(0) + \int_0^x \left(\frac{P}{AE} + \alpha\,\Delta T \right) dx. \tag{16.4-3}$$

Equation (16.4-3) is the counterpart of Eq. (11.1-6). The other equations in that section, specifically Eq. (11.1-7), are valid in the presence of thermal strain.

LATERAL EFFECT

Lateral displacements caused by temperature changes are specifically caused by the changes in temperature in the y (lateral) direction. Let a structural member that was originally at a uniform temperature T_0 undergo a temperature change $\Delta T_1 = T_1 - T_0$ at the bottom surface and a temperature change $\Delta T_2 = T_2 - T_0$ at the top surface. The distance between the top and bottom surfaces is denoted by h. Assuming that the temperature throughout the structural member varies linearly with y, the **thermal curvature** of the member is

$$\left(\frac{1}{\rho} \right)_t = \frac{\alpha(T_1 - T_2)}{h}. \tag{16.4-4}$$

Like the mechanical curvature, the thermal curvature is small, and both effects are added to obtain the total curvature. From Eq. (13.5-3),

$$\frac{1}{\rho} = \left(\frac{1}{\rho}\right)_m + \left(\frac{1}{\rho}\right)_t = \frac{M}{EI_z} + \frac{\alpha(T_1 - T_2)}{h}. \tag{16.4-5}$$

Notice that the thermal curvature is independent of $T_{avg} = \frac{1}{2}(T_1 + T_2)$, the average change in temperature. The temperature change $T_1 - T_2$ effects the beam laterally and T_{ave} effects the beam longitudinally. From Eqs. (13.5-4) and (16.4-5),

$$\frac{d^2w}{dx^2} = \frac{M}{EI_z} + \frac{\alpha(T_1 - T_2)}{h}. \tag{16.4-6}$$

Integrating Eq. (16.4-6) with respect to x and noting that for small angles $dw/dx = \tan\theta \approx \theta$, the bend angle as a function of x is

$$\theta(x) = \theta(0) + \int_0^x \left[\frac{M}{EI_z} + \frac{\alpha(T_1 - T_2)}{h}\right] dx. \tag{16.4-7}$$

Equation (16.4-7) is the counterpart of Eq. (13.5-7c). The other equations in that section, specifically Eqs. (13.5-7a,b,d), are valid in the presence of thermal strain.

The analysis of a structural member in the presence of thermal strain differs from the analysis of a structural member in the absence of thermal strain by the introduction of the two thermal effects given in Eqs. (16.4-3) and (16.4-7). Longitudinal changes in temperature cause expansions and contractions in the longitudinal direction that are accounted for in Eq. (16.4-3), and lateral changes in temperature cause changes in curvature that are accounted for in Eq. (16.4-7). The examples provide further clarification.

Examples

| 16.4-1 | A bar of length L is fixed at both ends as shown when the temperature of the bar is raised from T_0 to T_1. Determine an expression for the stresses that build up in the bar as a result of the temperature change.

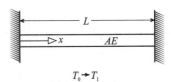

Example 16.4-1

Solution

The boundary conditions and the external force per unit length are

$$u(0) = 0, \quad u(L) = 0, \quad f_x = 0.$$

From Eq. (11.1-7), $P = P(0)$, and, from Eq. (16.4-3),

$$u = \frac{P(0)}{AE} x + \alpha(T_1 - T_0)x.$$

From the right boundary condition,

$$P(0) = -AE\alpha(T_1 - T_0).$$

The normal stress in the bar as a result of the temperature change is

$$\sigma = -E\alpha(T_1 - T_0).$$

The normal stress is compressive. Also notice that the displacement is zero everywhere.

| **16.4-2** | A beam with a rectangular cross section is pinned at both ends as shown when the temperature of the lower surface is raised from T_0 to $T_1 = T_0 + \Delta T$ and the temperature of the upper surface is lowered from T_0 to $T_2 = T_0 - \Delta T$. Determine the bending displacement of the beam.

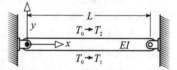

Example 16.4-2

Solution

The boundary conditions and the external force per unit length are

$$u(0) = 0, \quad M(0) = 0, \quad u(L) = 0, \quad M(L) = 0, \quad f = 0.$$

From Eqs. (13.5-7a,b), $V(x) = V(0)$ and $M(x) = M(0) - V(0)x$. From the boundary conditions, $0 = -V(0)L$, so $V(0) = 0$, and then $V(x) = 0$

and $M(x) = 0$. No shear forces or bending moments build up in the beam. From Eqs. (16.4-7) and (13.5-7d),

$$\theta = \theta(0) + \frac{2\alpha\,\Delta T}{h}x, \qquad u = \theta(0)x + \frac{\alpha\,\Delta T}{h}x^2$$

From the right boundary condition,

$$0 = \theta(0)L + \frac{\alpha\,\Delta T\,L^2}{h},$$

so

$$\theta(0) = -\frac{\alpha\,\Delta T\,L}{h}.$$

The displacement of the beam is

$$u = \frac{\alpha\,\Delta T}{h}x(x - L).$$

Notice that, internally, the beam undergoes no normal stresses and no shear stresses. Whereas the longitudinal temperature change in Example 16.4-1 caused non-zero internal stresses and no displacements, the lateral temperature change in this example caused no internal stresses and non-zero displacements.

Problems

16.4-1 The temperature of a bar increases from 100°F to 200°F while it is subjected to a longitudinal force F as shown. Determine the displacement of the tip of the bar. Let $F = 500\,\mathrm{N}$, $A = 0.1\,\mathrm{m}^2$, $L = 0.8\,\mathrm{m}$, $E = 250\,\mathrm{kN/m^2}$, $k = 30\,\mathrm{kN/m}$, and $\alpha = 0.002°\mathrm{F}^{-1}$.

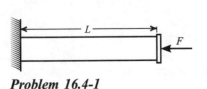

Problem 16.4-1

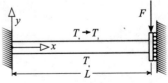

Problem 16.4-2

16.4-2 The temperature of the beam shown increases from 100°F to 120°F on the upper surface and there is no change in temperature on the lower surface. The cross section of the bar is square, having sides of length a. The temperature distribution is expressed as

$$T = 100 + \frac{20}{a}(y + \tfrac{1}{2}a) = 110 + \frac{20}{a}y \,°\text{F},$$

in which the 110°F is the longitudinal distribution of the temperature over the cross section and the $20y/a$ °F is the lateral distribution of the temperature. (The change in the longitudinal component of temperature was zero in Example 16.4-2.) Determine the normal stresses at the neutral axis at the right side of the bar and determine the displacement of the right side of the bar. Let $F = 500\,\text{N}$, $a = 0.2\,\text{m}$, $L = 0.8\,\text{m}$, $E = 250\,\text{kN/m}^2$, and $\alpha = 0.002°\text{F}^{-1}$.

APPENDIX

Tables

Table A.1 Material Properties

	Modulus of elasticity	Shear modulus of elasticity	Poisson's ratio	Yield stress	Ultimate stress	Coefficient of thermal expansion	Specific weight	Mass density
	E (ksi)	G (ksi)	ν	σ_y (ksi)	σ_x (ksi)	α $(10^{-6}/°C)$	ρ_g (kN/m^3)	ρ (kg/m^3)
Aluminum	10000	3700	0.33	2,900	10,000	23	26.6	2710
Glass	7000–12000	2700–5000	0.20–0.27		10,000	5–11	24–28	2400–2800
Steel	27000–30000	11000	0.27–0.30			10–18	77.0	7850
Rubber	0.10–0.60	0.03–0.015	0.45–0.50			130–200	9–13	960–1300
Wood	1400–2000						5–7	480–720

Table A.2 Mass Integrals and Area Integrals

Parallel axis theorems

$I_{Oxx} = I_{Cxx} + m(y_C^2 + z_C^2)$	$I_{Oxy} = I_{Cxy} - mx_C y_C$	$I_{Ozx} = I_{Czx} - mz_C x_C$
	$I_{Oyy} = I_{Cyy} + m(x_C^2 + z_C^2)$	$I_{Oyz} = I_{Cyz} - my_C z_C$
		$I_{Ozz} = I_{Czz} + m(x_C^2 + y_C^2)$

Line bodies

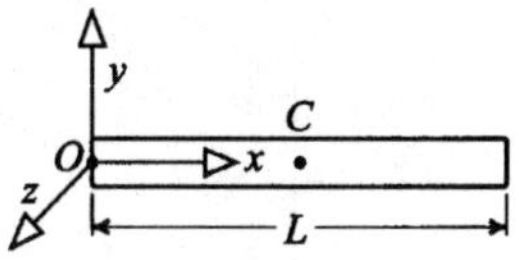	$I_{Oxx} = 0$ $I_{Oyy} = \frac{1}{3}mL^2$ $I_{Ozz} = \frac{1}{3}mL^2$	$I_{Cxx} = 0$ $I_{Cyy} = \frac{1}{12}mL^2$ $I_{Czz} = \frac{1}{12}mL^2$
$m = \rho_L L$ $x_C = \frac{1}{2}L$ $y_C = 0$ $z_C = 0$	$I_{Oxy} = 0$ $I_{Oyz} = 0$ $I_{Ozx} = 0$	$I_{Cxy} = 0$ $I_{Cyz} = 0$ $I_{Czx} = 0$

Table A.2   (*continued*)

Line bodies (continued)

$$m = \rho_L R \beta$$
$$x_C = R_0 \cos \tfrac{1}{2}\beta$$
$$y_C = R_0 \sin \tfrac{1}{2}\beta$$
$$z_C = 0$$
$$R_0 = \left(\frac{\sin \tfrac{1}{2}\beta}{\tfrac{1}{2}\beta} \right) R$$

$$I_{Oxx} = \frac{mR^2}{2}\left(1 - \frac{\sin 2\beta}{2\beta} \right)$$
$$I_{Oyy} = \frac{mR^2}{2}\left(1 + \frac{\sin 2\beta}{2\beta} \right)$$
$$I_{Ozz} = mR^2$$

$$I_{Oxy} = -\frac{mR^2}{2}\frac{\sin^2 \beta}{\beta}$$
$$I_{Oyz} = 0$$
$$I_{Ozx} = 0$$

$$I_{Cxx} = mR^2\left[\frac{1}{2} - \frac{\sin 2\beta}{4\beta} - \left(\frac{2\sin^2 \beta}{\beta} \right)^2 \right]$$
$$I_{Cyy} = mR^2\left[\frac{1}{2} + \frac{\sin 2\beta}{4\beta} - \left(\frac{\sin \beta}{\beta} \right)^2 \right]$$
$$I_{Czz} = mR^2\left[1 - 2\left(\frac{1 - \cos \beta}{\beta^2} \right) \right]$$

$$I_{Cxy} = mR^2\left[-\frac{\sin^2 \beta}{2\beta} + \frac{\sin \beta(1 - \cos \beta)}{\beta^2} \right]$$
$$I_{Cyz} = 0$$
$$I_{Czx} = 0$$

Table A.2 (*continued*)

Surface bodies

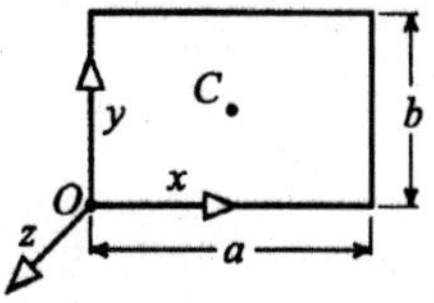	$I_{Oxx} = \frac{1}{3}mb^2$ $I_{Oyy} = \frac{1}{3}ma^2$ $I_{Ozz} = \frac{1}{3}m(a^2 + b^2)$	$I_{Cxx} = \frac{1}{12}mb^2$ $I_{Cyy} = \frac{1}{12}ma^2$ $I_{Czz} = \frac{1}{12}m(a^2 + b^2)$
$m = \rho_A ab$ $x_C = \frac{1}{2}a$ $y_C = \frac{1}{2}b$ $z_C = 0$	$I_{Oxy} = -\frac{1}{4}mab$ $I_{Oyz} = 0$ $I_{Ozx} = 0$	$I_{Cxy} = 0$ $I_{Cyz} = 0$ $I_{Czx} = 0$
$A = ab$	$I_{Oy} = \frac{1}{3}a^3b$	$I_{Cy} = \frac{1}{12}a^3b$

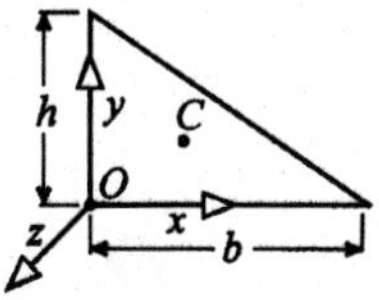	$I_{Oxx} = \frac{1}{6}mh^2$ $I_{Oyy} = \frac{1}{6}mb^2$ $I_{Ozz} = \frac{1}{6}m(b^2 + h^2)$	$I_{Cxx} = \frac{1}{18}mh^2$ $I_{Cyy} = \frac{1}{18}mb^2$ $I_{Czz} = \frac{1}{18}m(b^2 + h^2)$
$m = \frac{1}{2}\rho_A bh$ $x_C = \frac{1}{3}b$ $y_C = \frac{1}{3}h$ $z_C = 0$	$I_{Oxy} = -\frac{1}{12}mbh$ $I_{Czx} = 0$ $I_{Ozx} = 0$	$I_{Cxy} = \frac{1}{36}mbh$ $I_{Cyz} = 0$ $I_{Czx} = 0$
$A = \frac{1}{2}bh$	$I_{Oy} = \frac{1}{12}hb^3$	$I_{Cy} = \frac{1}{36}hb^3$

Table A.2   (*continued*)

Surface bodies (continued)

$$m = \tfrac{1}{2}\rho_A R^2 \beta$$
$$x_C = R_0 \cos\tfrac{1}{2}\beta$$
$$y_C = R_0 \sin\tfrac{1}{2}\beta$$
$$z_C = 0$$
$$R_0 = \frac{2}{3}\left(\frac{\sin\tfrac{1}{2}\beta}{\tfrac{1}{2}\beta}\right)R$$
$$A = \tfrac{1}{2}R^2\beta$$

$$I_{Oxx} = \frac{mR^2}{4}\left(1 - \frac{\sin 2\beta}{2\beta}\right)$$
$$I_{Oyy} = \frac{mR^2}{4}\left(1 + \frac{\sin 2\beta}{2\beta}\right)$$
$$I_{Ozz} = \frac{mR^2}{2}$$

$$I_{Oxy} = -\frac{mR^2}{4}\frac{\sin^2\beta}{\beta}$$
$$I_{Oyz} = 0$$
$$I_{Ozx} = 0$$

$$I_{Oy} = \frac{R^4\beta}{8}\left(1 + \frac{\sin 2\beta}{2\beta}\right)$$

$$I_{Cxx} = \frac{mR^2}{4}\left[1 - \frac{\sin 2\beta}{2\beta} - \frac{16}{9}\left(\frac{1-\cos\beta}{\beta}\right)^2\right]$$
$$I_{Cyy} = \frac{mR^2}{4}\left[1 + \frac{\sin 2\beta}{2\beta} - \frac{16}{9}\left(\frac{\sin\beta}{\beta}\right)^2\right]$$
$$I_{Czz} = \frac{mR^2}{2}\left[1 - \frac{16}{9}\left(\frac{1-\cos\beta}{\beta^2}\right)\right]$$

$$I_{Cxy} = mR^2\frac{\sin\beta}{\beta}\left[-\frac{\sin\beta}{4} + \frac{4}{9}\left(\frac{1-\cos\beta}{\beta^2}\right)\right]$$

$$I_{Cyz} = 0$$
$$I_{Czx} = 0$$

$$I_{Cy} = \frac{R^4\beta}{8}\left[1 + \frac{\sin 2\beta}{2\beta} - \frac{16}{9}\left(\frac{\sin\beta}{\beta}\right)^2\right]$$

Table A.2 (*continued*)

Volume bodies

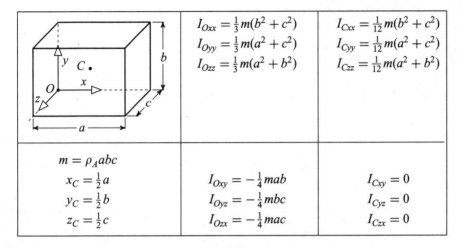

	$I_{Oxx} = \frac{1}{3}m(b^2 + c^2)$	$I_{Cxx} = \frac{1}{12}m(b^2 + c^2)$
	$I_{Oyy} = \frac{1}{3}m(a^2 + c^2)$	$I_{Cyy} = \frac{1}{12}m(a^2 + c^2)$
	$I_{Ozz} = \frac{1}{3}m(a^2 + b^2)$	$I_{Czz} = \frac{1}{12}m(a^2 + b^2)$
$m = \rho_A abc$		
$x_C = \frac{1}{2}a$	$I_{Oxy} = -\frac{1}{4}mab$	$I_{Cxy} = 0$
$y_C = \frac{1}{2}b$	$I_{Oyz} = -\frac{1}{4}mbc$	$I_{Cyz} = 0$
$z_C = \frac{1}{2}c$	$I_{Ozx} = -\frac{1}{4}mac$	$I_{Czx} = 0$

	$I_{Oxx} = \frac{1}{2}mR^2$	$I_{Cxx} = \frac{1}{2}mR^2$
	$I_{Oyy} = m(\frac{1}{4}R^2 + \frac{1}{3}a^2)$	$I_{Cyy} = m(\frac{1}{4}R^2 + \frac{1}{12}a^2)$
	$I_{Ozz} = m(\frac{1}{4}R^2 + \frac{1}{3}a^2)$	$I_{Czz} = m(\frac{1}{4}R^2 + \frac{1}{12}a^2)$
$m = \rho\pi R^2 a$		
$x_C = \frac{1}{2}a$	$I_{Oxy} = 0$	$I_{Cxy} = 0$
$y_C = 0$	$I_{Oyz} = 0$	$I_{Cyz} = 0$
$z_C = 0$	$I_{Ozx} = 0$	$I_{Czx} = 0$

INDEX

Solutions to Problems

Chapter 1

1.2-1 27.5 ± 1.01
1.2-2 62.5 ± 2.75
1.2-3 15.625 ± 0.1875
1.3-1 $\mathbf{r}_3 = -13\mathbf{i} + 2\mathbf{j}$
1.3-2 $\mathbf{F}_3 = 12\mathbf{j}, \mathbf{F}_4 = 44\mathbf{i} + 19\mathbf{j}$
1.3-3 $\mathbf{F}_3 = 4\mathbf{i} + 3\mathbf{j} + 7\mathbf{k}$
1.3-4 $\mathbf{F}_3 = 24\mathbf{i} + 33\mathbf{j} + 8\mathbf{k}, \mathbf{F}_4 = 12\mathbf{i} + 11\mathbf{j} - 18\mathbf{k}$
1.3-5 $6.708, 116.5°$
1.3-6 $5.385, 68.2°$
1.3-7 9.487
1.3-8 $-0.385\mathbf{i} + 0.923\mathbf{j}$
1.3-9 $-0.308\mathbf{i} - 0.231\mathbf{j} + 0.923\mathbf{k}$
1.4-1 $71.2°$
1.4-2 $-138.$
1.4-3 $0°$
1.4-4 $2.828, 3.162$
1.4-5 $-17\mathbf{j}$
1.4-6 $90°$
1.4-7 $\pm(0.577\mathbf{i} + 0.024\mathbf{j} - 0.817\mathbf{k})$

Chapter 2

2.1-1 $0.633i + 6.737j$ N
2.1-2 $0.2i + 136.8j$ lb
2.1-3 $0.305i - 50.848j$ lb, 50.849 lb at 0.3°
2.1-4 $20.48i + 10.60j + 29.26k$ N
2.1-5 $-164.9i + 247.2j + 779.9k$ N
2.2-1   2014 lb·ft
2.2-2   2014 lb·ft
2.2-3   2014 lb·ft
2.2-4 $308.3j + 738.4k$ N·m
2.2-5 $-202.5i - 1582.9j - 848.5k$ N·m
2.2-6 $-158i + 144j + 25.5k$ N·m
2.2-7 $35i - 315j$ N·m
2.3-1 $4.485i - 6.413j$ lb, $304.527k$ lb·ft, 529.0k lb·ft
2.3-2 $8i - 13.9j$ lb, $-487k$ lb·ft, $-623k$ lb·ft
2.3-3   1450 lb, 7450 lb·ft
2.3-4 $34.3i + 21.3j$ N, $144k$ N·m, $93.7k$ N·m
2.3-5 $-10.5i - 5j + 7.5k$ N, $-152.5i - 240.0j - 216.5k$ N·m, $-265i - 240j - 374k$ N·m
2.3-6   0 lb, $-212i - 212j - 212k$ lb·ft, $-212i - 212j - 212k$ lb·ft
2.3-7 $25i + 30j + 35k$ N, $1090i + 400j - 80k$ N·m, $1215j$
2.3-8 $188i - 43.7j - 161k$ lb, $398i - 1711j$ lb·ft, $398i - 377j + 568k$ lb·ft

Chapter 3

3.1-1 -51.7 N, 110.4 N, $-8.95°$
3.1-2   114.7 lb
3.1-3   140.8°
3.1-4   22.2°, 0°
3.1-5   25.4 N
3.1-6   809 N, 704 N, 991 N, 77.7 kg
3.1-7 $-28.868i + 50j$ lb
3.2-1 $1585.8i + 185.8j$ lb, $-890.9j$ lb
3.2-2 $-74.5i + 233j$ N, 41°
3.2-3 $-144.9i - 173.3j$ N, -134.5 N
3.2-4 $75i + 120j$ lb

3.2-5 $-70\mathbf{k}$ N, $129.9\mathbf{j} - 75\mathbf{k}$ N·m

3.2-6 $56.25\mathbf{j} + 60\mathbf{k}$ lb, $-188\mathbf{i} - 120\mathbf{j}$ lb·ft

3.2-7 $-170\mathbf{j} + 102\mathbf{k}$ N·m

3.2-8 $-34.2\mathbf{j} + 15.7\mathbf{k}$ N, 78.3 N, $75.2\mathbf{j} + 8.6\mathbf{k}$ N·m

3.3-1 10.585 lb

3.3-2 $22.1\mathbf{i}$ N, $46.1\mathbf{k}$ N·m

3.3-3 $-18.7\mathbf{i} + 41.6\mathbf{j}$ lb, $-6.0\mathbf{i} + 49.3\mathbf{j}$ lb, $-12.7\mathbf{i} - 40.9\mathbf{j}$ lb, $37.3\mathbf{i} + 33.3\mathbf{j}$ lb

3.3-4 $\theta = \arccos\left(\dfrac{\sqrt{R^2 - 4r^2}}{r4\sqrt{2}}\right) = 35.9°$

Chapter 4

4.1-1   No, $n = 7, m = 6, p = 6$

4.1-2   No, $n = 9, m = 19, p = 9$

4.1-3   No, $n = 7, m = 11, p = 7$

4.1-4 $R_F = F\mathbf{j}, R_K = F\mathbf{i}$

4.1-5 $n = 6, m = 8, p = 4, 2(6) = 8 + 4, T_{DF} = \frac{91}{75}mg$,
       $T_{EF} = \frac{26}{75}mg, T_{EC} = \frac{8}{25}mg, T_{BD} = -\frac{157}{100}mg$,
       $T_{AC} + \frac{77}{100}mg, R_{Ax} = -\frac{231}{500}mg = -0.462mg$,
       $R_{Ay} = -\frac{308}{500}mg = -0.616mg, R_{Bx} = 0.462mg$,
       $R_{By} = \frac{731}{500}mg = 1.616mg$

4.1-6 $R_{1x} = 0, R_{1y} = F, T_{12} = R_y - R_x, T_{13} = R_y\sqrt{2}$,
       $T_{23} = \frac{1}{2}F, T_{25} = R_y - R_x, T_{34} = -R_y\sqrt{2} + F/2\sqrt{2}$,
       $T_{35} = -F/2\sqrt{2}, T_{45} = 2R_y - \frac{1}{2}F$, the tensions are symmetrical in the structure.

4.1-7 $R_{1x} = 0, R_{1y} = \frac{25}{2}(4 + 3\sqrt{2}) = 103.1$ lb, $T_{12} = \frac{25}{4}(7 + 5\sqrt{2}) = $
       87.9lb, $T_{13} = -\frac{25}{4}(10 + 7\sqrt{2}) = -124.4$ lb, $T_{23} = \frac{25}{4}3 = $
       18.75 lb, $T_{25} = \frac{25}{4}(7 + 5\sqrt{2}) = 87.9$ lb, $T_{34} = $
       $-\frac{25}{4}(7 + 5\sqrt{2}) = -87.9$ lb, $T_{35} = -\frac{25}{4}(3 + 2\sqrt{2}) = -36.4$ lb,
       $T_{45} = \frac{25}{4}(12 + 8\sqrt{2}) = 1.45.7$ lb

4.2-1 2705 N

4.2-2 $P(x) = W(1 - 3x/4h)$ for $0 \leqslant x \leqslant h, P(x) = $
       $\frac{1}{4}W(2 - x/h)$ for $0 \leqslant x \leqslant 2h$

4.2-3 20 N·m

4.2-4　$V_{A,\text{right}} = -70$ lb, $M_{A,\text{right}} = 70$ lb-m, $V_{A,\text{left}} =$
-35 lb, $M_{A,\text{left}} = 70$ lb·m, $V_B = -35$ lb, $M_B = -122.5$ lb-m

4.2-5　$V_A = -82.5$ lb, $M_A = 61.875$ lb·ft, $V_B = -181.5$ lb,
$M_B = -605.458$ lb·ft

4.2-6　$V_A = -200$ N, $M_A = 800$ N·m

4.2-7　$V_{A^-} = 116.667$ N, $M_{A^-} = 800$ N·m, $V_{A^+} = 0$, $M_{A^+} = 800$ N·m

4.3-1　$R_{Ax} = -288.7$ lb, $R_{Ay} = 500$ lb, $R_{Bx} = 288.7$ lb, $R_{By} = 100$ lb

4.3-2　$\mathbf{R}_E = 192\mathbf{i}$ N, $\mathbf{R}_O = 192\mathbf{i} + 3400\mathbf{k}$ N, $\mathbf{M}_O = 6157\mathbf{i} +$
$575\mathbf{j}$ N·m, $P_y = 0$ N, $\mathbf{V}(y) = -191.5\mathbf{i} - (3400 + 850y)\mathbf{k}$ N,
$\mathbf{M}(y) = -(6800 + 425y^2)\mathbf{i} - (766 - 191.5)\mathbf{k}$ N·m

4.3-3　$-15\mathbf{i} - 25.98\mathbf{j}$, $15\mathbf{i} + 25.98\mathbf{j}$

4.4-1　$F = \dfrac{\mu \cos \theta + \sin \theta}{\cos \theta - \mu \sin \theta} mg \ (\mu < \cot \theta)$

4.4-2　$\theta = 180° - \left[\arccos\left(\dfrac{-m_2 \mu}{m_1 \sqrt{1 + \mu^2}} \right) - \arctan\left(\dfrac{1}{\mu} \right) \right] = 173.15°$

4.4-3　$F = \dfrac{\mu \cos \theta + \sin \theta}{\cos \theta - \mu \sin \theta} m_1 g \ (\mu < \cot \theta)$

4.4-4　$\arctan[(\mu^2 - 2)/(1 - 3\mu - \mu^2)]$ radians

4.4-5　1.9

4.4-6　$Wp/(4\pi \tan \theta)$

4.4-7　8 turns

4.4-8　46.67 N·m

4.5-1　487.9 N

4.5-2　$T_1 = 757$ N, $T_2 = 766$ N, $T_3 = 985$ N, $\theta = 5.07°$

Chapter 5

5.2-1　$\mathbf{r}_1 = 0.507\mathbf{i} + 1.088$ ft, $\mathbf{r}_2 = 0.983\mathbf{i} - 0.688$ ft,
$\mathbf{r}_3 = -1.2\mathbf{i}$ ft, $m = 72$ slug, $\mathbf{r}_C = 0.16\mathbf{i} + 0.13\mathbf{j}$ ft

5.2-2　$\mathbf{r}_1 = -1.2\mathbf{i}$, $\mathbf{r}_2 = 0.983\mathbf{i} - 0.688\mathbf{j}$, $\mathbf{r}_3 = 0.688\mathbf{i} + 0.983\mathbf{j}$,
$I_O = 104$ slug·ft^2

5.2-3　$I_{12} = 2.99$ slug·ft^2, $I_{13} = I_{23} = 0$

5.2-4　$\mathbf{r}_1 = 0$, $\mathbf{r}_2 = d\mathbf{j}$, $\mathbf{r}_3 = d(s\mathbf{i} + c\mathbf{j}) = d(0.866\mathbf{i} + 0.5\mathbf{j})$,
$\mathbf{r}_4 = d\{[(2s^2 - 1)/2s]\mathbf{i} + \tfrac{1}{2}\mathbf{j} + [(4s^2 - 1)^{1/2}/2s]\mathbf{k}\} =$

$d(0.289\mathbf{i} + 0.5\mathbf{j} + 0.816\mathbf{k})$, $M = 4m$, $\mathbf{r}_C = 0.288d\mathbf{i} + 0.5d\mathbf{j} + 0.294d\mathbf{k}$

5.2-5 $I_{x0} = \frac{13}{6}md^2$, $I_{y0} = \frac{3}{2}md^2$, $I_{z0} = \frac{7}{3}md^2$

5.2-6 $I_{xy} = -0.577\,d^2m$, $I_{yz} = -0.408\,d^2m$, $I_{xz} = -0.236\,d^2m$

5.3-1 $\mathbf{r}_C = d\mathbf{j}$, $I_{x0} = Md^2$, $I_{y0} = \frac{1}{12}ML^2$, $I_{z0} = M(\frac{1}{12}L^2 + d^2)$, $I_{xy} = I_{yz} = I_{zx} = 0$

5.3-2 $\mathbf{r}_C = \frac{2}{3}L\mathbf{i}$, $I_{x0} = 0$, $I_{y0} = I_{z0} = \frac{1}{4}\rho A_0 L^3$, $I_{xy} = I_{yz} = I_{zx} = 0$

5.3-3 $I_{x0} = 1.5mR^2$, $I_{y0} = 0.5mR^2$, $I_{z0} = 2mR^2$, $I_{xy} = I_{yz} = I_{zx} = 0$

5.3-4 $\mathbf{r}_C = (R/\theta)\sin\theta\,\mathbf{i} + (R/\theta)(1 - \cos\theta)\mathbf{j} + d\mathbf{k}$,
$I_{x0} = (MR^2/2\theta)(\theta + \frac{1}{2}\sin 2\theta)$, $I_{y0} = (MR^2/2\theta)(\theta - \frac{1}{2}\sin 2\theta)$,
$I_{z0} = d^2M$, $I_{xy} = (MR^2/4\theta)(\cos 2\theta - 1)$,
$I_{yz} = (MRd/\theta)(\cos\theta - 1)$, $I_{xz} = -(MRd/\theta)\sin\theta$

5.4-1 0.6 kg

5.4-2 $\mathbf{r}_C = 0.267\mathbf{i} + 0.1\mathbf{j}$ m

5.4-3 $I_{x0} = 0.027$ kg·m^2, $I_{y0} = 0.048$ kg·m^2, $I_{z0} = 0.075$ kg·m^2

5.4-4 $I_x = 0.0009$ m^4

5.4-5 $I_{xy} = -0.028$ kg·m^2, $I_{yz} = I_{zx} = 0$

5.4-6 $M = \frac{1}{2}\rho_A aR\pi$

5.4-7 $\mathbf{r}_C = \frac{1}{2}a\mathbf{i} + (2R/\pi)\mathbf{j} + (2R/\pi)\mathbf{k}$

5.4-8 $I_{x0} = \frac{1}{2}\rho_A aR^3\pi$, $I_{y0} = \rho_A aR\pi(\frac{1}{6}a^2 + \frac{1}{4}R^2)$,
$I_{z0} = \rho_A aR\pi(\frac{1}{6}a^2 + \frac{1}{4}R^2)$

5.4-9 $I_{xy} = \frac{1}{2}\pi Mat$, $I_{yz} = -\frac{1}{2}aR^3\rho t$, $I_{zx} = \frac{1}{2}\pi Mat$

5.5-1 ρabc, $\frac{1}{3}a\mathbf{i} + \frac{1}{3}c\mathbf{k}$

5.5-2 $I_{x0} = \rho(\frac{1}{6}abc^3 - \frac{1}{3}ab^3c)$, $I_{y0} = \frac{1}{6}\rho abc(c^2 + a^2)$,
$I_{z0} = \rho(\frac{1}{6}a^3bc - \frac{1}{3}ab^3c)$

5.5-3 $I_{xy} = I_{yz} = 0$, $I_{zx} = -\frac{1}{12}ra^2b^2c$

5.5-4 $M = \frac{1}{6}\rho abc$, $\mathbf{r}_C = \frac{1}{4}a\mathbf{i} + \frac{1}{4}b\mathbf{j} + \frac{1}{4}c\mathbf{k}$

5.5-5 $I_{x0} = \frac{1}{60}\rho ab^3c + \frac{1}{60}\rho abc^3$, $I_{y0} = \frac{1}{60}\rho a^3bc + \frac{1}{60}\rho abc^3$,
$I_{z0} = \frac{1}{60}\rho a^3bc + \frac{1}{60}\rho ab^3c$

5.5-6 $I_{xy} = -\frac{1}{120}\rho a^2b^2c$, $I_{yz} = -\frac{1}{120}\rho ab^2c^2$, $I_{zx} = -\frac{1}{120}\rho a^2bc^2$

5.5-7 $m = \frac{1}{3}\rho\pi R^2h$, $\mathbf{r}_C = \frac{1}{4}h\mathbf{k}$

5.5-8 $I_{x0} = I_{y0} = \frac{1}{20}\rho\pi R^4h + \frac{1}{30}\rho\pi R^2h^3$, $I_{z0} = \frac{1}{10}\rho\pi R^4h$

5.5-9 $I_{xy} = I_{yz} = I_{zx} = 0$

5.5-10 $M = \frac{3}{16}R^3\rho\pi$, $\mathbf{r}_C = \frac{7}{45}R\mathbf{k}$

5.5-11 $I_{xx} = \frac{7}{128}\rho\pi R^5$, $I_{yy} = \frac{7}{128}\rho\pi R^5$, $I_{zz} = \frac{5}{48}\rho\pi R^5$

5.5-12 $I_{xy} = I_{yz} = I_{xz} = 0$,

5.6-1 $\mathbf{r}_C = 0.311r$, $I_{xx} = I_{yy} = 0.703mr^2$, $I_{zz} = 1.406mr^2$,
$I_{xy} = I_{yz} = I_{zx} = 0$

5.6-2 $\mathbf{r}_C = -0.815\mathbf{i} + 1.596\mathbf{j}$

5.6-3 $\mathbf{r}_C = 2.094\mathbf{i} + 2.267\mathbf{j} - 0.399\mathbf{k}$ m, $I_{xx} = 76.419$ kg·m^2,
$I_{yy} = 65.814$ kg·m^2, $I_{zz} = 125.071$ kg·m^2, $I_{xy} = I_{yz} = 0$,
$I_{zx} = 13$ kg·m^2

5.6-4 $\mathbf{r}_C = \frac{1}{4}\sqrt{2}R\pi\mathbf{i} + \frac{1}{4}\sqrt{2}R\pi\mathbf{j}$, $I_{xx} = I_{yy} = \dfrac{R^6}{M}\left(\frac{1}{8}\pi + \frac{1}{3}\right)$,

$I_{zz} = \dfrac{R^6}{M}\left(\frac{1}{4}\pi + \frac{2}{3}\right)$, $I_{xy} = \dfrac{R^6}{M}\left(\frac{1}{3} - \frac{1}{4}\pi\right)$, $I_{yz} = I_{zx} = 0$

5.6-5 $0.039\mathbf{i} + 0.05\mathbf{j} + 0.086\mathbf{k}$ m, $I_{xx} = 0.00377$ kg·m^2,
$I_{yy} = 0.00300$ kg·m^2, $I_{zz} = 0.00377$ kg·m^2,
$I_{xy} = -0.00367$ kg·m^2, $I_{yz} = -0.00410$ kg·m^2,
$I_{zx} = -0.00519$ kg·m^2

Chapter 6

6.2-1 $a_x = g(\mu \cos\theta + \sin\theta) = 6.604$ m/s^2
6.2-2 $\mathbf{a} = 130\mathbf{i} + 450\mathbf{j}$ ft/s^2
6.2-3 $0.16\mathbf{i} - 0.0693\mathbf{j}$ m, $-0.125\mathbf{i} - 0.130\mathbf{j}$ m/s^2
6.2-4   (a) $\theta < \arctan\mu_s$, (b) $\arctan\mu_k < \theta < \arctan\mu_s$

Chapter 7

7.1-1 -11.94 ft, 9.56 ft
7.1-2 $\mathbf{F} = 1768\mathbf{i} + 1768\mathbf{j}$ N, $\mathbf{v} = 24.9\mathbf{i} + 2.18\mathbf{j}$ m/s
7.2-1 $v = 4t^3 - 40t + 6$ m/s, $a = 12t^2 - 40$ m/s^2
7.2-2 $v(x) = \sqrt{0.4x^3 - 5.562x + 6.886}$
7.2-3 $-\frac{3}{2}t^2 + 17t - 8$ m/s, $17 - 3t$ m/s^2
7.2-4 $x_{A/B} = -7 + 2t - 3.89t^2$ ft, $v_{A/B} = 2 - 7.77t$ ft/s,
$a_{A/B} = -7.77$ m/s^2
7.2-5 20.19 ft
7.2-6 $\mathbf{F} = -380\sin t\,\mathbf{i} + 152\mathbf{j}$
7.2-7 $\mathbf{r}_{A/B} = (3t^2 - 4t\cos t - 16)\mathbf{i} + (5t + 2t^3)\mathbf{j}$,
$\mathbf{v}_{A/B} = (6t - 4\cos t + 4t\sin t)\mathbf{i} + (5 + 6t^2)\mathbf{j}$,
$\mathbf{a}_{A/B} = (6 + 8\sin t + 4t\cos t)\mathbf{i} + 12t\mathbf{j}$
7.2-8 $\mathbf{v}_{A/B} = (3t^2 + 4t - 12)\mathbf{i} + (4t^2 + 2t)\mathbf{j}$
7.2-9 $\mathbf{r} = (0.2t^4 - 0.5333t^3 + 3.2t^2)\mathbf{i}$
$+0.1333t^3\mathbf{j} + 3600(1 - \cos\frac{1}{30}t)\mathbf{k}$ m, $\mathbf{v} = (0.8t^3 - 1.6t^2$
$+6.4t)\mathbf{i} + 0.4t^2\mathbf{j} + 120\sin\frac{1}{30}t)\mathbf{k}$ m

7.2-10 $\mathbf{r}_{R/ED} = (200 - 50 \sin t)\mathbf{i} + (5t - 50 \cos t)\mathbf{j} +$
$[800 - 32(t - 5)^2]\mathbf{k}$, $\mathbf{v}_{R/ED} = -50 \cos t\,\mathbf{i} + (5 + 50 \sin t)\mathbf{j} -$
$64(t - 5)\mathbf{k}$

7.2-11 $(-2000 - 70t)\mathbf{i} - 40\mathbf{j} + 18.333t^2\mathbf{k}$ ft

7.3-1 $\mathbf{n}_r = \dfrac{12 \cos t^2\mathbf{i} + (t - 3)\mathbf{j}}{\sqrt{144 \cos t^2 + (t - 3)^2}}$, $\mathbf{n}_\theta = \dfrac{-(t - 3)\mathbf{i} + 12 \cos t^2\mathbf{j}}{\sqrt{144 \cos t^2 + (t - 3)^2}}$

7.3-2 $\mathbf{F} = 23.0\mathbf{n}_r$ lb

7.3-3 $\frac{1}{2}\pi + 0.039t$ rad, 120.5 s

7.3-4 $\mathbf{r}(5) = 2.964\mathbf{i} + 5.306\mathbf{j}$

7.3-5 $(-128t^2 - 1200)\mathbf{n}_r + 128t\mathbf{n}_\theta$ lb

7.3-6 1.28 rad/s

7.3-7 1.188 ft/s

7.3-8 $v_r = 5\pi\sqrt{3}$, $v_\theta = 10\pi$ ft/s

7.4-1 $\mathbf{n}_r = 0.837\mathbf{i} + 0.547\mathbf{j}$, $\mathbf{n}_\theta = -0.547\mathbf{i} + 0.837\mathbf{j}$, $\mathbf{n}_z = \mathbf{k}$

7.4-2 $\mathbf{v} = 192\pi\mathbf{n}_\theta + 16/7\mathbf{n}_z$ in/s

7.4-3 26.95 ft/s

7.5-1 $\mathbf{n}_t = \dfrac{5}{\sqrt{(x - 40)^2 + 25}}\mathbf{i} + \dfrac{40 - x}{\sqrt{(x - 40)^2 + 25}}\mathbf{j}$,

$\mathbf{n}_n = \dfrac{40 - x}{\sqrt{(x - 40)^2 + 25}}\mathbf{i} - \dfrac{5}{\sqrt{(x - 40)^2 + 25}}\mathbf{j}$

7.5-2 $v = 8.02$ m/s

7.5-3 $a_t = \dfrac{256t}{\sqrt{4 + 256t^2}}$, $a_n = \dfrac{2048t^2 + 64}{\sqrt{4 + 256t^2}}$.

7.5-4 $\mathbf{v}_{A/B}(10) = 36\mathbf{i} - 440\mathbf{j} = -422\mathbf{n}_t + 130\mathbf{n}_n$,
$\mathbf{a}_{A/B}(10) = 2\mathbf{i} - 84\mathbf{j} = -78.74\mathbf{n}_t + 29.34\mathbf{n}_n$

7.6-1 $\phi = 147.1°$, $\theta = 127.7°$

7.6-2 $\mathbf{r} = -0.493\mathbf{i} + 1.838\mathbf{j} + 0.888\mathbf{k}$, $\mathbf{v} = -5.54\mathbf{i} + 5.96\mathbf{j} - 3.60\mathbf{k}$,
$\mathbf{a} = -24.7\mathbf{i} - 19.1\mathbf{j} - 35.2\mathbf{k}$

7.6-3 $\sqrt{5gR}[(2/\sqrt{5})\mathbf{n}_\phi + (2/\sqrt{5})\mathbf{n}_\theta]$

Chapter 8

8.1-1 3.01 m/s

8.1-2 $v_0 = 8.028$ ft/s

8.1-3   65.1 ft/s

8.1-4 $2P_2T_2 = P_1T_1$, the difference could be $P_2 = \frac{1}{2}P_1$, or
$T_2 = \frac{1}{2}T_1$, or a compromise

8.2-1   2.87°

8.2-2   2.1 ft

8.2-3   Setting no. 3

8.2-4   4.01 rad/s

8.2-7   58.3°

8.3-1   63.3 m/s

8.3-2   0.840 s

8.3-3   0.11 m

8.3-4 $3.92h_1$

8.3-5   11.36 m/s

8.4-1 $-1530\mathbf{k}$ kg·m^2/s

8.4-2 $v_3 = 36.65$ Gmile/year, $v_1 = 0.724$ Gmile/year

8.4-3   0.30 m

8.4-4 $\dot{\theta} = 1350$ rpm, $H = 14.14 \times 10^{-6}$ kg·m^2/s

Chapter 9

9.1-1 $1023\mathbf{i} - 988\mathbf{j} - 77\mathbf{k}$

9.1-2 $210\mathbf{i}_1 - 7542\mathbf{i}_2$

9.2-1 $-24\mathbf{i} + 72\mathbf{j}$ in/s, $-852\mathbf{i} - 324\mathbf{j}$ in/s^2

9.2-2 $v_A = (v_O - \omega d \sin \theta)\mathbf{i} + \omega d \cos \theta \mathbf{j} = 1.55\mathbf{i} - 2.06\mathbf{j}$ m/s
$a_A = (a_O - \alpha d \sin \theta - \omega^2 d \cos \theta)\mathbf{i} + (2\omega v_O + \alpha d \cos \theta - \omega^2 d \sin \theta)\mathbf{j} = 1.37\mathbf{i} - 45.66\mathbf{j}$ m/s^2

9.2-3 $-79.4\mathbf{i} + 53.4\mathbf{j}$ m/s, $-468.4\mathbf{i} - 734.4\mathbf{j}$ m/s^2

9.2-4 $v = 105\mathbf{i} + 27.2\mathbf{j}$ m/s

9.2-5 $966.1\mathbf{i} - 850.5\mathbf{j} + 272.5\mathbf{k}$

9.2-6 $v_A = -\omega R\mathbf{i} = -12\mathbf{i}$ m/s, $a_A = (\omega^2 R^2/\sqrt{L^2 - R^2})\mathbf{i} = 127\mathbf{i}$ m/s^2

9.2-7   3.214 rad/s, -19.151 ft/s, 53.97 rad/s^2, 116.05 ft/s^2

9.3-1   16.26 rad/s^2

9.3-2 $a_{m1} = \dfrac{2m_1 g}{m_2 + 2m_1}$, $t_{m1} = \sqrt{\dfrac{h(m_2 + 2m_1)}{m_1 g}}$, $t_{m1g \text{ force}} = \sqrt{\dfrac{hm_2}{gm_1}}$

9.3-3 $R_y = \dfrac{3m_1 + m_2}{2m_1 + m_2} m_2 g$, $R_x = 0$, $R_{y(\text{static})} = (m_1 + m_2)g > R_y$,
$R_{x(\text{static})} = 0$

9.3-4 $a_1 = -R_2\alpha, a_2 = R_1\alpha,$ where $\alpha = \dfrac{g(m_1 R_2 - m_2 R_1)}{mr_g^2 + R_2^2 m_1 + R_1^2 m_2}$

9.3-5 $10.06\mathbf{i}$ ft·s, 0 rad/s^2

9.3-6 $\mathbf{R}_O = \frac{3}{4}kx\mathbf{j} + \frac{5}{4}mg\mathbf{k} = 11.25\mathbf{j} + 18.93\mathbf{k}$ N, $\boldsymbol{\alpha}_{\text{horizbar}} = (3g/2d)\mathbf{j} = 36.8\mathbf{j}$ rad/s^2, $\boldsymbol{\alpha}_{\text{vertbar}} = -(3kx/2md)\mathbf{i} = -37.5\mathbf{i}$ rad/s^2

9.4-1 $\omega = \sqrt{\dfrac{3g}{l}}$

9.4-2 $w_{\max} = \sqrt{\dfrac{k}{2m}}$

9.4-3   11 m, 5.61 rad/s

Chapter 10

10.1-1 $R_1 = 0.282$ inch, $R_2 = 0.230$ inch, $\Delta_1 = 2.8$ inches, $\Delta_2 = 4.0$ inches

10.1-2 $F_{ll} = 1200$ lb, $F_{yp} = 1800$ lb

10.2-1   3.98 kN/m^2

10.2-2 $D_{\text{for strength}} = 0.1262$ m, $D_{\text{for stiffness}} = 0.1193$ m

10.3-1   0.27 ft

Chapter 11

11.1-1   1.27 cm

11.1-2 $u(h) = 0.119$ m

11.1-3   0 cm, 0.175 cm

11.1-4 $u(6) = 2.29$ mm

11.2-1 $u(x) \begin{cases} 0, & 0 < x < 0.75, \\ -x/1000, & 0.75 < x < 2.25, \\ -2.25/1000, & 2.25 < x < 3, \end{cases}$

$\varepsilon = 0$ for $0 < x < 0.75$ and $2.25 < x < 3$,

$\varepsilon_{\max}$ for $0.75 < x < 2.25$, $u_{\max}$ for $2.25 < x < 3$

11.2-2 $u(L) = 0.02$ m

11.2-3   0.37 cm

Chapter 12

12.1-1 $2.72x$ rad, 1.044 rad at $x = 0.2^+$

12.1-2 $\phi(h) = 0.05071$ rad $= 2.91°$

12.1-3 5.73 rad

12.2-1 $\phi = \begin{cases} 99.5x, & 0 < x < 0.2 \text{ rad}, \\ 14.92 + 24.87x, & 0.2 < x < 0.5 \text{ rad} \end{cases}$

12.2-2 $\phi = \begin{cases} Tx, & x < \frac{1}{2}L, \\ Tx - \dfrac{kTL}{JG}\left(\dfrac{JG + kL}{2JG - kL}\right)(x - \frac{1}{2}L), & x > \frac{1}{2}L \end{cases}$

12.2-3 $\phi = \dfrac{4T}{JG(\alpha + 8)} \begin{cases} \frac{1}{2}x, & 0 < x < \frac{1}{4}L, \\ \frac{1}{2}L - x, & \frac{1}{4}L < x < \frac{3}{4}L, \\ x - L, & \frac{3}{4}L < x < L \end{cases} \quad (\alpha = KL/JG)$

12.3-1 0.4 rad, 11.55 kN/m^2

12.3-2 $\phi(L) = 0.48$ rad

12.3-3 $t_1 = t_2 = 0.1$ in

Chapter 13

13.1-1 $2500y(20 - x)$ lb/in^2

13.1-2 $\sigma(y) = -173.6y$ lb/in^2

13.2-1 $-7500(x - 8)^2$ lb, $9000(x - 8)^3$ lb·ft

13.2-2 $V(x) = F = 300$ N, $M(x) = F(L - x) = 750 - 300x$ N

13.3-1 $V = \begin{cases} 16.67 \text{ N}, & 0 < x < \frac{1}{2}L, \\ -83.33 \text{ N}, & \frac{1}{2}L < x < L, \end{cases}$

$M = \begin{cases} -16.67x \text{ N·m}, & 0 < x < \frac{1}{2}L, \\ -16.67x + 100(x - 0.75) - 50 \text{ N·m}, & \frac{1}{2}L < x < L, \end{cases}$

13.3-2 $V = \begin{cases} Fa/L, & 0 < x < \frac{1}{2}(L - a), \\ F(a - L)/L, & \frac{1}{2}(L - a) < x < \frac{1}{2}(L + a), \\ Fa/L, & \frac{1}{2}(L + a) < x < L \end{cases}$

$$M = \begin{cases} -Fax/L, & 0 < x < \tfrac{1}{2}(L-a), \\ -\tfrac{1}{2}Fax/2 + F(x - \tfrac{1}{2}(L-a)), & \tfrac{1}{2}(L-a) < x < \tfrac{1}{2}(L+a), \\ -Fax/L + Fa, & \tfrac{1}{2}(L+a) < x < L \end{cases}$$

13.4-1 $$\tau = \begin{cases} \dfrac{24F\{y^2 - [\tfrac{1}{4}(3a+t)]^2\}}{(a+3t)^3(a+t) - (a-t)^4}, & -c < t < a-c, \\[4mm] \dfrac{24F\{[(a+3t)]^2 - y^2\}}{(a+3t)^3(a+t) - (a-t)^4}, & a-c < t < a-c+t \end{cases}$$

13.4-2 $\tau = 3F(b^2 - 4y^2)/(2ab^3) = 312.5 - 1250y^2 \text{ lb/in}^2,$
$V = f_0(\tfrac{1}{2}L - x), M = f_0[(x - \tfrac{1}{2}L)^2 - \tfrac{1}{12} - L^2],$

13.5-1 $\theta = \dfrac{2f_0}{a^4 E}x(L-x)(\tfrac{1}{2}L - x), w = \dfrac{f_0}{2a^4 E}x^2(x-L)^2$

13.5-2

$$V = \begin{cases} -\tfrac{1}{2}F, & x < L, \\ \tfrac{1}{2}F, & x > \tfrac{1}{2}L, \end{cases} \qquad M = \begin{cases} -\tfrac{1}{8}FL + \tfrac{1}{2}Fx, & x < \tfrac{1}{2}L, \\ \tfrac{3}{8}FL - \tfrac{1}{2}Fx, & x > \tfrac{1}{2}L, \end{cases}$$

$$\phi = \dfrac{2}{3\pi R^4}\begin{cases} -\tfrac{1}{8}FLx + \tfrac{1}{4}Fx^2, & x < \tfrac{1}{2}L, \\ \tfrac{3}{8}FLx - \tfrac{1}{4}Fx^2, & x > \tfrac{1}{2}L, \end{cases}$$

$$w = \dfrac{2}{E\pi R^4}\begin{cases} -\tfrac{1}{16}FLx^2 + \tfrac{1}{12}Fx^3, & x < \tfrac{1}{2}L, \\ \tfrac{3}{16}FLx^2 - \tfrac{1}{12}Fx^3, & x > \tfrac{1}{2}L \end{cases}$$

13.5-3 $V = \dfrac{3B}{2L}, M = B\left(1 - \dfrac{3x}{2L}\right), \theta = \dfrac{BL}{4EI}\left(1 - \dfrac{x}{L}\right)\left(3\dfrac{x}{L} - 1\right),$

$w = -\dfrac{BL^2}{4EI}\dfrac{x}{L}\left(\dfrac{x}{L} - 1\right)^2$

Chapter 14

14.1-1 -43.9 ksi, 92.26 ksi
14.1-2 $\sigma' = 93.3$ klb/in, $\tau' = 25$ klb/in, $\tau_{\max} = 50$ klb/in at $\theta = 0°$
14.2-1 $-0.0439, 0.1852$
14.2-2 $\varepsilon' = 0.0933, \gamma' = 0.05, \gamma_{\max} = 0.1$ at $\theta = 0°$
14.3-1 -0.013 in
14.3-2 $\gamma_{\max} = 0.02$
14.3-3 $v = \tfrac{1}{2}$

Chapter 15

15.1-1 $\sigma_x = \dfrac{F}{\pi R^2} - \dfrac{4B}{\pi R^4 L} xy, \quad \tau = -\dfrac{BR^3}{3L}\left[1 - \left(\dfrac{y}{R}\right)^2\right]$

15.1-2 $\sigma_x = \dfrac{-2f_{y0}x(x-L)y}{\pi R^4}, \quad \tau_{xy} = \dfrac{2f_{y0}(L-2x)[1-(z/R)^2]}{3\pi R^2}$

$-\dfrac{T_{x0}(L-2x)z}{\pi R^4}, \quad \tau_{xz} = \dfrac{T_{x0}(L-2x)y}{\pi R^4}$

15.1-3

$\sigma_{\max} = \dfrac{2c + 6(a/L)s}{4c^2 + (a/L)^2 s^2}\dfrac{F}{a^2} \quad$ at $(0, \pm\tfrac{1}{2})$ (at the top and bottom),

$\tau_{\max} = \dfrac{\tfrac{3}{4}s(a/L)^2}{4c^2 + (a/L)^2 s^2}\dfrac{F}{a^2} \quad$ at $y = 0$ (along the middle)

$(s = \sin b, c = \cos b)$

15.1-4

$$\mathbf{P}_1 = 0, \ \mathbf{M}_1 = -B\mathbf{k}, \ \boldsymbol{\phi}_1 = \dfrac{-4Bx_1}{E\pi R^4}\mathbf{k}, \ \mathbf{u}_1 = \dfrac{-2Bx_1^2}{E\pi R^4}\mathbf{j},$$

$$\sigma_{1x} = \dfrac{4By_1}{\pi R^4}, \ \tau_{1xy} = 0, \ \tau_{1xz} = 0,$$

$$\mathbf{P}_2 = 0, \ \mathbf{M}_2 = B\mathbf{i}, \ \boldsymbol{\phi}_2 = \left(\dfrac{2Bx_2}{G\pi R^4} - \dfrac{4BL}{E\pi R^4}\right)\mathbf{i},$$

$$\mathbf{u}_2 = \dfrac{-2Bx_2}{E\pi R^4}\mathbf{j}, \ \sigma_{2x} = 0, \ \tau_{2xy} = \dfrac{2Bz}{\pi R^4},$$

$$\tau_{2xz} = \dfrac{2By}{\pi R^4}$$

Chapter 16

16.1-1 $w_r = \sin(r\pi x/L), \ P_r = EI_2(r\pi/L)^2 \ (r = 1, 2, \ldots)$

16.1-2 $w = \left\{\left[1 - \left(\dfrac{\alpha}{\beta L}\right)^2\right](1 - \cos\beta x) + \left(\dfrac{\alpha}{\beta L}\right)^2\left(\dfrac{x}{L} - \dfrac{\sin\beta x}{\beta L}\right)\right\}w(L),$

$F = (\beta L)^2\dfrac{EI}{L^2},$

When solving for buckling load, you must end up numerically solving for βL with the equation $-\beta L \sin \beta L + \cos \beta L = 1$.

16.2-1 31.5 ksi

16.2-2 $p = 0.111$ ksi

16.3-1 $\sigma = \dfrac{3F(L-x)y}{D} \begin{cases} E_1, & -c < y < a - c, \\ E_2, & a - c < y < 2a - c, \end{cases}$

where $c = \dfrac{a\,E_1 + 6E^2}{2\,E_1 + 2E_2}$ and

$D = a^2 E_1(a^2 + 3c^2 - 3ac) + 2a^2 E_2[a^2 + 3(a-c)(2a-c)]$

16.4-1 0.176 m

16.4-2 $\sigma = -E\alpha\Delta T + y\left[\dfrac{6F(2x-L)}{L^4} - \dfrac{\alpha(T_1 - T_2)E}{h}\right]$

ABOUT THE AUTHORS

LARRY SILVERBERG is a professor of Mechanical and Aerospace Engineering at North Carolina State University. He has been teaching engineering college students for almost 20 years, and has received many awards for teaching and research excellence.

JAMES P. THROWER is a very talented Ph.D. graduate student who is working under the direction of Dr. Larry Silverberg at North Carolina State University. Both authors reside in Raleigh, North Carolina.